III International Conference on Cognitive Robotics 2018

IOP Conference Series: Materials Science and Engineering
Volume 516

Tomsk, Russia
22 - 24 November 2018

ISBN: 978-1-5108-8634-6
ISSN: 1757-8981

Printed from e-media with permission by:

Curran Associates, Inc.
57 Morehouse Lane
Red Hook, NY 12571

Some format issues inherent in the e-media version may also appear in this print version.

Printed by Curran Associates, Inc. (2019)

Institute of Physics
Dirac House, Temple Back
Bristol BS1 6BE UK

Phone: 44 1 17 929 7481
Fax: 44 1 17 920 0979

techtracking@iop.org

Additional copies of this publication are available from:

Curran Associates, Inc.
57 Morehouse Lane
Red Hook, NY 12571 USA
Phone: 845-758-0400
Fax: 845-758-2633
Email: curran@proceedings.com
Web: www.proceedings.com

Table of contents

Volume 516

III International Conference "Cognitive Robotics"

22–24 November 2018, Tomsk, Russian Federation

Accepted papers received: 12 March 2019
Published online: 26 April 2019

Preface

Papers

IOP Conf. Series: Materials Science and Engineering **516** (2019) 011001 doi:10.1088/1757-899X/516/1/011001

1. Abstract

This issue of the journal is dedicated to the scientific and applied research presented at the III International Conference "Cognitive Robotics". The conference was held on November 22-24 2018 in Tomsk (Russia).

The conference brought together experts from different countries and organizations who received a great opportunity to share knowledge in sessions, to initiate discussions on topics of interest to all participants.

The key topics of the conference were:

1. Cognitive technologies and artificial intelligence in control systems for land and air-based robotic systems, and robotic submarines.
2. Sensing technology for land and air-based robotic systems, and robotic submarines.
3. Design of robotic systems (UAVs, AUVs).
4. Psycho-emotional and socio-humanitarian aspects of robotics.
5. Educational robotics

CONFERENCE THEMES:

- Practical application of cognitive technologies;
- Land and air-based robotic systems, and robotic submarines;
- Design of robotic systems;
- Psycho-emotional and socio-humanitarian aspects of robotics;
- Intelligent reconfigurable systems for control, navigation and image processing for autonomous mobile robots;
- Algorithms and mathematical models of high-speed processing and analysis of images;
- Educational robotics.

Program committee

Chairman: Alexey V. Yurchenko – D.Eng.Sc., professor, Senior Researcher of the international laboratory Vision Systems;

Co-chairman: Vladimir I. Syryamkin, D.Eng.Sc., professor, Head of the Department of Quality Control of the Faculty of Innovative Technologies, National Research Tomsk State University; academician of International Academy of Sciences of Higher Education and Russian Academy of Natural Sciences; Director of TSU scientific and educational center «Recognition: navigation, diagnostics, mechatronics»

Co-chairman: Stanislav V. Shidlovskiy - D.Eng.Sc., Dean of the Faculty of Innovative Technologies of National Research Tomsk State University; Head of the Laboratory of high-performance reconfigurable systems (NR TSU); Professor of Tomsk Polytechnic University.

IOP Conf. Series: Materials Science and Engineering **516** (2019) 011001 doi:10.1088/1757-899X/516/1/011001

MEMBERS OF THE PROGRAM COMMITTEE

V.S. Titov, D.Eng.Sc., professor, honored scientific worker of the Russian Federation, Head of the Department of Computer Science of Southwest State University (Kursk, Russia)

A.V. Lapota, Director-Chief constructor of the Russian State Scientific Center for Robotics and Technical Cybernetics, Saint-Petersburg, Russia

Anatoliy N. Soldatov – Dr. Sci. in Physics and Mathematics, professor, Head of the Regional Center of the Russian Association of Laser, Honored Inventor of the Russian Federation, dean of the FIT TSU

R. V. Mescheryakov, Vice-Rector of TUSUR for Research and Innovation, Chairman of the Russian National Committee of RoboCup

Nikola Sobutonov – Professor, Academician of the Bulgarian Academy of Sciences

Valery L. Vengrinovich – D.Eng.Sc., professor, Minsk, Belarus

Sergey B. Suntsov – JSC Information Satellite Systems named after Academician MF Reshetnev. (Krasnoyarsk);

Ali D. Mehtiyev – Ph.D. in Eng.Sc., Head of the Department of Technologies and Communication Systems of Karaganda State Technical University

Alexey G. Yakunin – Professor, Head of the Department of Computer Science, Computer Engineering and Information Security, Polzunov Altai State Technical University

Eduard Siemens- professor, Anhalt University of Applied Sciences (Germany);

Jurgen Schreiber CEO German-Korean Company «NUGA LAB GmbH», Prof. Fraunhofer Institute of Ceramic Technology and Systems (Germany)

E. S. Shandarov, head of the laboratory of robotics and artificial intelligence TUSUR, executive secretary of the Russian National Committee RoboCup.

Organizing committee

Conference chairman: Vladimir I. Syryamkin, D.Eng.Sc., professor, Head of the Department of Quality Control of the Faculty of Innovative Technologies, National Research Tomsk State University; academician of International Academy of Sciences of Higher Education and Russian Academy of Natural Sciences; Director of TSU scientific and educational center «Recognition: navigation, diagnostics, mechatronics».

Conference co-chairman - Eduard V. Galazhinsky, Doctor of Science (Psychology), Professor, academician of the Russian Academy of Education, Rector of Tomsk State University, Professor.

Conference co-chairman – Stanislav V. Shidlovskiy - D.Eng.Sc., Dean of the Faculty of Innovative Technologies of National Research Tomsk State University; Head of the Laboratory of highperformance reconfigurable systems (NR TSU); Professor of Tomsk Polytechnic University.

MEMBERS OF THE ORGANIZING COMMITTEE:

Ivan V. Ivonin, TSU Vice-Rector for Academic Research, Doctor of Physics and Mathematics, Professor

Eduard Siemens- professor, Anhalt University of Applied Sciences (Germany)

Ali D. Mehtiyev – Ph.D. in Eng.Sc., Head of the Department of Technologies and Communication Systems of Karaganda State Technical University

Alexander Yu. Verkhovskiy – Head of the Laboratory of Biochemistry of Transport Systems, Faculty of Innovative Technologies, National Research Tomsk State University

IOP Conf. Series: Materials Science and Engineering **516** (2019) 011001 doi:10.1088/1757-899X/516/1/011001

Anatoliy N. Soldatov – Dr. Sci. in Physics and Mathematics, professor, Head of the Regional Center of the Russian Association of Laser, Honored Inventor of the Russian Federation, dean of the FIT TSU

Maxim V. Syryamkin – director of the interacademic educational and scientific production center Technology Management, National Research Tomsk State University

Artem A. Ananin – Head of the scientific and educational projects of National instruments

Alexey A. Poguda – deputy Dean of the FIT TSU, coordinator of the conference site;

Denis N. Kuznetsov – FIT TSU, Conference Secretary, engineer of the Laboratory of highperformance reconfigurable systems.

3. The Organizing Committee of the III International Conference "Cognitive Robotics" wishes to thank the IOP Conference Series Materials Science and Engineering-Team for collaboration.

IOP Conf. Series: Materials Science and Engineering **516** (2019) 011002 doi:10.1088/1757-899X/516/1/011002

Peer review statement

All papers published in this volume of *IOP Conference Series: Materials Science and Engineering* have been peer reviewed through processes administered by the proceedings Editors. Reviews were conducted by expert referees to the professional and scientific standards expected of a proceedings journal published by IOP Publishing.

IOP Conf. Series: Materials Science and Engineering **516** (2019) 012001 doi:10.1088/1757-899X/516/1/012001

Quantitative characteristics of high-resolution spectra to the benefit of astrophysics: The case of an ethylene molecule and its isotopologues

Yu V Konova, N V Kashirina, A G Ziatkova
National Research Tomsk Polytechnic, Tomsk, Russia
E mail: ch yuliya28@mail.ru

Abstract. Study of deuterated isotopologues is significant for the expansion of information quantity about "mother" modification of molecule. The spectra of cis-ethylene-d_2 ($C_2H_2D_2$-cis) and ethylene-d_3 (C_2HD_3) molecules have been recorded with a Bruker IFS 120 HR Fourier spectrometer in the region of the ν_3 bands location with spectral resolution of 0.0021 cm^{-1}. As a result of the experimental spectrum analysis, more than 1200 and 1000 transitions have been assigned for the ν_3 bands of $C_2H_2D_2$-cis and C_2HD_3, respectively. Study of these bands for both investigated molecules was made at first time.

1. Introduction

For the first time, the composition of the atmosphere of Titan was remotely studied by the Voyager missions in the 1980s [1], then studies were carried out by the Infrared Space Observatory in 1998 and 2004 [2, 3]. Currently, there is actual information obtained from ground-based observatories and from the Cassini-Huygens mission [4]. The mentioned above and many other studies have shown that the atmosphere of Titan not only demonstrates the rich content of organic molecules and a peculiar thermal structure, but also shows impressive new meteorological phenomena.

The first studies in the field of infrared spectroscopy made it possible to obtain information only about the temperatures and chemical structures of the studied planets (for example, for the planet Titan [1]; the spectral resolution was 4.3 cm^{-1}). Further studies, which were carried out already from the surface of the Earth, made it possible to register and determine the lines of methane in the spectrum, first to detect the mesosphere on Titan in the infrared range [5], and also to register the high-resolution spectra of ethylene on Titan (with spectral resolution is equal to 0.53 cm^{-1}) [6].

As a result of the previous remote missions and research of the obtained data, the study of the spectral characteristics of the C_2H_4 molecule is in demand, due to the fact that it is included in the list of molecules found in the atmosphere of the giant - planets. In addition, the presence of deuterated modifications of the detected molecules was shown in [7], which leads to the need for researching the spectra of isotopologues for the completion of databases (such as HITRAN and GEISA).

The study of different ethylene isotopologues is of great interest as in the spectra of such kinds of molecules there are additional transitions, which are forbidden in the basic modification. This is because it has a lower symmetry as compared with molecule of ethylene, which symmetry group is isomorphic to D_{2h} point symmetry group and has 8 irreducible representations. For comparing, C_{2v} point symmetry group has 2 times less irreducible representations (namely, 4).

This paper is devoted to the study of the ν_3 fundamental bands for the two different isotopologues of ethylene: $C_2H_2D_2$-cis and C_2HD_3.

IOP Conf. Series: Materials Science and Engineering **516** (2019) 012001 doi:10.1088/1757-899X/516/1/012001

2. Experimental details

The experimental spectra have been recorded in the Braunschweig Infrared Laboratory (Germany) using IFS120HR Fourier transform infrared (FTIR) spectrometer based on the principle of the Michelson interferometer in combination with stainless steel White cell with a base length of 1 m.

The gas $C_2H_2D_2$-cis sample was purchased from the Cambridge Isotope Laboratories. The chemical and isotopic purity of the sample was estimated to be 99.9%.

The gas sample of the C_2HD_3 molecule was obtained via the exchange reaction between C_2D_4 and H_2 on a nickel-wire catalyst which was electrically heated to the temperature range of 120–180°C. As the result of this procedure, seven different isotopologues have been registered: C_2H_4, C_2H_3D, $C_2H_2D_2$ (include the –trans, -as and -cis variants), C_2HD_3 and C_2D_4. The ratio of these components is presented in Table 1.

Table 1. Maximum concentration of different isotopologues listed on the gas sample.

Name of isotopolog	**Max. concentration, %**
C_2HD_3	42.2
C_2H_4	0.4
C_2H_3D	4.7
$C_2H_2D_2$	21.1
C_2D_4	31.6

For FTIR spectrometer a Globar IR source, a KBr beamsplitter, a mercury – cadmium – telluride semiconductor detectors have been used. In total four spectra have been analyzed for the $C_2H_2D_2$-cis and C_2HD_3 molecules. More experimental details are presented in Table 2. Survey spectra of both investigated molecules in the region of the ν_3 bands centers location are demonstrated in the left parts of Figures 1 and 2.

The nominal instrumental resolution defined as $1/d_{MOPD}$ (maximum optical path difference) was 0.0021 cm^{-1} for all spectra. The spectra were calibrated by using spectral lines of the N_2O and H_2O molecules.

Table 2. Experimental setup.

Molecule name	**No. of spectr.**	**Region, cm^{-1}**	**Resolution, cm^{-1}**	**No. of scans**	**Optical path length, cm^{-1}**	***T*, K**	***P*, mbar**
$C_2H_2D_2$-cis	I II	1060-2000	0.0021	400 470	4 16	300	2
C_2HD_3	III IV	640-1150	0.0021	320 500	4	300	2 0.2

IOP Conf. Series: Materials Science and Engineering **516** (2019) 012001 doi:10.1088/1757-899X/516/1/012001

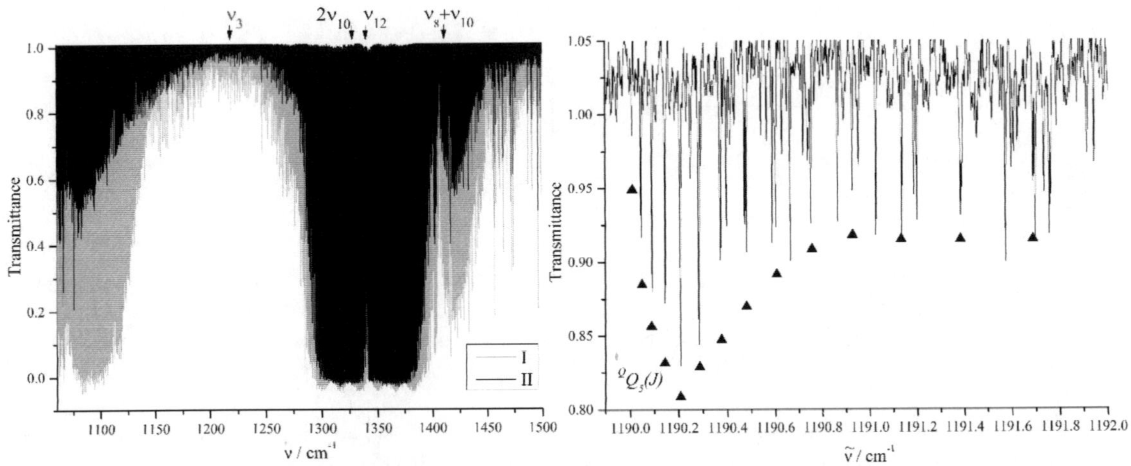

Figure 1. The survey spectra of $C_2H_2D_2$-cis in region 1060–1500 cm^{-1} (left); a part of of its Q – branch (right).

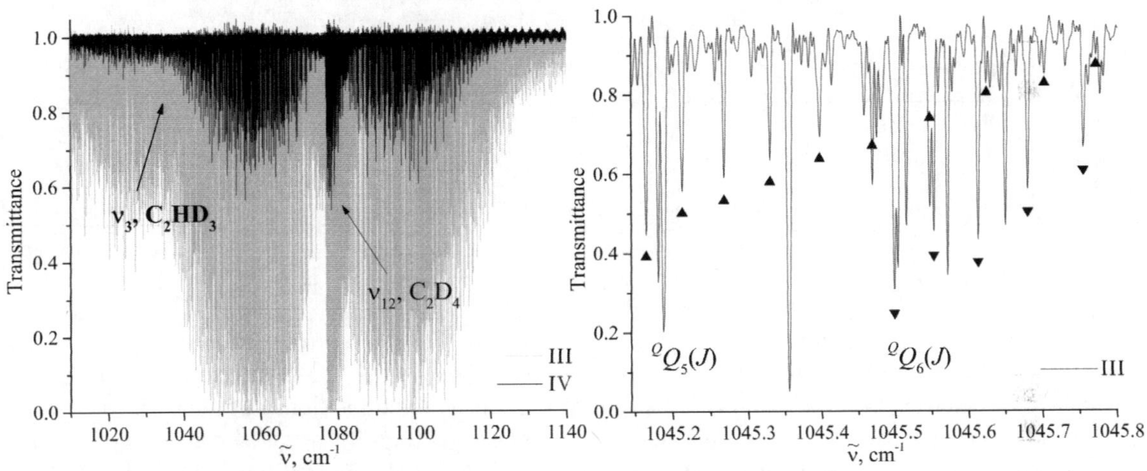

Figure 2. Experimental spectra of the C_2HD_3 molecule in the region of the ν_3 band (left) and part of its Q - branch (right).

3. Theoretical background

3.1. General information about $C_2H_2D_2$-cis molecule

Cis-ethylene-d_2 is an asymmetric top molecule and the symmetry of this molecule is isomorphic to the C_{2v} point symmetry group (see Figure 3). It has 12 vibrational modes and 8 irreducible representations. One can mention from Table 3 investigated band ν_3 is A_1 symmetry. It means that ν_3 is b-type band and selection rules $\Delta J = 0; \pm 1$ and ΔK_a = odd, ΔK_c = odd.

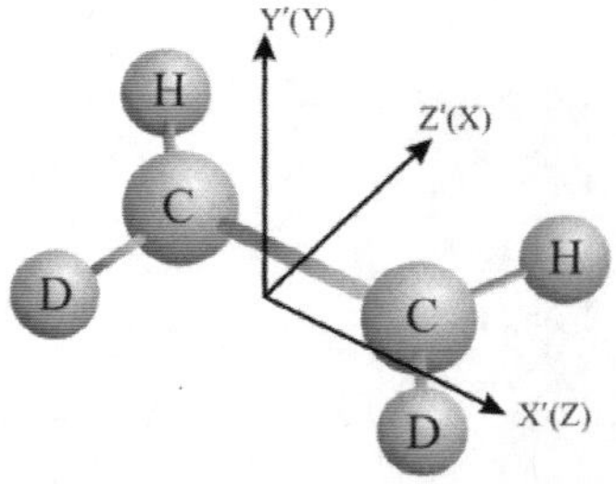

Figure 3. Axes definitions used in the present work for the $C_2H_2D_2$-cis molecule. The primed symbols refer to the axis definitions for the C_{2v} symmetry group used in the classification of the vibrational modes. The unprimed symbols refer to the Cartesian axis definitions of the I^r representation of Watson's *A* –reduced effective Hamiltonian.

For convenience of the reader, the symmetry properties of $C_2H_2D_2$-cis are presented in Table 3. Table 3 includes the q_λ vibrational coordinates in accordance with the list of irreducible representations and table of characters of the C_{2v} symmetry group.

Table 3. Symmetry types and characters of irreducible representations of the C_{2v} group (application to the $C_2H_2D_2$-cis molecule)

Representation	E	C_2	$\sigma_y(xz)$	$\sigma_y(yz)$	Vibrational coordinates
A_1	1	1	1	1	$q_1, q_2, q_3, q_9, q_{10}$
A_2	1	1	-1	-1	q_4, q_8
B_1	1	-1	1	-1	q_5, q_6, q_{11}, q_{12}
B_2	1	-1	-1	1	q_7

3.2 General information about C_2HD_3 molecule

C_2HD_3 molecules are asymmetric top and have 12 vibrational modes too (see Figure 4). Symmetry group of the molecule is isomorphic to C_s point symmetry group, which has only two irreducible representations (A' and A'', the symmetry properties of the molecule C_2HD_3 are presented in Table 4). As the consequence, the $A' \leftarrow A'$ bands are the hybrid bands, and transitions of both *a* and *b* – types are allowed in such bands. In the $A'' \leftarrow A'$ bands only *c*-type transitions are allowed. Using Table 4 one can determine, that the ν_3 investigated band can be identified as hybrid band with allowed both type of transitions – *a* and *b* - types. For such bands following selection rules corresponds: $\Delta J = 0, \pm 1$; $\Delta K_a =$ any; $\Delta K_c =$ odd.

Table 4. Symmetry types and characters of irreducible representations of the C_s group (application to C_2HD_3).

Representation	E	$\sigma_{x'z'}$	Vibrational coordinates
A'	1	1	$q_1, q_2, q_3, q_5, q_6, q_9, q_{10}, q_{11}, q_{12}$
A''	1	-1	q_4, q_7, q_8

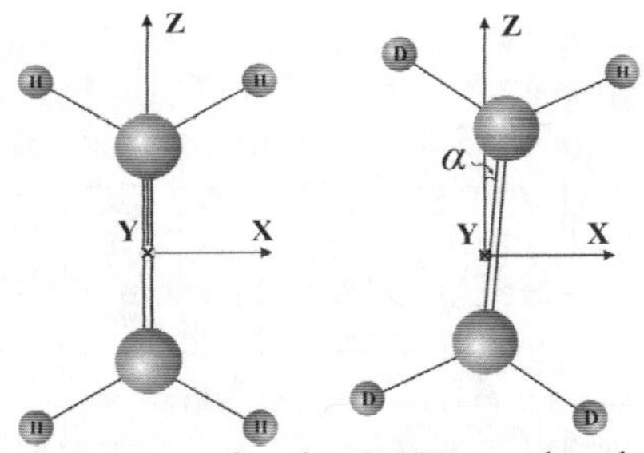

Figure 4. Axes definitions for the C_2HD_3 molecule (right part).
For comparison, the "mother" C_2H_4 molecule is presented at the left part of the figure.

3.3 Effective Hamiltonian model

Model of effective Hamiltonian is used for theoretical analysis of experimental data in following form:

$$H^{vib-rot} = \sum_{v,\tilde{v}}^{n} | v >< \tilde{v} | H^{v\tilde{v}}, \quad (1)$$

where the sum is taken over all the vibrational states. The diagonal blocks in the Hamiltonian expression $H^{v,\tilde{v}}$, describing the unperturbed rotational structure of vibrational states, have a form of reduced effective Hamiltonian (Watson's operator) [8]:

$$H^{v,\tilde{v}.} = E^v + \left[A^v - \frac{1}{2}(B^v + C^v) \right] J_z^2 + \frac{1}{2}(B^v + C^v) J^2 + \frac{1}{2}(B^v - C^v) J_{xy}^2 - \\ - \Delta_K^v J_z^4 - \Delta_{Jk}^v J_z^4 J^2 - \varDelta_J^v J^4 - \delta_K^v \left[J_z^2, J_{xy}^2 \right] - 2\delta_J^v J^2 J_{xy}^2 + H_K^v J_z^6 + \\ + H_{KJ}^v J_z^4 J^2 + H_{JK}^v J_z^2 J^4 + H_J^v J^6 + \left[J_{xy}^2, h_K^v J_z^4 + h_{JK}^v J^2 J_z^2 + h_J^v J^4 \right] + \ldots \quad , \quad (2)$$

where E is the center of the band; A, B, C are rotational constants; $\varDelta_K$, $\varDelta_{JK}$, $\varDelta_J$, etc., are the different order centrifugal distortion coefficients; $J_{xy}^2 = J_x^2 - J_y^2$.

4. Results of assignment of transitions

Assignment of transitions was made on the basis of the Ground State Combination Differences (GSCD) method. To implement this method, it is required to know the exact values of the parameters of the ground state for the investigated molecules. The ground state rotational energies have been calculated with the parameters from [9] and [10] papers for $C_2H_2D_2$-cis and C_2HD_3 molecules, respectively. For convenience of the reader, these parameters are represented in the Table 5.

Table 5. Spectroscopic parameters of ground vibrational states for the $C_2H_2D_2$-cis and C_2HD_3 molecules.

Parameter	**$C_2H_2D_2$-cis**	**C_2HD_3**
A	3.32454129	2.8446778
B	0.847825953	0.78580090
C	0.673768816	0.61415904
$\varDelta_K \times 10^4$	0.49105	0.31568
$\varDelta_{KJ} \times 10^5$	0.36722	0.34767
$\varDelta_J \times 10^5$	0.117725	0.09472
$\delta_K \times 10^5$	0.64619	0.5023
$\delta_J \times 10^6$	0.28756	0.24149
$H_K \times 10^7$	0.172	-
$H_{KJ} \times 10^9$	-0.377	-
$H_{JK} \times 10^{10}$	0.373	-
$H_J \times 10^{12}$	0.95	-

$C_2H_2D_2$-cis. As mentioned above, we have determined the energy transitions from the ground vibrational state to the upper vibrational states ($v_3 = 1$) of $C_2H_2D_2$-cis. The number of energy transitions belonging to the *b* - type v_3 band is about 1200. For demonstartion of spectrum analysis on the right part of Figure 1 one series of *Q* – branch corresponding to $K_a = 5$ is shown. The maximum values of quantum numbers J^{max}/K_a^{max} are 25/10, respectively.

C_2HD_3. As a result of the study of the experimental spectra of C_2HD_3 more than 1000 ro-vibrational transitions have been assigned for the v_3 fundamental band. In the ground of this amount of transitions, about 400 energies of upper excited state have been calculated. Obtained information corresponds to the maximum values of quantum number $J^{max}/K_a^{max} = 26/9$. By way of illustration, in Figure 2 (right part) two series for $^{Q}Q_5(J)$ and $^{Q}Q_6(J)$ are represented

5. Summary

In this study, we made the analysis of the high-resolution IR spectrum of the v_3 bands of the $C_2H_2D_2$-cis and C_2HD_3 molecules. Experimental spectra for investigated molecules have been recorded using FTIR spectrometer Bruker IFS 120 HR. The experimental resolution is 0.0021 cm^{-1}. More than 1200 transitions belonging to the v_3 of the $C_2H_2D_2$-cis molecule were assigned with the maximum values of the upper quantum numbers $J^{max} = 25$ and $K_a^{max} = 10$ for the first time. Also for C_2HD_3 more than 1000 transitions with $J^{max} = 26$ and $K_a^{max} = 9$ were assigned for the first time too.

Acknowledgments

The research was executed with the financial support of the Russian Foundation for Basic Research (Project No. 18-32-00116 mol_a).

References

[1] Hanel R., Conrath B., Flasar F. *et al.* 1981 *Science* **212** 192–200 doi: 10.1126/Science.212.4491.192
[2] Coustenis A., Salama A., Lellouch E. *et al* 1998 *Astron. Astrophys.* **136** 85–89 doi: 1998A&A...336L..85C
[3] Vervack R., Sandel B., Strobel D. 2004 *Icarus* **170** 91–112 doi: 10.1016/j.icarus.2004.03.005
[4] Coustenis A. *et al* 2007 *Icarus* **189** 35–62 doi: 10.1016/j.icarus.2006.12.022
[5] Penteado P., Griffith C., Greathouse T. *et al* 2005 *Astrophysical Journal Letters* **629** L53–L56 doi: 10.1086/444353
[6] Roe H., de Pater I., McKay C. 2004 *Icarus* **169** 440—461 doi: 10.1016/j.icarus.2004.01.002
[7] Lellouch E. *et al.* 2001 *Astronomy and Astrophysics* **670** 610–622 doi: 10.1051/0004-6361:20010259
[8] Watson J. 1967 *The Journal of Chemical Physics* **46** 1935—1967 doi: 10.1063/1.1840957
[9] Ulenikov O., Gromova O., Bekhtereva E.S. *et al.* 2016 *Quant. Spectrosc. Radiat. Transf.* **170** 69–82 doi: 10.1016/j.jqsrt.2015.10.011
[10] Ng L., Tan T., Gabona M., Godfrey P. et, McNaughton D. 2015 *Journal of Molecular Spectroscopy* **316** 79–83

IOP Conf. Series: Materials Science and Engineering **516** (2019) 012002 doi:10.1088/1757-899X/516/1/012002

Development of a tracked platform control system

A V Pilipenko, A S Vetrov, A Y Pilipenko, N V Goncharova

OGU named after I.S. Turgenev, Oryol, Russia
E-mail: a@pilipenko.info

Abstract. In this paper, the authors describe in detail the technology of development and creation of a tracked robot, presenting sketches of parts and components. In addition to the mechanical part, the paper presents the development technology of an automated control system based on the ARM STM32, taking into account the bounce of buttons and other engineering subtleties. The authors presented a number of documents developed in accordance with State Standards, such as a functional automation diagram and a circuit diagram. The authors devote a large amount of material to photographic materials demonstrating the efficiency of development.

1. Introduction

On the basis of a tracked platform, it is possible to build both combat robotic mechanisms with a portable ammunition and all-terrain transporters that can transport cargo in terms of production or the external environment [1]. Periodically, the enterprises need to remotely manage various objects. As a rule, this need arises in several situations: when it is necessary to protect a person from the negative effects of the environment or the loads to which the object is exposed; when the characteristics of the device do not allow direct human participation in its management; when full autonomy of the system is required. A very urgent task is the development of systems for manual control of tracked platforms.

The task of managing tracked platforms comes down to controlling 2 DC motors. Today, there are several different approaches to controlling DC motors. The simplest of these is to connect the motors through a relay to a power source. There is also an approach to control such motors by including the supply of a rheostat into the break, which allows you to make a smooth adjustment of the current in the circuit. And, the third, last method is control with the help of a current driver and a power MOSFET transistor through a microcontroller by generating a PWM signal of different duty cycle [2]. In this paper, the authors consider an approach to the development of an automated control system for a tracked platform with the inclusion in the power supply circuit of the relay and the use of a wired console.

2. Conception

When creating a track, its potential carrying capacity, overall dimensions, availability of a platform for the location of batteries, electromechanical drives, a contact relay group and a control board are taken into account. Overall dimensions of the platform 475×350×75 mm, such dimensions allow to accommodate all the above elements. Figure 1 and Figure 2 show a general view of the trolley.

IOP Conf. Series: Materials Science and Engineering **516** (2019) 012002 doi:10.1088/1757-899X/516/1/012002

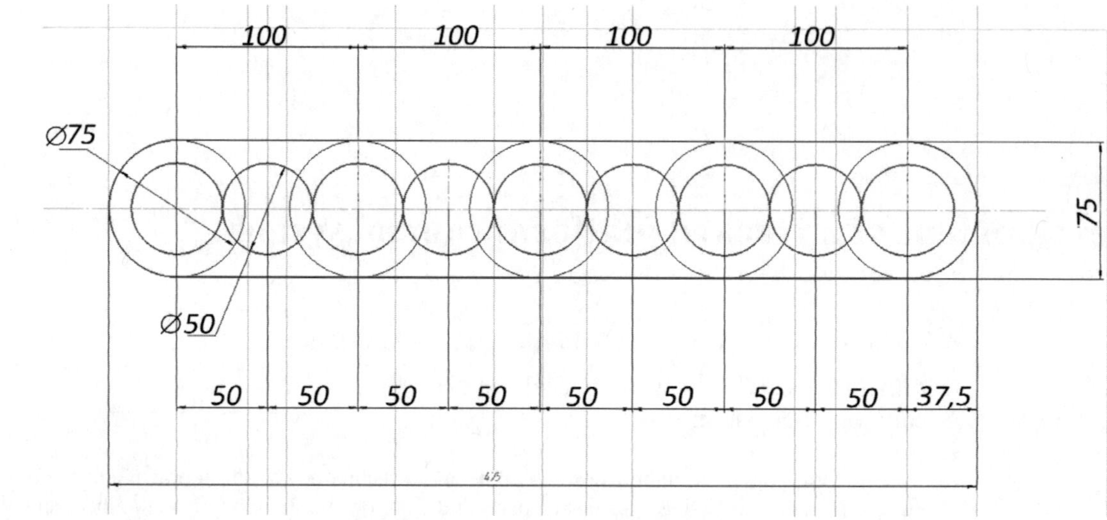

Figure 1. Sketch trolley, left view.

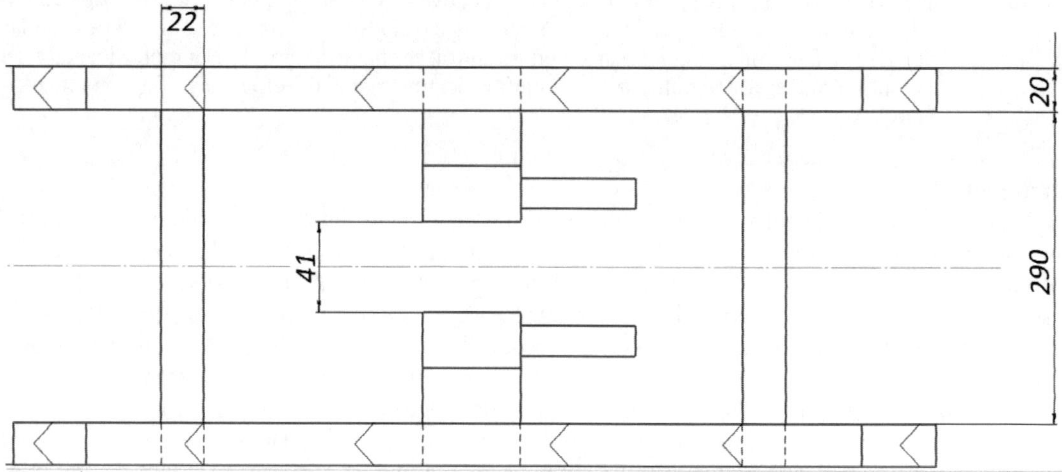

Figure 2. Sketch trolley, top view.

The design is made of two symmetrical sides. The side is an electric tracked truck. There is also a set of gears for transmitting the rotation to the tracks. The sides are connected by a metal profile by the method of electric arc welding. Such a connection will give the proper strength and reliability in operation. The motor transmits rotation through a gearbox to a gear wheel, which is mounted on a rectangular axis (Figure 3). This allows you to transfer almost all the energy of rotation in a rigid hook. Through the central hole with the help of a stud fastens this wheel to the frame of the cart. If it is decided to remove the cart from the tracks, there are three small holes that are used to connect to a conventional wheel.

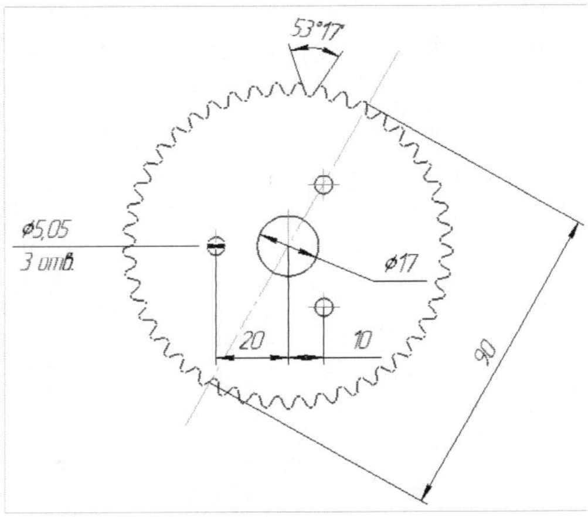

Figure 3. Gear.

The rotation is transmitted to the tracks by means of a drive gear, which is fitted onto the rectangular spindle with a square hole. The wheel taking the rotation from the spindle transmits it to other wheels and a tracked track. The material of these gears - organic sheet glass State Standart 10667-90 (GOST 10667-90). This material is easily processed by cutting, drilling and milling. The control object is shown in Figure 4, the maximum working load is 67 kg.

Figure 4. Tracked platform mockup.

3. Management system development

The motor control is based on the idea of alternately switching the relay using the stm32f0 discovery microcontroller [3]. Commands to the controller are given using a remote control with direct galvanic coupling. Since the selected microcontroller has an output voltage of 3V and low current, it will not be possible to connect the relay directly to the microcontroller's leg, so it was decided to design the product, which will be called "discrete logical amplifier". The schematic diagram of this product is provided in Figure 5 [4]. From pin, a protective resistor R1 1K is installed, and the gate contact is pulled to the

IOP Conf. Series: Materials Science and Engineering **516** (2019) 012002 doi:10.1088/1757-899X/516/1/012002

ground after 10 KΩ. The protective diode VD1 protects the pin from current surges, during switching on and off of the relay. This approach allows using modularity to achieve ease of connection.

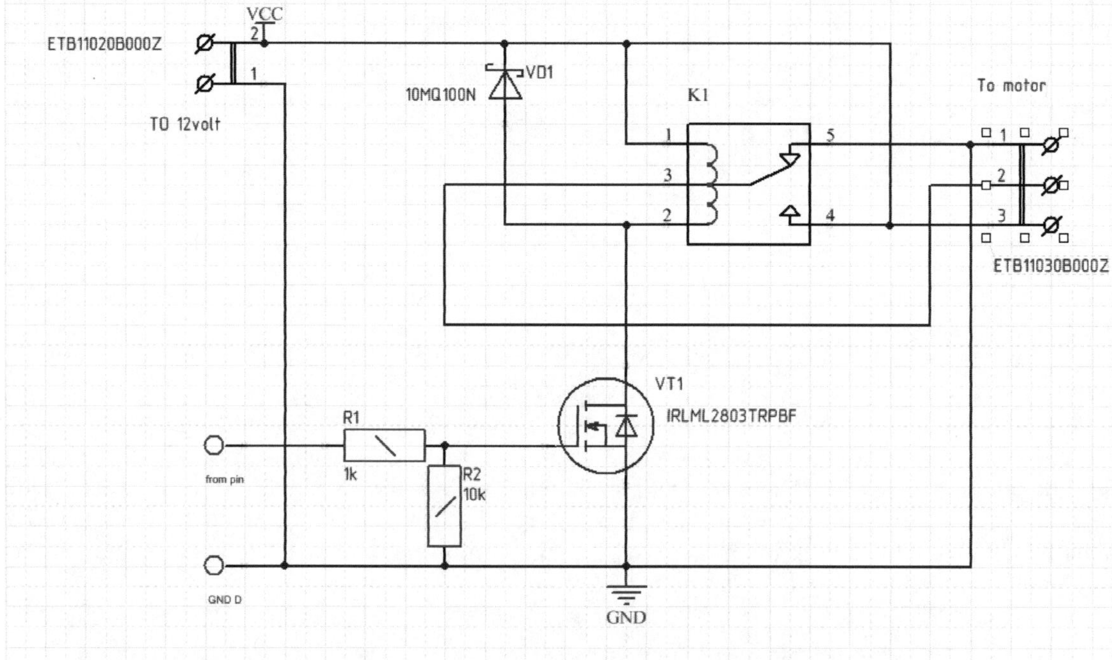

Figure 5. Schematic diagram of the product.

Consider the features of connecting buttons on the diagram in Figure 6.

When using buttons with mechanical contact inevitably, an effect called "contact bounce" occurs. At first glance, there are no bounce protection for the SB1 and SB2 buttons, and there is no pull-up to power or to ground, which is considered to be a wrong connection of the contacts, but, however, the microcontroller implements a hardware pull-up to the ground. Bounce also cleaned programmatically.

For buttons SB3, SB4, the connection is different from SB1 and SB2. There is a circuitry tightening to the power supply, as well as an RC circuit added parallel to the buttons, which filters the bounce of the contacts. Calculate the main characteristic of the RC circuit - the value of τ.

The calculation is carried out according to the formula $\tau = RC$. For our button, we take a resistor with a nominal value of R = 10 KΩ, and a capacitor with a capacity of C = 0.1 μF. Thus τ = 10,000 Ohm * 0.0000001 F = 0.0001 s. Our τ = 1ms. But it should be remembered that τ shows how much time the capacitor C is charged at 37% of the supply voltage. According to the transitional characteristics of the RC circuit, the full charge of the capacitor occurs over a time interval equal to 5 * τ. Then 5 * 1 ms = 5 ms, this time is enough for the natural disappearance of contact bounce, which means that the parameters of the RC circuit are selected correctly. The resistors are turned on directly to the power supply and to the pin of the microcontroller, and the capacitor is turned on parallel to the button, this creates an RC circuit symmetrical in all directions, with all the states of the buttons. For the buttons SB3, SB4, a different approach to the software processing of clicks occurs, so it was decided to install circuit protection against bounce.

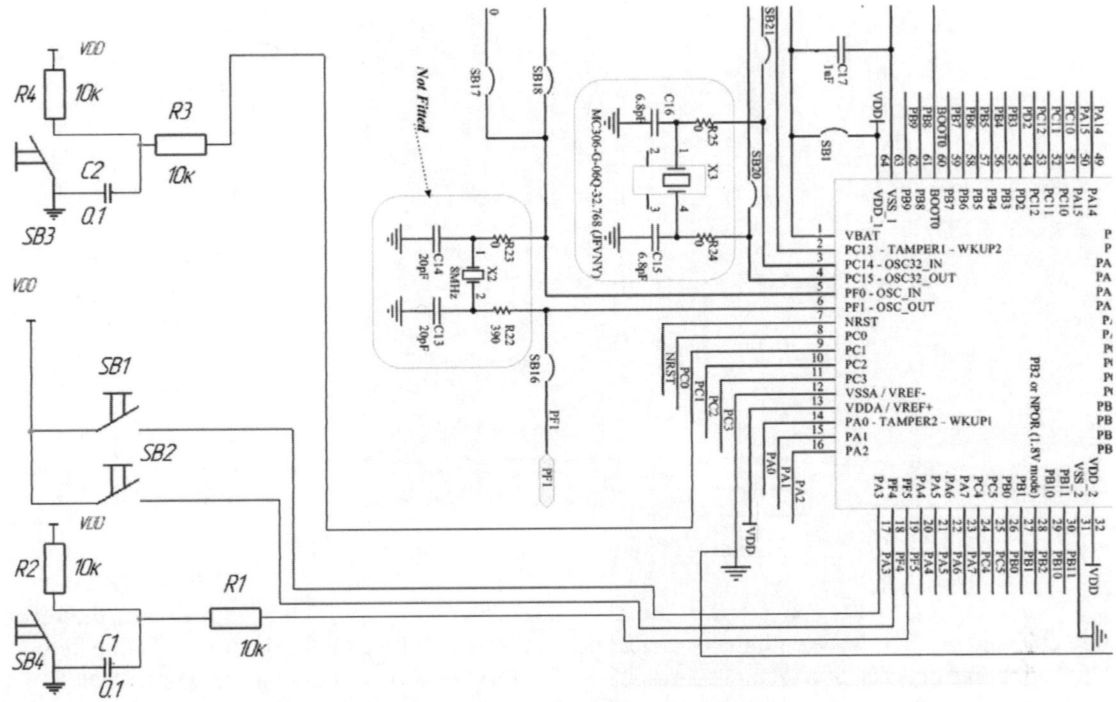

Figure 6. Control panel connection.

The design of the functional automation scheme (hereinafter FAS) was made in accordance with GOST (All-Union State Standard) 21.404–85 [5]. The main control device is a microcontroller [6]. Means of switching, control buttons are shown by the corresponding designation in accordance with GOST 21.404-85. The actuators of our system are geared motors. Trolley control is reduced to the management of engines. Referring to Figure 7 and Figure 8, consider the FAS system.

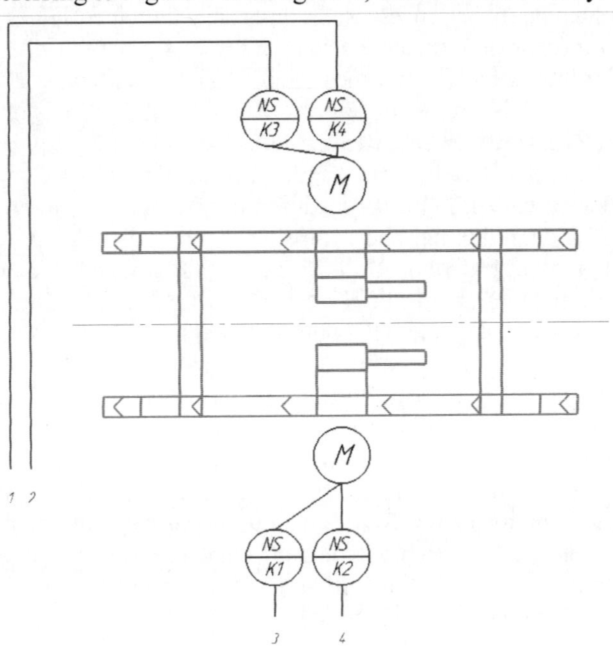

Figure 7. A fragment of FAS "in place".

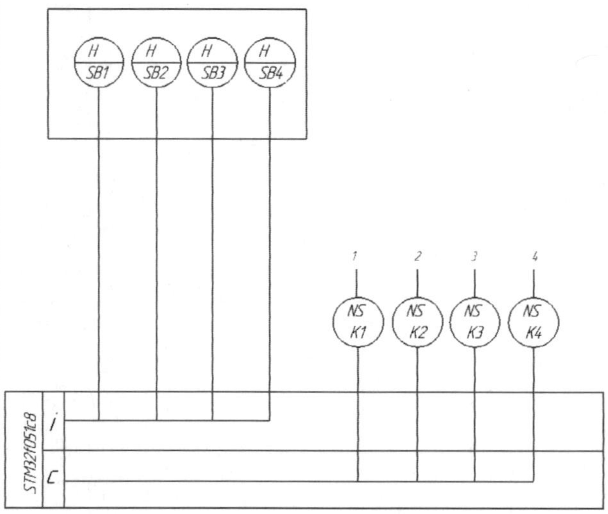

Figure 8. A fragment of FAS "On the shield".

The control line "c" is connected to the elements of the NS relay. The designation of the relay is supplemented in accordance with the electrical principle scheme both on the device and in the field. The registration line consists of a remote control of 4 H buttons, these buttons are also complemented by the designation according to the circuit diagram. The functional connection between the control panel, the controller and the actuators with the switches is clearly traced.

The basis of the algorithm is alternately switching the relay. The relay switches using GPIO, it follows that the system will be controlled by switching the state of the microcontroller's legs. The main part of the program is a cyclical polling of the buttons SB1, SB2, and then sending commands to the microcontroller's legs, but before that it is necessary to initialize the periphery of the microcontroller. According to the polling results of the button, the program either continues to poll other buttons further and execute other commands, or it processes the pressing, incrementing or decrementing the uint8_t numOfGear variable transmission indicator. Then follows a simple but effective bounce protection in the form of a delay of 100ms, which is 20 times longer than the maximum bounce time.

According to the results of the survey of buttons, the controller decides on the choice of the mode of operation of the engine. Since the implementation of the code takes place in the C language, these actions are best carried out with switch - case - break statements. If the numOfGear variable is zero, then the setNeutralGear procedure is called, in this procedure all the legs of the microcontroller are switched to the logical zero mode. By default, when the controller is turned on, this program will be set. Consider the other modes of operation of the robot. If the value numOfGear = 1, then the setFirstGear procedure is called. This procedure includes engines in the forward driving mode. Similarly, the trolley is moving backwards, we simply invert the state of all the output legs. The setReverseGear procedure is responsible for this. Let's now consider a special type of movement - torsion around its axis. In order for it to appear it is necessary that the variable numOfGear be equal to 2. At this point in time, one track moves forward and the second backward, thereby creating a circular motion in place.

Considered all modes of movement forward, backward, parking and turning on the spot. Now we describe the control of turning right and left. To do this, we will take a completely different approach to controlling the pressing of buttons SB3, SB4. This approach does not consist in cyclic polling, but in handling external interrupts. Set up interrupts for triggering on the falling edge. We describe the function of the external interrupt handler. There is a software, handwritten flag - the variable uint8_t sideFlag, it records the number of button presses. The external interrupt system in the STM32 family is divided into lines.

The SB3 button is on line 1, and the SB4 button is on line 5. By interrupting, when entering the handler, we parse the interrupt on which line, and then increment or decrement the sideFlag variable.

There is a common global practice to make interrupt handlers as short and as less resource intensive as possible so as not to miss another interrupt and not fall into a hardfault.

4. Results

Figure 9 shows testing a system with a microcontroller, a relay, and a geared motor. Checking the display of high emissions during engine shutdowns, and therefore it is recommended that low-voltage circuits be made as short as possible and kept as far as possible from the engines in order to avoid interference.

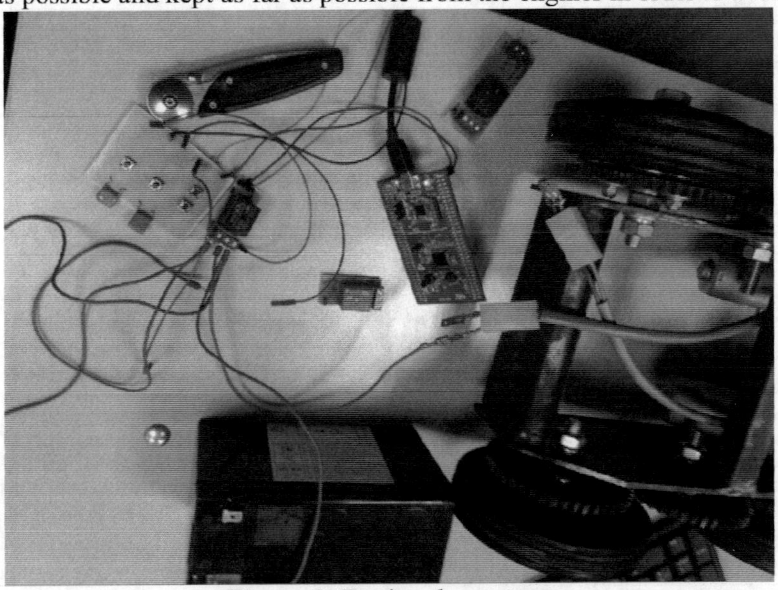

Figure 9. Testing the system.

In this work, a robot with a tracked course was designed, manufactured and programmed. The management program was written in IDE IAR Workbench 8.20, and initialized in the Cube MX program [7]. The drawings were made in the program Compass 16, and the design of the circuitry of the additional product and its printed circuit board in the program Altium Designer 17 [8]. This robot can be easily modified for a wide range of tasks.

References

[1] Kamaev V.A. *et al.* 2013 *Proceedings of the Volgograd State Technical University* **24** (127)
[2] Hu C., Tam S. C., Hsu F. C., Ko P. K., Chan T. Y., Terrill K. W. 1985 *IEEE Transactions on Electron Devices* **32(2)** 375–385
[3] STM32F0, S. ARM-based 32-bit MCU with 16 to 64-KB Flash, timers. ADC, communication interfaces, 2-4.
[4] Apkhudov T.M., Shekikhacheva L. Z. 2015 *Symbol of science* **(7-1)**
[5] Bourjault A. 1988 *Methodology of assembly automation: A new approach. Robotics and Factories of the Future'87* (Springer: Berlin, Heidelberg)
[6] Lan Gongjin, Kai He, Duru Xu. 2012 Research on motion controller for SCARA robot based on STM32. Recent Advances in Computer Science and Information Engineering (Springer: Berlin, Heidelberg)
[7] Düll M. *et al.* 2015 *Designs, Codes and Cryptography* **77(2–3)** 493-514
[8] Rahman M. S. A., Subhas C. M., Pak-Lam Yu. 2014 *Novel Sensors for Food Inspection: Modelling, Fabrication and Expe rimentation* (Springer:Cham)

IOP Conf. Series: Materials Science and Engineering **516** (2019) 012003 doi:10.1088/1757-899X/516/1/012003

Robotic quality control system for the components with complex shape

Y A Salchak, V U Zhvyrblya, I O Bolotina, A M Lider, D A Koneva, D A Sednev

Tomsk Polytechnic University, Tomsk, Russia
E-mail: vzhrb3001@gmail.com

Abstract. This paper represents the concept of an intelligent system for manufacturing components with the complex shape that combines automated technological operation and non-destructive testing modules. The engineering solution for development of the automated ultrasound tomography module is discussed, which is the key element for quality assessment of the components at both early and final stages of the manufacturing process. The proposed solution includes a robotic manipulator to perform scanning of products of any size and shape, as well as the use of a specialized positioning table to achieve accurate testing results. The proposed automation solution for the ultrasound tomography module is performed as a three-level system, including development of an automated workplace for the operator, which provides interconnection between every module included in the manufacturing process.

1. Introduction

Taking into account growing manufacturing demands and industrial design complication the technological processes require constant improvement. One of such advances is considered to be automation of different technological stages. It is vital for different industries such as textile, automotive, chemical and others [1–3]. Automation is considered to be most pressing for the mechanical and metalworking manufacturing industries. In addition to the increase of the technological process efficiency by reducing number of the manufacturing stages, automation also allows to improve the quality of the final components via implementation of advanced testing techniques [4].

Metalworking manufacturing industry is a segment of mechanical engineering complex which plays the crucial role in Russian manufacturing market, embracing fabrication of components for application in various industrial fields, including critical areas such as military industry, mining, nuclear energy, and others. This conditions toughening the quality requirements for the produced components.

At the same time, components with specific structure, such as large dimensions and complex shape, are often used in these critical areas. This not only affects the manufacturing process, but also complicates the following quality assessment performance.

Thus, a new concept of smart manufacturing was introduced as a response to these challenges [5]. It provides for industrial automation by implementation of advanced technological and information technology solutions that allow to optimize the manufacturing procedure.

One of the most important steps to ensure the effectiveness of the proposed automation of the manufacturing is the initial determination of the factors that may lead to the loss of quality of the final components. For the mechanical manufacturing these factors include: low initial quality of the workpiece; inaccurate settings of the final component characteristics; and failures in the technological process.

In this paper we discuss manufacturing of the critical components with complex shape. In particular, shutoff and regulating pipeline components are considered. Manufacturing of these products is primarily associated with the use of metalworking equipment, such as milling machines. The object of study is made of steel by casting, and further complex geometry is achieved by milling.

Milling process has improved significantly over the past decades. At present, application of computer numerical control (CNC) equipment is quite common. It is implemented so that the setting of the final component's characteristics is generated via software and represented as a digital file. Further, this file is used to determine the parameters for the mechanical treatment. At that stage the accuracy of the final component characteristics is vital, as it may influence the reliability of further mechanical treatment.

For this purpose, this paper suggests development of an automated workplace for the operator (AWO), which will enable the use of special software. To achieve the highest accuracy, it is suggested to determine the characteristics of the final products by creating appropriate CAD models and further CNC based on these models. It should be noted that the important function of the developed software is to provide the option of manufactured items registration and storage of all the information regarding them, including specified CAD models. Thus, further necessary adjustments may be possible.

The main advantage of the proposed solution is enabling the quality control of the components throughout the entire manufacturing process, as well as providing the option to adjust the parameters before the start of manufacturing process. In order to achieve high quality and reliability of the manufactured pipeline components, it is proposed to create modules for non-destructive testing (NDT) at various stages of the production process. Each of them is responsible for ensuring the reliability of a particular stage and will be based on the implementation of physical principles. Further to that, within the framework of smart manufacturing concept implementation, integration of these modules by the means of the developed AWO is considered. Therefore, it will be possible to conduct integrated automation of the entire manufacturing process.

2. The concept of implementation of advanced measuring and information technology systems into the manufacturing process of complex shaped components

As part of the study, modernization and improvement of the existing manufacturing process of complex shaped components is proposed. When creating an automated system, it is necessary to take into account the existing features of technological processes to enable gradual automation, while developing a new smart manufacturing system starting not from zero but rather performing smooth step-by-step transformation. This approach is considered to be more realistic, rather than development of a new design, in which technological processes are developed entirely anew, and not on the basis of existing industrial facilities.

At that, as mentioned earlier, one of the main aspects is effective quality assessment of the components. Inspections are to be conducted at the final stage of the technological process as well as at the initial and early stages of the process to assess the quality of the workpieces thus enabling the quality monitoring during treatment to eliminate appearance of any possible defects caused by machining processing. Such approach requires structuring the entire process into several modules, while providing the option for interaction among them for further optimization and acceleration, at the same time minimizing human intervention and applying the principles of digitalization.

There are the following modules proposed:

— X-ray testing module.

— Ultrasonic testing module (UT testing module).

— Optical testing module.

— Automated machining module.

— Automated workplace for the operator (AWO).

The existing manufacturing cycle stages with the modules integrated are represented in Figure 1.

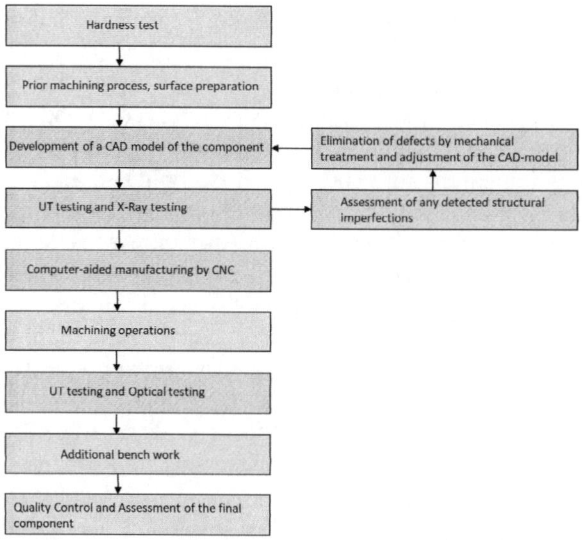

Figure 1. Modified manufacturing cycle diagram.

Each block of the diagram requires direct human intervention which significantly increases labour costs and consequently the total production costs as well as the probability of defects occurrence. The first step to transition to smart manufacturing is development of an automated workplace for managing all manufacturing stages concurrently.

The X-ray and UT testing modules are designed to assess the component in terms of the presence of any defects in its internal structure. Moreover, the UT testing module allows to control the quality of the final component. The implementation of these two modules will enable to determine the size and location of a defect with very high accuracy. Further, this information will enable CNC parameters adjustment to provide better settings of the machining path for the efficient use of the workpiece while avoiding the detected structural defects. More detailed description of the proposed solutions for the discussed modules is given in [6].

This paper discusses in more detail the engineering solutions of the regulating system and the automated workplace of the UT testing module, as the most important part of quality assurance of the component during the entire manufacturing process. The design of the UT testing module, its technical characteristics, and the structure of the its regulation system are described further; prospective implementation of information technology solutions is discussed.

3. Automated ultrasonic testing module

The structure of the automated regulation system (ARS) of the UT testing module, based on ultrasound tomography, includes:

— Ultrasound tomography system, including the robotic manipulator.
— Electrical actuator for the positioning.
— Controller for collecting, transmitting and managing data.
— The control system for the coupling water control.
— Automated work place for the operator.

Figure 2 represents the block diagram of the ARS.

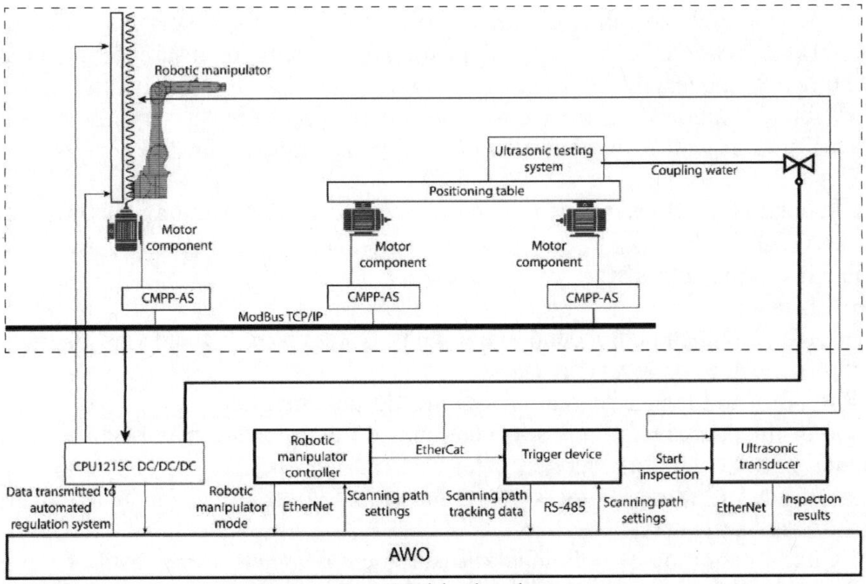

Figure 2. ARS block scheme.

The interacting parts of the UT testing module are the following: a turning table with the device for center adjustment and positioning of the controlled object; robotic manipulator with fixed ultrasonic transducer and acoustic coupling system; robotic manipulator positioning module; coupling water handling system.

The structural diagram of the interaction of these parts is represented in Figure 3.

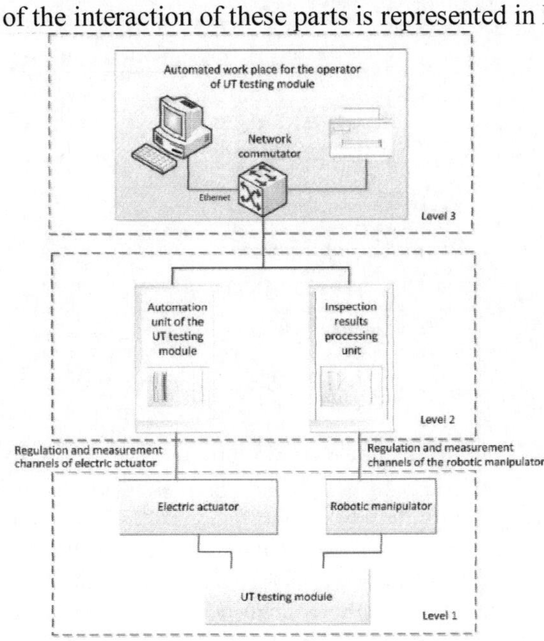

Figure 3. Three-level interaction system of the ARS of the UT testing module.

Operation process regulation, control of the technological equipment and quality assessment performance regulation are carried out at three levels:

1) Local automation (positioning table and robotic manipulator controllers) provides specified table rotations with a strictly fixed component positioned in the center as well as vertical movement of an

industrial manipulator with an ultrasonic transducer and an acoustic coupling system placed on it. Automated control of rotations of the table and the robotic manipulator ensure the movements according to the control program sequence.
2) Controllers for data collecting, transmitting and managing connect servodrives of the electrical actuator with the PLC, as well as the controller of the robotic manipulator. They are used for transmitting the data via:
- industrial Ethernet (TCP / IP exchange protocol), which forms the data transmission channel between controllers and AWO;
- control channels of the manipulator;
- drive control channels.
3) The regulation of AWO and ultrasound tomography system is performed via Ethernet network.
Interaction of these regulation levels provides:
- high reliability of the automated regulation systems during operation;
- simplification of the procedures for assessment and self-assessment, maintenance of most important specific channels in hot operational mode;
- development of automated regulation systems individually due to the modular replacement at each level of software and hardware;
- compatibility and integration of individual measurement channels at the level of the programmable logic controller in the ARS.

The operational mode of the ARS is determined by the setting of the quality inspection program. The operational diagram of the automated regulation system of the UT testing module is presented in Figure 4.

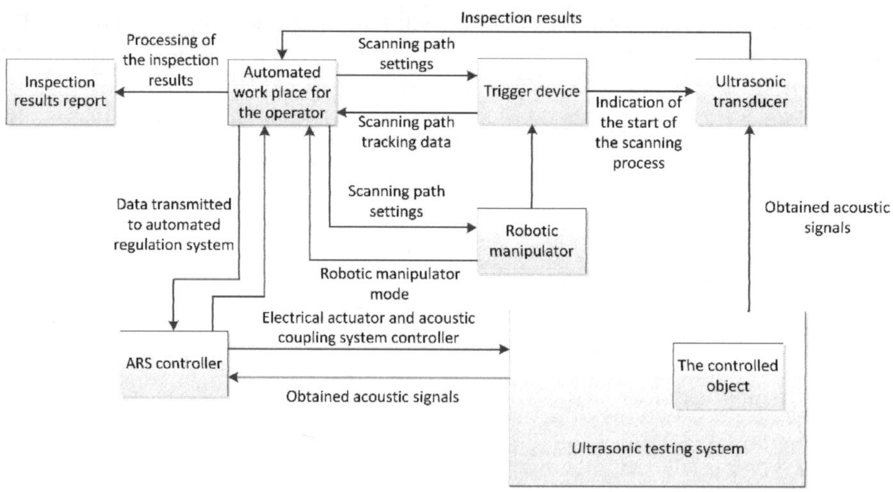

Figure 4. Operational diagram of the automated regulation system of the UT testing module.

The controller of the external automated regulation system of the automated ultrasound testing module for the components operates the electric drives of UT testing module using the data of the external automation. The selected electric drives for regulation of the mobile components of the UT testing module perform positioning of the robotic manipulator with specified accuracy. The scanning path settings are provided by the robotic manipulator movements, the latter is regulated by the robotic manipulator controller. The electric drives of the positioning table enable rotations of the component.

The control of the movements is regulated through the encoder signals. The operator's automated workplace receives the data about the positioning and its correspondence to the control program sequence settings. The control and measurement units are synchronized with the special Ethernet communication channel. The AWO allows processing of the scanning results in accordance with the defect qualifications and generating the inspection protocol. The scanning results are also provided in real time as three-dimensional reconstruction of the internal structure of the component, which allows the operator to conduct overall assessment of the quality. The reconstruction and visualization is performed as ultrasonic tomography based on digital focused array technique (DFA) developed by TPU in collaboration with Fraunhofer Institute [7]. Figure 5 represents the UT testing module, as well as an example of the obtained inspection results.

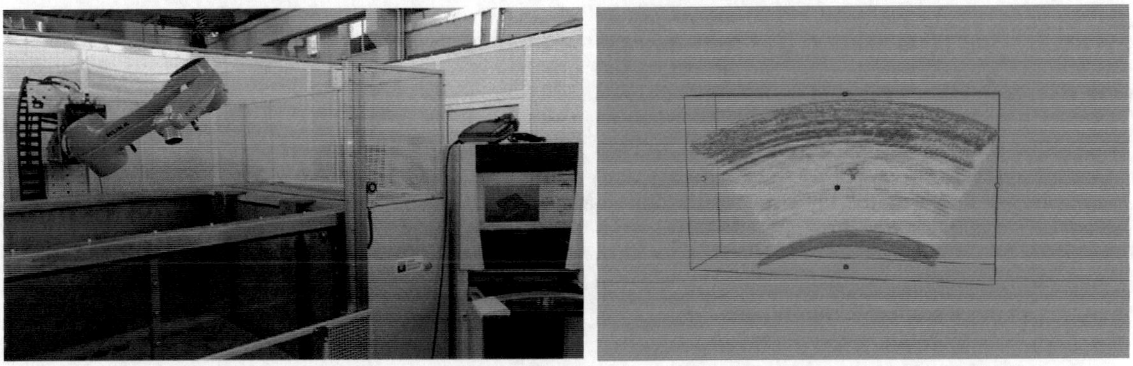

Figure 5. An example of ultrasound tomography results using a robotic manipulator.

4. Summary

The discussed concept allows to create smart manufacturing system which increases the quality of the fabricated components with complex shape by implementation of advanced hardware and software solutions for non-destructive testing and regulation of the technological process stages. The automation solution for the ultrasonic testing module, which is a part of the system, is described in more detail. This module enables to conduct ultrasound tomography inspections of the workpiece, as well as of the final component. The use of a robotic manipulator, while providing interaction through the automated regulation channels and enabling processing of the obtained results and further adjustment of machining of the component has been discussed in this paper. This will significantly improve the technological processes efficiency, minimize the risk of failures, and reduce the probability of manufacturing defects.

Acknowledgments

The study is financially supported by the Ministry of Science and Higher Education of the Russian Federation within the Federal Target Program entitled "R&D in Priority Areas of Development of Russia's Scientific and Technological Complex for 2014–2020" (unique project ID: RFMEFI57817X0251).

References

[1] Tsai W. 2018 *Energies* **11(8)** 2072
[2] Peng T., Zhou B. 2018 *Assembly Automation* **38(3)** 347–360
[3] Naghib S., Di Maio F., De Bartolo L. *et al* 2018 *Journal of Chemical Technology & Biotechnology* **93(3)** 710–719
[4] Schmidt M., Spieth H., Haubach C. *et al* 2019 *100 Pioneers in Efficient Resource Management* (Springer Spektrum: Berlin, Heidelberg)
[5] Kang H., Lee J., Choi S. *et al* 2016 *International Journal of Precision Engineering and Manufacturing-Green Technology* **3(1)** 111–128
[6] Salchak Y., Kotelnikov A., Sednev D. *et al* 2018 *IOP Conference Series: Materials Science and Engineering* **327(2)** 022091
[7] Bolotina I., Bulavinov A., Lider A. *et al* 2015 *IOP Conference Series: Materials Science and Engineering* **81(1)** 012073

IOP Conf. Series: Materials Science and Engineering **516** (2019) 012004 doi:10.1088/1757-899X/516/1/012004

Investigation of additional losses in optical fibers under mechanical action

A V Yurchenko[1], A D Mekhtiyev[1,2], F N Bulatbaev[2], Y G Neshina[1], A D Alkina[1], P Sh Madi[1]

[1] National Research Tomsk Polytechnic University, Tomsk, Russia
[2] Karaganda State Technical University, Karaganda, Kazakhstan
E-mail: niipp@inbox.ru

Abstract. The relevance using of high-performance information and measurement systems based on fiber optic sensors to create systems for monitoring the geotechnical state of the rock mass is an extremely promising issue. This is a new generation of high-speed measurement monitoring systems that are able to function effectively in conditions that are dangerous due to a sudden explosion of methane gas as they do not transmit electrical signals and underground mines are one of the most difficult and tough environments for people's work. The authors substantiate the idea of using the optical fiber of the ITU-T G.652.D standard to create systems for monitoring the geotechnical state of a rock mass since the fiber-optic sensors developed on its basis have a sufficiently high accuracy, measurement speed and have a good linearity of characteristics. For testing control methods and measurement of geotechnical parameters of output the imitation laboratory standing on the basis of fiber-optical sensors was developed. Quartz single-mode optical fiber 9/125 µm (OS2) Corning SMF-28e+® was used. The authors conducted a series of experiments to study additional losses due to mechanical action on optical fiber.

1. Introduction

Safety of mining operations is an important aspect and priority of the complex mine system.

The first stage of development of information-measuring system based on fiber-optic sensors (IIS FOS) is an analytical study of existing achievements in this area based on a systematic approach and the solution of the problem as a whole. The results of the literature analysis are allowed to use the accumulated experience for the development of information-measuring system based on fiber-optic sensors that can work effectively in mines dangerous for sudden release of methane and dust.

During the operation of optical cables there are certain technical aspects associated with the occurrence of mechanical loads that can lead to bending of the optical waveguide which in turn causes signal attenuation and additional losses.

Laying fiber-optic communication lines there are often situations associated with the influence of external influences that affect the technical characteristics of the optical cable and the bandwidth of the telecommunication channel. The most important fact as practice has shown is the influence of the bending parameters of optical fibers and the occurrence of additional losses in the radiation power of the signal source in the optical cable.

These losses are especially high at such a bending radius, which is critical when the fiber conductor is on the verge of mechanical damage. This critical radius can be very small (only a few millimeters) for fibers with a high numerical aperture, while the allowable bending radius is much larger (often tens of centimeters) for fibers in single-mode mode with a large cross-mode area.

It was completed the scientific analysis of similar works of foreign authors who work with optical fiber and develop fiber-optic sensors [1–3].

This is a new generation of high-speed measurement monitoring systems that are able to function effectively in conditions that are dangerous due to a sudden explosion of methane gas, as they do not transmit electrical signals, and underground mines are one of the most difficult and tough environments for people's work. Safety of mining operations is an important aspect and priority of the complex system of the mine.

The principal difference between the idea and existing analogues in the world is the integration of functions in one device with the possibility of automated monitoring, in real-time technical condition of mine workings on the basis of fiber-optic sensors (FOS) with which a multifunctional measurement system with high metrological accuracy. This system based on optical fiber provides wireless transmission of information from the sensors to the module of the automated control system has no analogues. A fiber-optic sensor is used which is a small device in size and allows detecting changes in various values of electrical and non-electrical parameters, which allows real-time tracking of deviations of the object under study.

2. Laboratory tests

For working out of methods of control and measurement of geotechnical parameters of developments the imitation laboratory stand on the basis of fiber-optical sensors which is presented in Figure 1 was developed.

Quartz single-mode optical fiber 9/125 μm (OS2) Corning SMF-28e+® with low "water peak" (ITU-t standard-652) was used.(D). It is not desirable to use fiber with Ultra series as it has a lower sensitivity to bending. The fiber has a primary coating of 245 microns (with an outer shell). To determine the values of optical radiation power and losses the VIAVI optical power meter (JDSU) SmartPocket OLP-38 was used. It operates in the dynamic range from -50 to +26 dBm with a wavelength range of 780-1650 nm. As a radiation source was used SmartPocket OLS-34/35/36 with built in options Auto-λ and Multi-λ, SmartPocket OLP-38 can automatically measure the power level and insertion loss in single-mode and multimode fiber optic cable. The connection to the optical fiber was carried out using a universal 2.5 mm UPP adapter and optical connectors FC.

The length of the compensation coil is 2 km of optical fiber (ITU-t standard 652.(D).

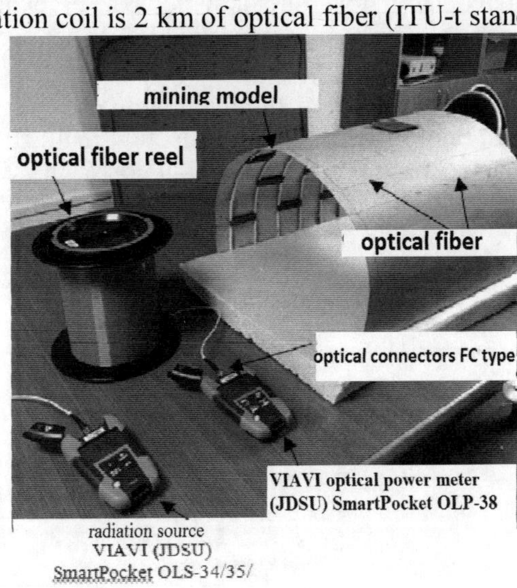

Figure 1. Simulation laboratory stand.

With the help of the developed laboratory stand, a number of experiments were carried out to determine the optical fiber losses at different pressure values. Quartz single-mode optical fiber 9/125 μm (OS2) Corning SMF-28e+®was used. Also, modeling and calculation were performed with the help of finite element analysis in ANSYSv18 program. The General view of the model is shown in Figure 2.

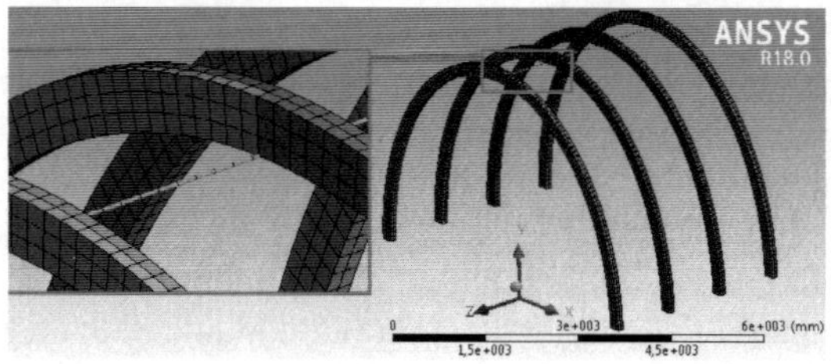

Figure 2. General view of the model performed in the ANSYSv18 software environment.

The frame structure is under the pressure created by the rock. The calculation is carried out in 10 steps, at each subsequent step the pressure increases by 1 MPa, taking the initial value of 1 MPa and the final -10 MPa.

Fiber optic sensors have a number of advantages: the possibility of multiplexing, remote measurement, resistance to electromagnetic interference, the lack of electricity at the point of measurement, long-term stability.

We present a number of technical advantages of FOS in relation to traditional electronic systems of measurement and monitoring, through which FOS is planned to be used in mining IMS:

- safe use in explosive environment, complete intrinsic safety;
- elasticity with a minimum bending radius of 2 mm;
- wide signal bandwidth and the ability to multiplex signals;
- low weight and overall dimensions, high reliability and corrosion resistance;
- excellent parameters of noise immunity and non-inductance of channels, low energy costs for operation [4–6].

3. Full-scale modeling

With the help of the developed laboratory stand, a number of experiments were carried out to determine the optical fiber losses at different pressure values. When exposed to the fiber optic meter changes the output form of the light spot.

The numerical aperture (NA) is the sinus of the maximum angle of incidence of the light beam at the fiber end, at which the condition of total internal reflection is fulfilled. This parameter determines the number of modes propagating in the optical fiber [7–9]. Also the value of the numerical aperture affects the accuracy with which the optical fibers should be docked with each other and with other components of the line. Figure 3 shows a diagram of the change in the deformation of the light spot forms.

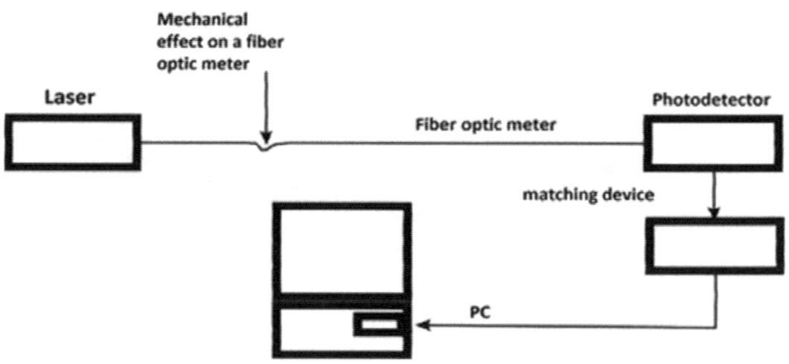

Figure 3. Diagram of deformation change in the shape of a light spot.

The aperture before and after the application of pressure on the optical fiber can be seen in Figure 4, which presents a diagram of the deformation change in the shape of the light spot. The camera captures the change in the shape of the spot and transmits the information to the computer where the program compares with the samples.

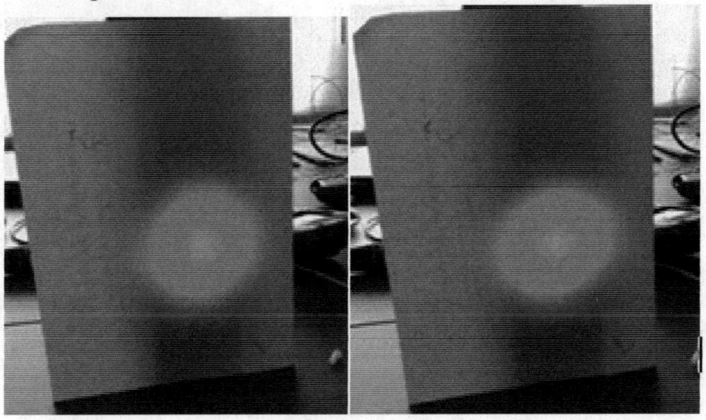

Figure 4. Aperture before and after applying pressure to the optical fiber.

The aperture shows that when the optical cable is deformed the light spot changes its shape.

The value of additional losses and optical refractive indices in the optical fiber depends on the level of mechanical action and temperature which can be expressed through the following dependence:

$$\Delta q/q=1/q\ (dq/qT)\ k\ \Delta T + \delta q/q$$

where q – additional loss of the optical fiber;

Δq – change of additional losses;

$(dq/qT)k$– is a partial derivative of the temperature dependence of optical fiber losses;

δq – is the variation of additional losses due to the photoelasticity method.

$1 / q\ (dq/qT)$ the first part of the equation takes into account the changes in the additional losses arising in the optical fiber under mechanical action, and part δq/q photoelasticity effect due to fiber deformation [10–14].

Change of additional losses due to photoelasticity δq. We use the famous formula using coefficients of Pockels R:

$$\Delta q= -q^2/2\ (\rho_{11} \cdot\sigma_1 + \rho_{21} \cdot\sigma_2 + \rho_{12} \cdot\sigma_z),$$

where σ_1 and σ_2 – the value of the relative deformation occurring in the cross section of the optical fiber;

σ_z-– the value of the relative deformation occurring in the cross section d along the fiber axis;

$$\sigma_z = \Delta l/l,$$

where Δl – change in the length of the fiber segment with length l.

$$\sigma_1 = \sigma_2 = -\nu\ \sigma_z,$$

where ν - is the value of the Poisson's ratio.

To register the excess of a given level of deformation, and hence the appearance of additional losses during deformation of the rock mass and the structure of the shaft support within 0.5%, a certain value of the change in the refractive index and the change in the cross section of the optical fiber, namely its core, due to a slight deformation due to photoelasticity equal to δq=9.6·10-4. With the known speed of propagation of the optical signal:

$$V=c/n,$$

where v– is the propagation velocity of the optical signal in the fiber core;

s – speed of light in vacuum;

n – is the refractive index of the optical fiber core.

IOP Conf. Series: Materials Science and Engineering **516** (2019) 012004 doi:10.1088/1757-899X/516/1/012004

If you take the length of the optical fiber used for measurement, conditional length of 1 km, with the deformation within 0.5%, will cause a change in the optical signal delay. For example if the length of the measuring transducer section is l=1 km, then due to the deformation of 0.3%, the change in the optical signal delay in this section will be about 3 NS. By increasing the length of the control measurement area using a fiber-optic light guide with a length of 5 km with the already self-oscillating mode of operation can be 5 μs. This allows you to get the difference in the accumulation of pulses. The pulse accumulation time is 30 seconds. The change in the delay time in the optical fiber within 0.16 NS will be allowed to obtain a difference in the number of accumulated pulses equal to 2.

To increase the sensitivity of the measurement system, it is possible to increase the length of the optical fiber and the pulse accumulation time. The dynamic range of the strain measurement system will depend on: the sensitivity of the photodetector the power of the radiation source, the optical length of the measuring part, the value of the reflected power, namely the level of mechanical action and the formed radius of local bending.

It is possible to determine the dynamic range of the meter, with a minimum backscatter signal. The backscattering will be the reverse optical pulse reflected from the bending zone under mechanical action on the optical fiber:

$$q > q_m + \alpha_l \cdot l,$$

where q–is the reflection loss of one optical pulse;

q_m – losses per reflection of one optical pulse at the output end of a fiber optic conductor;

l– is the length of the fiber optic meter.

The dynamic range W can be calculated by the formula:

$$W = W_0 - 2n \cdot l \cdot \alpha_l - n\, q,\ dB,$$

where W_0 – is the energy reserve of the sensor.

$$R_i/P_{range};$$

R_i – pulse power of optical radiation source;

P_{range} – sensitivity digital image sensors, dynamic range;

n – number of optical connectors in the meter circuit;

α_{oc} – optical losses introduced by the meter;

L – is the length of the fiber optic meter;

α_l – value of linear attenuation of the pulse in the fiber-optic meter;

q – reflection loss per optical connection.

4. Discussion and conclusion

Under mechanical action on the optical fiber there are micro and macro bends, leading to additional losses of the optical signal in the fiber. These losses can be measured and pressure values can be set on the optical fiber, and the displacement value can also be determined. An important advantage of this measurement system is its complete spark and fire safety. The energy passivity of the sensors is allowed to exclude power sources that are directly in the area of coal mining. Information-measuring systems of fiber-optic sensors are able to monitor mining operations around the clock.

One of our experiments was the appearance of additional losses due to mechanical action on the optical fiber. When exposed to the fiber optic meter changes the output form of the light spot.

Measurement of rock pressure and rock displacements - are two important parameters that can be used to quantify the effectiveness of support for the roof of mining in a given state of geoengineering.

The use of optical fiber standard ITU-T G. 652.D for the control of the coal mine rock mass is very promising since the fiber-optic sensors developed on its basis have a sufficiently high accuracy, speed of measurement and have a good linearity of characteristics. It is not desirable to use fiber with Ultra series as it has a lower sensitivity to bending. Taking into account the above mentioned advantages of FOS our work is aimed at the creation of high-performance IMS based on fiber-optic sensors with improved metrological and operational characteristics for subsequent implementation at various enterprises of Kazakhstan.

References

[1] Liu X. *et al* 2009 *ACTA Photonica Sinica* **38** 112–114 doi:10.3390/s16101759
[2] Kumar A. *et al* 2011 *International Journal of Control and Automation* **3** 63–70 doi: 10.1007/s12404-010-0401-3
[3] Naruse H. *et al* 2007 *Measurement Science and Technology* **18**(10) 3202–3210 doi: 10.1088/0957-0233/18/10/S23
[4] Yurchenko A.V. *et al* 2018 *Russian Journal of Nondestructive Testing* **7** 61–66 doi: 10.1134/S1061
[5] Gorlov N., Eyrikh V.I. 2012 *Proceedings of the International scientific conference* **2** 68-–70
[6] Yurchenko A.V. *et al* 2016 *IOP Science. Journal of Physics* **671** doi:10:1088/1742-6596/671/1/012001
[7] Chotchayev H.O. 2016 *Geology and Geophysics of Southern Russia* **3** 129–140
[8] Buimistryuk G.Ya. 2011 *Foton-Express* **6(43)** 38–39
[9] Buimistryuk G. 2013 *Control engineering Russia* **3(45)** 34–40
[10] Kim S.T. *et al* 2015 *Sensors (Switzerland)* **15(1)** 1060–1070 doi: 10.3390/s150101060
[11] Liu T. *et al* 2013 *Proceedings of the 2013 International Conference on Microwave and Photonics* 102–111
[12] Yiming Zh. *et al* 2016 *Sensors (Switzerland)* **16(10)** 112–117 doi: 10.3390/s16101759
[13] Kovalev P.V. *et al* 2016 *Metalurgija* **55(4)** 715–718
[14] Issagulov A.Z. *et al* 2016 *Metalurgija* **55(3)** 426–428

IOP Conf. Series: Materials Science and Engineering **516** (2019) 012005 doi:10.1088/1757-899X/516/1/012005

Numerical modelling of levitating particles in air

D Sukhanov, S Roslyakov

Tomsk State University, Tomsk, Russia
E-mail: sdy@mail.tsu.ru

Abstract. A numerical model for the particles levitation in an ultrasonic field is developed on the basis of the finite difference time domain method. The force effect of the ultrasonic field on the particles is calculated via gradient of the Gor'kov potential, also the friction force in gases for spherical particles is taken into account. With the help of the OpenCL technology, a parallel computation of the trajectories of a set of particles is implemented.

1. Introduction

The control of particles in the ultrasound field has recently received considerable attention. The technology of acoustic levitation of particles in the air has prospects for use as contactless tweezers for manipulating small objects. Such methods will be used in 3D printing, medicines manufacturing, for the manufacture of complex chemical composites. In the work of Gor'kov, an expression for radial pressure on a small spherical particle is given [1]. In work [2] is presented a technology of manipulation by levitating particles using phased arrays of ultrasonic sources. Acoustic field can cause such effects as attraction and repulsion [3]. Acoustic effects on particles attract the attention of researchers [4–7]. Numerical simulation of acoustic levitation is important part preliminary to experimental research. In [8] has been proposed a single frequency numerical method to calculate potential for acoustic radiation force on the basis of matrix representation.

We propose a numerical model of levitating particles in gases optimized for parallel computing on the basis of finite difference time domain method. The simulated ultrasonic field is represented as a set of harmonic waves and stored in memory as a set of complex coefficients.

2. Mathematical model

To predict the behavior of particles in the ultrasonic field, it is necessary to conduct a numerical simulation of the effect of the ultrasonic field on the particles.

The calculation of the force acting on a particle in an ultrasonic field can be accomplished using the Gor'kov formula [1]:

$$\mathbf{F} = -\nabla U , \tag{1}$$

where $U = 2\pi R^3 \rho \left(\frac{\overline{p^2}}{3\rho^2 c^2} f_1 - \frac{\overline{v^2}}{2} f_2 \right)$ – Gor'kov potential, $f_1 = 1 - \frac{c^2 \rho}{c_0^2 \rho_0}$, $f_2 = 2\frac{\rho_0 - \rho}{2\rho_0 + \rho}$, ρ – density environment (air), ρ_0 – particle density, p – acoustic pressure, v – acoustic velocity, $\overline{p^2}$ – average of field pressure square, $\overline{v^2}$ – average of field velocity square.

The particle trajectory is modeled based on the equations of motion: $\mathbf{r}(t+\Delta t)=\mathbf{r}(t)+\mathbf{v}(t)\Delta t+\mathbf{a}\Delta t^2/2$ – coordinates of the particles in the next time, $\mathbf{v}(t+\Delta t)=\mathbf{v}(t)+\mathbf{a}\Delta t$ – particle velocity, $\mathbf{a}=\left(\mathbf{F}+\mathbf{F}_{fr}+\mathbf{F}_{gr}\right)/m$ – particle acceleration, m – particle mass, $\mathbf{F}_{fr}$ – friction force, $\mathbf{F}_{gr}$ – gravity force.

The speed potential of wave field is defined as a set of harmonic waves:

$$\Phi(\mathbf{r},t)=\sum_n A_n(\mathbf{r})\cos(\omega_n t+\varphi_n(\mathbf{r})), \tag{2}$$

where ω_n – cyclic frequency of nth harmonic, $A_n(\mathbf{r})$ – magnitude of wave field at point $\mathbf{r}$, $\varphi_n(\mathbf{r})$ – phase of wave field at point $\mathbf{r}$. For each wave defined magnitude $A_n(\mathbf{r})$ and phase $\varphi_n(\mathbf{r})$ in different points of media by solving Helmholz equation for frequency ω_n. To minimize computational efforts for field calculation we propose to calculate field pressure and velocity only at points of particles location. Pressure of acoustic field is calculated via speed potential by following operation: $p(\mathbf{r},t)=-\rho\frac{\partial\Phi(\mathbf{r},t)}{\partial t}=\rho\sum_n A_n(\mathbf{r})\omega_n\sin(\omega_n t+\varphi_n(\mathbf{r}))$. The speed is defined as: $\mathbf{v}=\nabla\Phi$, numerically we calculated it by expression:

$$\begin{aligned}\mathbf{v}(\mathbf{r},t)= \; & \mathbf{x}_0\frac{\Phi(\mathbf{r}+\mathbf{x}_0\Delta x,t)-\Phi(\mathbf{r}-\mathbf{x}_0\Delta x,t)}{2\Delta x}+\\ & \mathbf{y}_0\frac{\Phi(\mathbf{r}+\mathbf{y}_0\Delta y,t)-\Phi(\mathbf{r}-\mathbf{y}_0\Delta y,t)}{2\Delta y}+\\ & \mathbf{z}_0\frac{\Phi(\mathbf{r}+\mathbf{z}_0\Delta y,t)-\Phi(\mathbf{r}-\mathbf{z}_0\Delta y,t)}{2\Delta z},\end{aligned}$$

where $\mathbf{x}_0=(1,0,0)$, $\mathbf{y}_0=(1,0,0)$, $\mathbf{z}_0=(1,0,0)$ – orthogonal unit vectors, Δx, Δy, Δz – spatial steps of field calculation for x, y and z axis respectively. To calculate force acting on the particle via formula (1) it is required to evaluate gradient operator that requires definition of field in 6 points for numerical computation of symmetrical spatial derivative. Overall, to find acoustic force acting on the particle it is required to calculate $\Phi(\mathbf{r},t)$ at 19 different points, where $\mathbf{r}$ – is the coordinates of the particle.

We used the OpenCL technology for the software implementation of the algorithm for calculating the trajectory of levitating particles in the ultrasonic field. This technology allows you to perform parallel computing on the set of cores of the graphics card processor. Consider the possibility of launching M parallel threads; then, to calculate the trajectories of N particles, it will be necessary to calculate the positions of N/M different particles in each thread. Arrays containing data on the coordinates and velocities of particles are stored in the graphics card's memory to reduce data exchange with the central processor. The ultrasound field is defined as three-dimensional arrays of complex numbers for various frequencies, which are loaded into the memory of a video card before the simulation of particle motion begins.

3. Numerical modeling

For numerical test we propose to compute particles dynamics in wave field of special configuration (Figure 1). We consider that source field is generated by planar phased array at frequency 22 kHz and allows forming the focused wave field of the letter "R" shape at distance 20 cm.

Under the focusing plane we placed planar reflector to create standing wave. Particles should levitate at a quarter of a wavelength above the reflector and concentrate around local maximums of field.

Software was developed based on the OpenCL parallel programming technology for simultaneous calculation of the trajectory of many particles in an ultrasonic field. Considered source field is shown on Figure 2*a*, and resulting focused field above the reflector is shown on Figure 2*b*. The result of particle trajectories modeling show on Figure 3.

IOP Conf. Series: Materials Science and Engineering **516** (2019) 012005 doi:10.1088/1757-899X/516/1/012005

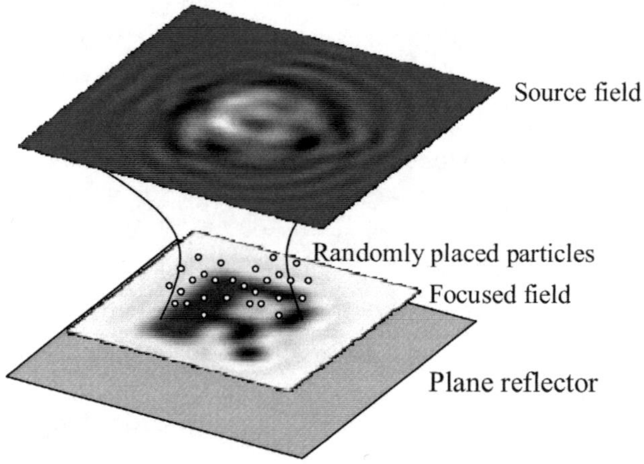

Figure 1. Considered numerical test of particle levitation.

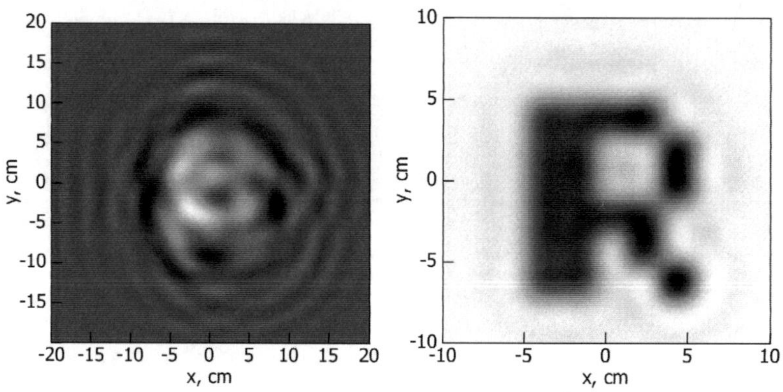

a b

Figure 2. Speed potential field of plane phased array source : a – at the source plane; b – at focus plane.

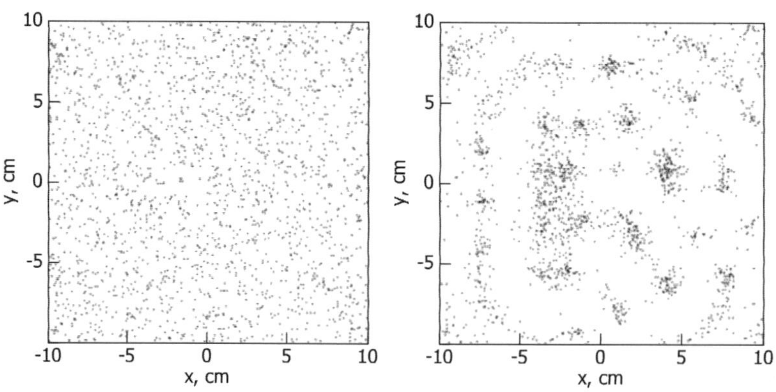

a b

Figure 3. The initial distribution of particles (a) and after 2 s (b) in an ultrasonic field with a frequency of 22 kHz and an amplitude of 1.2 kPa.

Modeling of levitation of plastic particles with density 1050 kg/m^3 and diameter of 100 to 1000 μm was carried out in the considered ultrasonic field with a frequency of 22 kHz in air with a speed of sound of 343 m/s. On the vertical axis, the field has a maximum corresponds to focusing distance that equal to 20 cm, which should provide levitation in the field of gravity. The developed program made it possible to calculate the positions of 2048 particles with a random initial coordinate distribution (Figure 3*a*) and to obtain the position of the particles after 2 s (Figure 3*b*).

We can observe concentration of particles into letter "R" shape. Particles tend to local field maxima, which does not allow the formation of images with homogeneous regions. From here we can conclude that a single array of emitters is not enough to form an arbitrary image. The behavior of the particles does not contradict to theoretical expectations.

4. Conclusion
A numerical model of particle levitation in an acoustic field has been developed suitable for parallel computation. We proposed to represent an ultrasonic field as a set of harmonic oscillations in the whole modeling space, and further calculate the time dependence of the field only at the points where the particles are located. This approach saves the memory required for storing a three-dimensional field at different points in time. The calculation of the acoustic force acting on the particles requires the determination of the field at 19 neighboring points. The field is found by summing the harmonic components of waves, taking into account their amplitudes and phases. The performed simulation has shown results that are consistent with theoretical expectations.

Acknowledgment
The scientific research was carried out with the support of the grant of the Russian Science Foundation No. 17-79-20051.

References
[1] Gor'kov L.P., *On the forces acting on a small particle placed int an acoustic field in an ideal liquid*, Sov. Phys. Dokl., 6(9), 773-775 (1962)
[2] Yoichi Ochiai *et al* 2014 PLOS ONE **9**(7) e102525 doi: 10.1371/journal.pone.0102525
[3] Mitri F.G. 2017 *Annals of Physics* **386** 1–14 doi: 10.1016/j.aop.2017.08.023
[4] Settnes M., Bruus H. 2012 *Physical review E.* **85** 016327 doi: 10.1103/PhysRevE.85.016327
[5] Diego Baresch *et al* 2013 *The Journal of the Acoustical Society of America* **133** doi:10.1121/1.4770256.
[6] Shilei Liu *et al Sensors* 2017 **17** 1664 doi: 10.3390/s17071664
[7] Muller P. B. *et al* 2012 The Royal Society of Chemistry doi: 10.1039/c2lc40612h
[8] Andrade M.A., Perez N., Buiochi F., Adamowski J. 2011 *IEEE Transactions on Ultrasonics Ferroelectrics and Frequency Control* **58(8)** 1674 – 1683 doi: 10.1109/TUFFC.2011.1995

IOP Conf. Series: Materials Science and Engineering **516** (2019) 012006 doi:10.1088/1757-899X/516/1/012006

A mechanism for the formation of unconditioned reflexes

V I Syryamkin, V N Shumilov

National Research Tomsk State University, Tomsk, Russia
E-mail: vnshumilov@rambler.ru

Abstract. The paper considers a possible mechanism for the formation of unconditioned reflexes in organisms with the nervous system before their birth, that is, before the environmental impact on a certain organism. The effect of the environment is detected by receptors. It generates signals coming from the receptors into the nervous system, which organizes the organism's reaction to the environmental effect. Inter-neuronal connections, which carry out unconditioned reflexes, realize input of signals about the environmental effect from the receptors in the nervous system.

1. Introduction

There are unconditioned and conditioned reflexes. Conditioned reflexes are produced by organisms during their own lives and may be replaced by new reflexes if necessary. Conductive connections are formed between excited neurons. Let us discuss the beginning of the formation of conditioned reflexes, when neurons appear in the developing organism and there are no connections between these neurons. Because of the lack of connections, neurons cannot be excited, and therefore, connection between them cannot be formed. The development of the nervous system (the formation of conditioned reflexes) begins with unconditioned reflexes that arise in the organism unconditionally regardless of the environmental effect on it [1].

Unconditioned reflexes are normally formed in all organisms of a given species before the birth of the organism. At the beginning of life, neurons of the nervous system still have no connections with other neurons, through which they could be excited. That is, neurons cannot be excited without connections at their input; they cannot react to events, and therefore, cannot record traces of events.

The formation and development of reflexes (chains of cooperative neurons) to a certain extent is similar to the appearance and development of frost patterns on a clean windowpane. These patterns begin to develop from some place (usually from an edge with favorable conditions to the pattern initiation) and grow along the edge of the pattern in the direction of cold air movement. On new places of free space (in isolation from the previous pattern), new patterns are formed rarely because of absence of initiating basis for them [2, 3].

How are unconditioned reflexes formed? They are formed in the course of evolution in all average organisms and are formed in each individual organism during its development in the mother's womb or in the egg.

Evolution is the result of two main processes - stability and variability. Stability is provided by replication (self-copying of body cells). Variability is caused by mutations. Evolution of the species is a natural development of organisms' line, their adaptation to the environment (there is always something to improve). In addition, the environment may change, and organisms must also adapt to these environmental changes. Mutations occur at the level of initial (maternal) sell when the life of each organism is initiated, at the moment of conception of each new embryo. If the mutation proves to be useful, its results contribute further to the survival and expansion of a growing and mature organism, that is, to the continuation of the organism and its descendants with the properties caused by mutations. This means that the successful mutation contributes to the survival and expansion of the species. An organism that carries a harmful mutation is more likely to die without producing descendants. Evolution occurs with the survival and expansion of organisms that are best adopted to the current environmental conditions (existing at the given time in a given location).

2. Unconditioned reflexes are the basis of the personal development

Organisms of sufficiently highly developed species are born or come out of the egg with a set of unconditioned reflexes, which form the basis for the further development of the nervous system of organisms [4].

In the organism, the variety of possible useful processes and mechanisms on their basis is limited because of the limited amount of information contained in the original maternal cell. Therefore, we can conclude that the mechanism for forming connections between neurons that make up the chain of unconditioned reflex cannot significantly differ from the mechanism of forming conditioned reflexes because such a difference would require the existence of additional mechanisms [5].

The information capacity of the initial cell is limited; therefore, processes occurring during the cell development in the embryo and during the further development of the embryo can only be the simplest and natural in the emerging conditions.

Unconditioned reflexes are short and simple chains of neurons from receptors to effectors. Connections are formed between neurons that realize the unconditioned reflex. To generate short reflex chains, less information (conditions) is required. The mechanism for forming connections that realize unconditioned reflexes cannot significantly differ from the mechanism of forming connections between neurons that were excited by signals from other excited neurons through their input connections. However, at first, neurons do not have input connections, therefore, they can be excited only by means of almost random self-excitation.

By conducting signals about the environmental effect on the receptors, unconditioned reflexes provide the possibility of excitation of neurons, between which associative connections are formed during the life of the organism. Random self-excitations of neurons in mature organisms do not happen very often. With such excitations, random connections can form. There is little benefit in these random connections between self-excited neurons. Such connections do not reflect anything; therefore, their benefit is only in expanding the possibilities for forming connections between a group of neurons (to which a neuron joins as a result of the neuron connection after its self-excitation) with other neurons. In the beginning, such connection does not correspond to any particular event.

For the further development of the nervous system (formation of associative connections), it is necessary that interneurons can be connected to the reflex chain [6, 7]. Because of the presence of connections between unconditioned reflexes, signals initiated in the receptors enter the nervous system, where neuronal chains begin to form; these chains reflect traces of events effecting the organism and fix the organism's experience. Figure 1 shows the processes of the nervous system functioning.

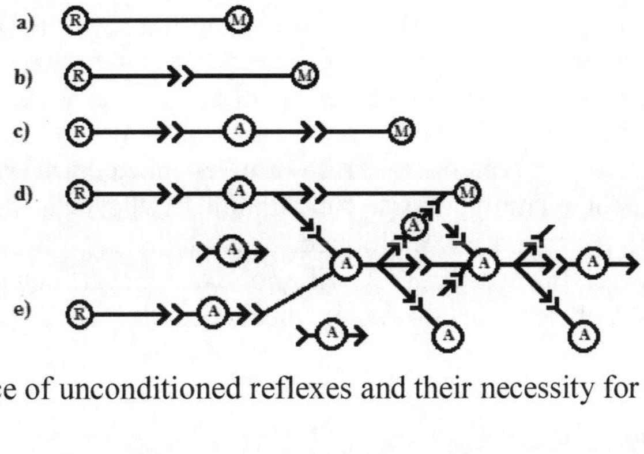

Figure 1. Appearance of unconditioned reflexes and their necessity for the nervous system.

IOP Conf. Series: Materials Science and Engineering **516** (2019) 012006 doi:10.1088/1757-899X/516/1/012006

In this Figure, *R* is receptors, *M* is manipulators (effectors), *A* is associative neurons.

a) Combined receptor and effector (manipulator) in the shortest chain of unconditioned reflexes.

b) Separate receptor and effector with increasing distance between them.

c) Appearance of interneurons allowing increasing distance from the receptor to the effector.

d) Connection of the advanced information processing systems that provide forming of complex conditioned reflexes to effector control.

e) Other receptors and still unused neurons A.

It is clear that the main condition for the formation of connections between neurons is their neighborhood in space and mechanical contact between them, which provide *an opportunity* for the formation of connections between them. The neighborhood of neurons is provided by the development of nerve tissues, when during the course of the organism's development, new cells are formed as a result of the division of the earlier formed cells starting from the original cell.

The mechanism for the formation of unconditioned reflexes is the same as the mechanism for the formation of conditioned reflexes. The difference is that self-excitations of neurons occur much more often at the age of the embryo than in an adult organism. The specific weight of neuron self-excitations is also much higher. This is because of the fact that the potentials and threshold values of the neuron excitation have slightly different values at the youngest age of the organism than in the adult organism. Moreover, during the development of the organism in the maternal womb, the excitation of various neurons occurs less frequently than in an adult organism with a large number of neuronal connections, and therefore the neuron's readiness for excitation increases to the maximum. In this period, the threshold level of neuron excitation is much lower because of the rarer neuron excitations. Therefore, during the period of intrauterine growth, self-excitation of neurons often occurs, which ensures the formation of connections between neurons realizing unconditioned reflexes, even in conditions of the absence of initiating neuronal stimuli.

The formation of traces of events results in accumulation of the organism's experience and formation of conditioned reflexes that provide an adequate response of the organism to current events by using various manipulators of the organism (limb muscles and glands producing necessary substances). Reflexes in the nervous system are realized as a set of neurons united in a single control sequence through information (associative) connections between neurons. These connections are formed under the influence of the environment in previous episodes.

During the period of intrauterine development, the organism cannot yet be fully affected by stimuli, which will appear only after the birth of the organism. At first, in the developing organism, there are still no neurons, connections between them, and connections of receptors with the subsequent neurons. Gradually, in the developing organism the number of neurons and connections between them grows. The question arises. How are associative connections between neurons formed if they do not yet have input connections for excitation of neurons by the initial stimuli?

Differences in the levels of neurons excitation in different organisms (along with other factors) determine the psychological type and the temperament of a person - some people are more excitable, others are more inert. Since the threshold levels affect the excitation of individual neurons, they also affect the formation of connections between neurons and the memory features of individuals.

The threshold of neuron excitation of specific organisms can be adjusted by appropriate nutrition or by taking special medications.

The set of unconditioned reflexes in different organisms can vary within some limits. Some reflexes can be expressed clearly, while others may be weak or nonexistent.

Thus, if for some reason (for example, because of unbalanced nutrition) the desired nutrients do not enter neurons in sufficient quantity, the self-excitation of the neurons will be less frequent than necessary for forming all necessary unconditioned reflexes. As a result, the embryo, which has already begun to develop (not yet born organism) will not have a sufficient basis for the formation of conditioned reflexes, which are necessary for the successful development and expansion of the organism.

IOP Conf. Series: Materials Science and Engineering **516** (2019) 012006 doi:10.1088/1757-899X/516/1/012006

The absence of any secondary unconditioned reflexes does not have a significant effect on the development of the child's organism. For example, conjunctival and scleral reflexes, which are often absent normally, do not have much significance for the development of the nervous system.

Functionality and the utility of the majority of unconditioned reflexes, which are formed during the intrauterine life of the organism, is obvious. At the same time, some unconditioned reflexes do not bring any visible benefit to the body. There are also other secondary effects (phenomena) outside the nervous system in various mature organisms. For example, the hair on various parts of the human body. Some of these signs are atavisms and are manifested through many generations from those times when these signs were useful. Some signs are purely secondary effects; they continue to persist in the species, although they are useless from them. At the same time, they are not harmful. More likely, lines of organisms with negative properties died because they did not stand up to collisions with the dangers during life. Therefore, organisms with negative features did not really advance into the future. Just secondary features did not facilitate but also did not impede the survival and expansion of the species [8, 9].

The mechanisms of both conditioned and hundreds of unconditioned reflexes are of the same nature. Therefore, unconditioned reflexes, which are not involved or used rarely in the postnatal period of the organism's life, fade away with time in the same way as conditioned reflexes, which do not receive frequent repetitions (reinforcements).

The absence of any standard unconditioned reflexes (observed in average statistical organisms of this species) indicates that in this organism there were no conditions for forming a complete set of unconditioned reflexes ensuring a normal successful development of the organism. What is the reason for insufficient development of unconditioned reflexes?

Sometimes the inadequate development of unconditioned reflexes is due to mutations in the organism when the organism is conceived, when the formation of a changed set of unconditioned reflexes is predetermined by genetic changes during the initiation of the organism (embryo). Another reason for the lack (underdevelopment) of certain unconditioned reflexes during the intrauterine development of the organism is the absence of conditions for forming missing connections between neurons. Such conditions may be, for example, in the lack of necessary nutrients and microelements resulting in an increase in the threshold excitation potential of the neuron, or in too high connecting potential even in case of contact between excited neurons.

The absence of any secondary unconditioned reflexes does not have a significant effect on the development of the child's organism. Moreover, even the complete absence of the sense organ (the provider of information to the brain), including organs of hearing or even sight, which are the most informative, does not deprive a person of the opportunity to become a full member of the human society. On the other hand, feral children who have lived isolated from human contact from a very young age (being raised by animals (monkeys or wolves)) [10], grew up with the psyche of animals and with a set of acquired reflexes inherent in the animal community that sheltered the child.

There are cases when people found such children ('Mowgli'), took them away from their already habitual environment and tried to accustom them to life in the human environment. However, this never led to the complete adaptation and socialization of such children because during the period of forming foundations for perception of the world ("the period of laying the foundation for an individual") such children lived with animals. Therefore, they grew up with habits, customs, and values of animal communities that sheltered them.

This proves that, although the person's personality is formed on the physiological basis of the available capabilities of the organism and its nervous system, which ensures the **delivery of information** to the brain, the resultant behavior of the organism and personality is influenced by the personal experience accumulated in the organism during its existence. The personal experience means the **information received** by the body from the outside world and acquired by it, that is, the values gained as a result of education.

IOP Conf. Series: Materials Science and Engineering **516** (2019) 012006 doi:10.1088/1757-899X/516/1/012006

3. Formation of conditioned reflexes in the prenatal period

Some scientific studies provide data that children of alcoholics are prone to alcoholism, but if such children are put with a good family, then they grow as normal people (Blakemore and Cooper, 1970). This proves that the mother's alcoholism during the period of intrauterine development results not in "alcoholic" genes, which are transferred to the newborn child, but in alcohol dependence, which develop in the alcoholic mother' womb. This dependence (the totality of reflexes) can be suppressed and replaced by education and by creating right reflexes in a good family.

4. Conclusion

The work considers the formation of unconditioned reflexes in the perinatal period (before the birth of the organism), which provide the simplest necessary reactions of the organism to the environmental effect. These reflexes are not a characteristic feature of a specific organism, but of the whole species of animals, to which this organism belongs. Reflexes are generated by the connections between the corresponding neurons that conduct signals about the environmental effect on the organism from receptors to effectors that perform an active response to the environmental effect. The number of connections forming unconditioned reflexes is relatively small because the amount of information about the structure of the future organism is small. This information is in a single initial maternal cell in a mediated form. In addition to the obvious function of transferring signals from receptors to effectors, chains of connections between neurons, which carry out an unconditioned reflex, provide input of signals from receptors to the nervous system consisting of a set of neurons that conduct signals and fix traces of events - the effects of the environment.

Acknowledgement

The paper was supported by the competitiveness improvement programme of National Research Tomsk State University and grant of the Russian Foundation for Basic Research No 16-29-04388.

The authors are grateful to Tatiana B. Rumyantseva from Tomsk State University for English language editing.

References:

[1] Shumilov V. 2015 *Principles of the brain functioning* (Publishing house of Tomsk State University: Tomsk)

[2] Shumilov V. 2016 *The way we think. Mechanisms of the brain functioning* (Lap Lambert)

[3] Voronin V., Shumilov V., Shumilov Yu. *et al* 2015 *Neurocomputers: development and application* **4** 16360

[4] Johnson M., Munakata Y., Gilmore R. 2002 *Brain Development and Cognition* (Blackwell Publishers Ltd: Oxford)

[5] Shumilov V., Shumilov I., Syryamkin V. 2015 *Neurocomputers: development and application* **4** 16390

[6] Lie D., Song H., Colamarino S. *et al* 2004 *Annual Review of Pharmacology and Toxicology* **44** 399—421 doi: 10.1146/annurev.pharmtox.44.101802.121631

[7] Hofman M., Boer G., Someren E. *et al* 2002 *Plasticity in the Adult Brain: From Genes to Neurotherapy* (Elsevier Science)

[8] Shumilov V. 2018 *MATEC Web Conf* **155** 01050 doi: 10.1051/matecconf/201815501050

[9] Shumilov V, Syryamkin V, Syryamkin M. 2015 *AIP Conference Proceedings* **1688** 040007 doi: 10.1063/1.4936040

[10] Daw N., Wyatt H. 1976 *J. Physiol* (257) 155–170 URL: https://physoc.onlinelibrary.wiley.com/doi/pdf/10.1113/jphysiol.1976.sp011361

IOP Conf. Series: Materials Science and Engineering **516** (2019) 012007 doi:10.1088/1757-899X/516/1/012007

Software tools for web mapping systems

A Kadochnikov[1], N Shaparev[1,2], A Tokarev[1], O Yakubailik[1]

[1]Institute of Computational Modelling SB RAS, Krasnoyarsk, Russia
[2]National Research Tomsk State University, Tomsk, Russia
E-mail: oleg@icm.krasn.ru

Abstract. Technologies and software components developed for geoinformation web systems and geoportal services are considered. The program and user interfaces, implementation features, the general architecture of the projected systems are discussed. In particular, we consider a subsystem for visualizing the web map data, a CSW directory service, a style editor for geoportal maps and layers. Examples of completed developments and experience of implementation are given.

1. Introduction

Development of geoportals and their various components, i.e. cartographic web interfaces, distributed storage systems for spatial data, associated metadata management systems – are one of the modern trends in the development of geoinformation technologies [1, 2]. The need to create a similar class systems is caused by the large number of distributed information resources having different formats, requiring different conditions for maintenance and access; all these factors complicate greatly their search, classification, and distribution [3, 4].

The geoportal of the Institute of Computational Modeling of the Siberian Branch of the Russian Academy of Sciences (ICM SB RAS) is the unity of software tools, a set of software, technology and information and computing support, focused on the rapid development of the applied geoinformation web systems. Created in the first edition about 10 years ago, it is constantly developing, getting improved in the framework of the basic fundamental and applied research, with partial support of integration projects of the SB RAS and various scientific foundations, in the implementation of commercial custom information systems. It has successfully proved itself as a software and technological basis for resource-intensive information and analytical systems of the regional level for tasks of various subjects, ex. sectoral management informational support (in the sphere of health and education), environmental monitoring and the state of the environment assessment, forecast of the region's social and economic development, centralized information support with cartographic data, and so on [5, 6].

The wide range of tasks performed by this system, which are related to processing of large volumes of cartographic information and provision of tools for information interaction between users, requires the system components not only to have a developed functional part, but also effective and high-quality user interfaces [7].

This scientific work represents a number of software and technology solutions implemented for the geoportal recently.

2. Geoportal system architecture

From a technological point of view, the geoportal includes the following components [8, 9]:

- a repository of information (thematic) resources, i.e. file server with geospatial data in popular GIS formats, geospatial data bases, etc.,
- a catalogue of the registered resources, i.e. database of metadata of all information resources of the created system, as well as a set of software libraries (APIs) for various processing operations,
- an administration system to register resources in a directory, to manage the access rights, to enter and edit data and metadata, to set up classifiers, etc.,
- an editor of style design, i.e. a specialized software for thematic coloring of layers and maps, etc.

Within the framework of the objective aimed at geoportal creation, the ICM SB RAS identified the task of designing and developing software tools and technological solutions, based on the web services of the ICM SB RAS geoportal and designed to administer the geoportal, including creation and edition of thematic style of individual layers and maps, as a composition of individual geospatial data layers [10].

The general characteristics of the developed toolkit functionality:

- connection to a geo-information web server, getting the list of cartographic resources,
- thematic design of the geospatial data layers available on the portal,
- saving styles in the portal database,
- formation of maps from individual geospatial data layers,
- creation of new cartographic resources,
- automation of thematic design,
- management of publications,
- editing the metainformation of the resources available on the portal.

A generalized scheme of geospatial information processing, which is implemented on the geoportal, is shown in Figure 1.

3. Subsystem for web map data visualization

The generated map of a geo-information web monitoring system can contain a set of raster and vector thematic geospatial data layers, additional reference layers of images from below (on background) and from above (on top) of the user's thematic data, such as the own cartographic base maps, maps of third-party services, annotations of geographical objects. The client software provides interactivity of the created maps in a standard web browser. The data presentation configuration provides the ability to select the geodatabase format on the client (vector or raster), to control various display options (interactive legend, list of fields displayed in attribute tables), to use custom templates for displaying the table information according to the layer objects, to manage the data access rights, from the ability to render a layer or map through the web browser, to access spatial data through the WMS / WFS cartographic web services, and editing via web interface, or downloading as ArcGIS shape files. An integral part of thematic web maps based on the geoportal is a customized rasterization and data caching service that provides a noticeable performance increase for "heavy" multi-layer maps. A good addition to maps on the geoportal is a set of cartographic base maps, i.e. map-schemas, mosaics of satellite images, digital terrain models, and geospatial data of the third-party web services.

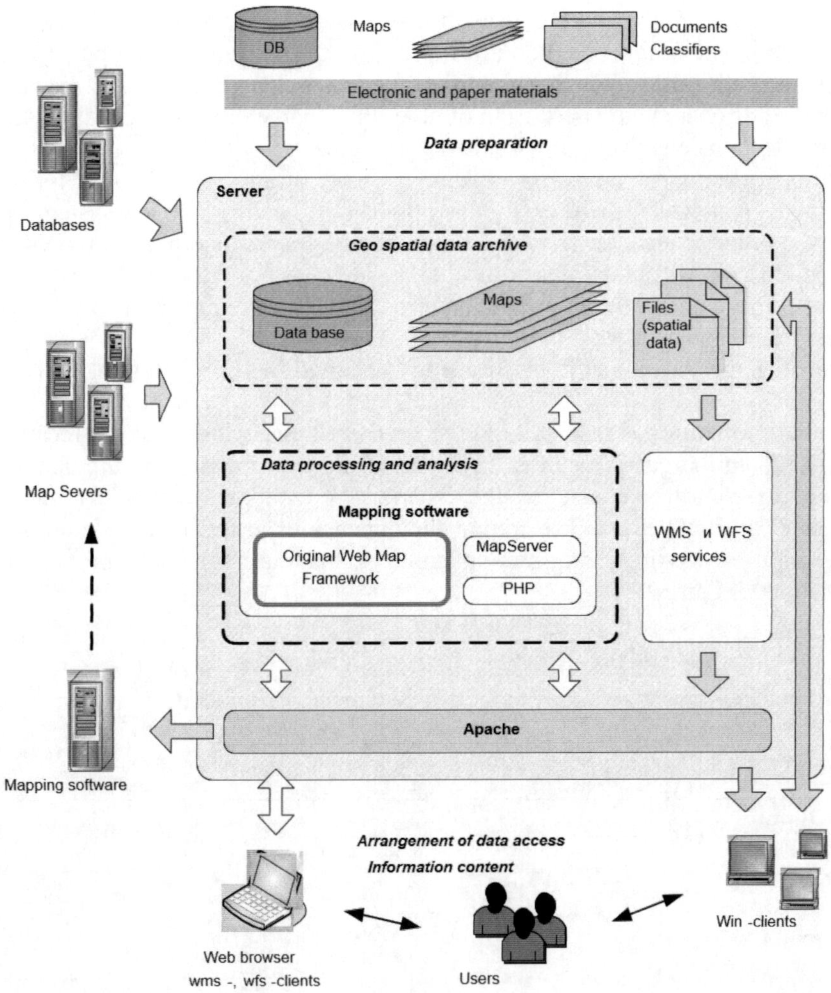

Figure 1. Software architecture of geospatial information processing system.

The first block of the modernized software, developed during the reporting period, is connected with the visualization of data on objects of an interactive web map. Geoportal tools allow to view information in a pop-up window in HTML format, when prompted on a web map with a mouse click. The information can be obtained:

- according to the objects of a separate geoportal layer;
- according to the objects of some or all geoportal map layers;
- according to the coordinate, with the help of the informational geoportal service "address search".

In early geoportal versions, the developer of the map web application controlled visualization of the object data when the user requested it. A universal form of an information pop-up window was created, which made it possible to display the attributive information of objects in an accessible form, regardless of its content [11]. A mechanism of the data visualization according to the interactive web map objects based on the template system and external services was added to the current geoportal version. The new system features allow the operator, creating geoportal maps and layers for the end user, to form the necessary type of result of information according to the layer objects. The first method allows the user to control the output of information about the objects with the help of TWIG templates (http://twig.sensiolabs.org/). TWIG is a compiling template handler with open source code written in the PHP programming language. Templates allow to change the order and form of the

IOP Conf. Series: Materials Science and Engineering **516** (2019) 012007 doi:10.1088/1757-899X/516/1/012007

attributive data output according to the objects in map layers, including different style design (colors, font parameters, etc.). The second method allows, during the information output, to use an external service via the specified link, to which the attributive data is transferred. Such services will help to create more complex answers with application of other third-party services and databases [12].

The program block mechanism for visualizing the data according to the objects is as follows. Executing an information query on vector objects, an array of attributive data is formed on the web map on its all layers (for which there exists this opportunity, defined by the system operator) in the form of an indexed or associative array. Then, depending on the cartographic web application settings, two following options are possible:

- the attributive information is transferred as an array to the web application, where the HTML-code of the request result is generated;
- based on the attributive data array, the result is generated in HTML format and sent to the web application.

As a result, the information is displayed in the user's browser window at the location, determined by the developer. When the query result is displayed in the pop-up window, the operator can use the additional option to control the size of the information window, which can be necessary and convenient for some attributive data. Previously, the dimensions were calculated automatically, taking into account the window size limits. For pop-up windows, the minimum and maximum allowed widths or heights are defined (Figure 2).

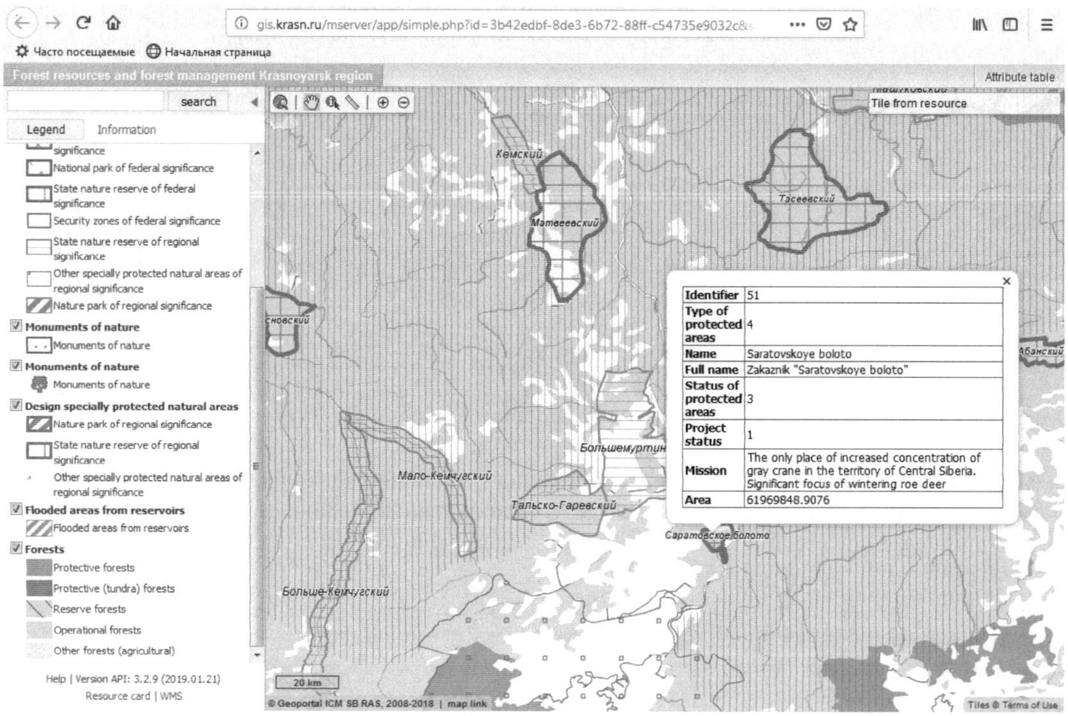

Figure 2. The web-mapping user interface of the geoportal with pop-up info window.

The web map visualization on the user's side is possible in two modes: as a raster image or a set of vector objects, with use of the Web Feature Service (WFS) protocol. The current mechanism of the data visualizing with the help of templates and external services is currently working only in raster image mode. At displayment in vector format, the information about the objects is visualized with the help of a universal form as tooltips.

IOP Conf. Series: Materials Science and Engineering **516** (2019) 012007 doi:10.1088/1757-899X/516/1/012007

4. CSW web directory service

The use of the CSW (Catalog Service for the Web) protocol of the OGC (Open Geospatial Consortium) will significantly expand the application of the ICM SB RAS geoportal, which already contains tools and services for storing, cataloguing, publishing, and downloading of spatial data, searching and filtering on metadata, and interacting web-visualization.

The CSW standard defines the mechanism for searching and publishing metainformation about the spatial data, services and objects associated with them in terms of information. This service provides information exchange in distributed systems and improves the efficiency of geospatial data search by client applications [13, 14]. The catalogue contains records that are encoded in XML format in accordance with one of the metadata storage standards, i.e. ISO 19139, Dublin Core, DIF, FGDC, and Atom. An example of individual metadata elements is the title, format, type (data, data set, service), bounding area on the map, projection, etc. Queries can be sent to the directory in several ways via the HTTP protocol: GET or POST with parameters, POST with XML document. The CSW 2.0.2 specification defines the following operations for the directory:

- *GetCapabilities* – to get metainformation about the service itself.
- *DescribeRecord* – to get a description of the elements of the information model of resources.
- *GetRecords* – to search for entries in the directory using filters.
- *GetRecordById* – to get the catalog entry by its unique identifier.
- *GetDomain* – to get a set of valid values for individual metadata fields.
- *Harvest* – to download metadata from another source (server).
- *Transaction* – to create / edit / delete directory entries.

After analyzing the existing libraries and individual applications for organizing CSW access to the directory, Pycsw solution was chosen for a basis, which meant the full implementation of the OGC CSW server in Python. Based on this, the software was created for the ICM SB RAS geoportal. Pycsw has open source code, it is also distributed under the MIT license and works on all major platforms (Windows, Linux, Mac OS X). Pycsw allows to publish and collect metadata through several API options (CSW, OpenSearch, OAI-PMH, SRU), providing standard-based implementations of the directory service. It supports several metadata formats with transparent conversion – ISO, Dublin Core, DIF, FGDC and Atom. To collect metadata from the external servers, the information model of the geoportal catalog was modified. A new element, i.e. "Catalogue", has been added, which specifies the source address (URL), the frequency of its polling and other settings. After the import is completed, the information element is created in the form of child nodes.

The key elements of the scheme for expanding the ICM SB RAS geoportal are as follows:

- *Resource Directory* is an internal catalogue of geoportal resources with its programming interface via SOAP – CatalogueAPI protocol.
- *Pycsw* is an expanded pycsw application with a database on PostgreSQL and 2 directories, both public and official. The public catalogue is accessible from outside via the CSW protocol and contains meta descriptions of geoportal resources.
- *The synchronization service* is a service providing filling of the public directory pycsw, launch of data collection tasks through the pycsw service directory.
- *External servers* are various sources of spatial data and metadata, with access via WMS, WFS, WMTS, CSW protocols.

Also, a separate data synchronization service was developed. The service runs on schedule and performs 2 main tasks:

1. *Filling the pycsw database*. This service is connected to the geoportal catalogue via the SOAP protocol and crawls all public resources. If any resources were added / changed / deleted, the corresponding operation is performed on a public copy of the pycsw directory. The process is performed incrementally, only those resources that were updated are modified. Metadata is sent in XML format, regulated by the standard ISO 19115: 2003/19139: 2007. If metadata was specified on the geoportal only in a simplified form, an automatic conversion is made.

IOP Conf. Series: Materials Science and Engineering **516** (2019) 012007 doi:10.1088/1757-899X/516/1/012007

2. *Organization of data collection (harvest)*. To accomplish this task, pycsw options are used for importing data from the external sources. A separate, service pycsw directory, unavailable from outside the geoportal, is used here. At next iteration, the synchronization service polls all created items of the "Catalogue" type on the geoportal via CatalogueApi. For new objects, or for those that have time for the next synchronization, the asynchronous query of Harvest to pycsw is executed. As a result, a service pycsw directory is filled with resource metadata from an external source. Further, the synchronization service creates corresponding information resources in the geoportal catalogue.

5. Editor of the style design of geoportal layers and maps

A new style editor of the geoportal thematic maps was created in the web interface. It forms an XML description of the style design of geoportal layers and maps, and stores it in the resource directory. Previously, this task was solved with the help of a separate windows application *GeoExpress*. The implemented new solution simplifies significantly the process of setting the parameters of the spatial data representation on the geoportal.

The implementation is made in the concept of a single-page web application (single-page application, SPA). A single HTML document is used as a shell for all web pages, with organization of interaction with the user through dynamically loaded HTML, CSS, JavaScript via AJAX. All visual interface elements are constructed directly in the browser using JavaScript by manipulating the document's DOM structure. The development was performed using the AngularJS client-side JavaScript framework. The framework works with HTML, containing additional custom attributes, that are described by the directives, and links the input or output of the page region to a model that is a regular JavaScript variables [15]. The values of these variables are set manually or extracted from the static or dynamic JSON data. The library is flexible for choosing design patterns, and models MVC (Model-View-Controller) or MVVM (Model-View-View Model) can also be chosen. The framework adapts and extends traditional HTML to provide two-way data binding for dynamic content, which automatically synchronizes the model and view. As a result, AngularJS reduces the role of DOM manipulations and improves testability [16, 17].

The main web application for geoportal setting up and managing, was updated in the part of styling setting up. Style editors for layers and maps were developed in the form of SPA-applications built into the main user interface [18, 19].

The layer style editor supports several types of design: thematic coloring, pie charts, bar charts, cumulative histograms. Style classes are defined for a range of scales and conditions for attributes of spatial objects. All changes can be immediately viewed on the map in the preview tab.

For the full-fledged functioning of the client part of the style editor, server support is required in the form of a set of services. The following tasks are solved:

- Navigation through the directory resource tree to select a layer. Getting the values of the specified attribute layer field.
- Getting the list of the available libraries of symbols and fonts, as well as their elements.
- Forming an image of a font table and individual characters.
- Preview of a layer or map with the current style settings.
- Projection transformation for an array of points.

6. Implementation experience based on geoportal technologies

The list of selected projects implemented on the basis of this approach:

- "Spatial Data Bank of the Krasnoyarsk Territory" for the Ministry of Informatization and Communication of the Krasnoyarsk Territory;
- "Automated system for monitoring the municipal entities" for the Ministry of Economy and Regional Development of the Krasnoyarsk Territory;

- "Geoinformation system for monitoring the state of the environment in the zone of operation of the oil and gas industry enterprises of the Krasnoyarsk Territory" for the Siberian Federal University and the Ministry of Natural Resources and Forestry of the Krasnoyarsk Territory;
- Dispatch and navigation system for vehicle monitoring based on GLONASS / GPS "REGNASS" satellite data for the Ministry of Transport and Communication of the Krasnoyarsk Territory;
- Geoinformation web system "The network of Krasnoyarsk educational services" for the Ministry of Education and Science of the Krasnoyarsk Territory;
- "The health map of the Krasnoyarsk Territory" for the Ministry of Health of the Krasnoyarsk Territory.

The modular architecture of the systems under consideration, the use of standard web services for data exchange between these modules, ensures the rapid adaptation of the existing software to the customer requirements, the replication of individual components, and their complementarity. In its turn, registration of created resources, i.e. spatial data in the centralized geoportal catalogue, provides the possibility of their simultaneous application in several developments. This service-oriented approach, based on the active implementation of web technologies into application information systems, is increasingly being used now [20].

7. Conclusion

For application geoinformation web systems, based on geoportal, a set of special software has been created. Web based visualization services for cartographic data have been implemented by mouse click on the basis of the compiler template handler, which allows to change the order and form of the attributive data output according to the objects in map layers, including different style design, i.e. color, font parameters, etc. New web services have been created to represent raster data, including data for remote sensing, which have tools for creating thematic maps, managing information requests, building a legend based on the palette. Implemented support by the geoportal of the embedded resources in vector format WFS, support for requests in POST format, created a single URL for accessing to services.

To support the integration of heterogeneous distributed geospatial data in the system of resource cataloguing, based on the ICM SB RAS geoportal, the CSW protocol is supported, which defines the mechanism for searching and publishing meta information about spatial data, services and information objects associated with them.

As a result, the spatial metadata catalogue is provided in a distributed mode, with information transmission (synchronization) between the distributed nodes in the network.

The software for creating and editing the style of thematic maps in the web interface has been created, which greatly simplifies the process of setting the parameters of the spatial data representation on the geoportal.

The above software and technology solutions implemented in a modular service-oriented architecture will help to ensure the effective implementation of new geospatial web applications and services.

References

[1] Smits P. *et al* 2005 *Computers, Environment and Urban Systems* **29** 15–31
[2] Wenwen Li *et al* 2015 *Computers, Environment and Urban Systems* **54** 314–325
[3] Gang Han *et al* 2015 *Journal of Photogrammetry and Remote Sensing* **103** 66–80
[4] Dimitrios Gkatzoflias *et al* 2013 *Computers & Geosciences* **52** 21-–33
[5] Kadochnikov A.A. *et al* 2008 *Journal of Siberian Federal University. Engineering & Technologies* **4** 377–386
[6] Yakubailik O.E. *et al* 2009 *Vestnik of SibGAU* **25** 61–66
[7] Kadochnikov A.A. 2007 *Computational Technologies* **2** 70–78
[8] Rautenbach V. *et al* 2013 *Computers, Environment and Urban Systems* **37**107–120

IOP Conf. Series: Materials Science and Engineering **516** (2019) 012007 doi:10.1088/1757-899X/516/1/012007

[9] Chin-Te Jung, Chih-Hong Sun, May Yuan 2013 *Computers, Environment and Urban Systems* **38** 45–57
[10] Yakubailik O. *et al* 2015 *15th International Multidisciplinary Scientific GeoConference SGEM* 2015 **1** 487–494
[11] Yakubailik O.E. *et al* 2009 *Computational Technologies* **6**116–126
[12] Matveev A.G. *et al* 2013 *Fundamental research* **10-15** 3358–3362
[13] Liping Di *et al* 2016 *Computers & Geosciences* **92** 1–8
[14] Wang H.*et al* 2014 *Expert Systems with Applications* **41** 2362–2371
[15] Michener W. K. 2015 *Ecological Informatics* **29** 33–44
[16] Stefano S. *et al* 2012 *Information Systems* **37** 539–556
[17] Blackburn M. R. *et al* 2015 *Procedia Computer Science* **61** 141–146
[18] Zhang J. 2012 *Ecological Informatics* **8** 68–77
[19] Blower J.D. *et al* 2013 *Environmental Modelling & Software* **47** 218–224
[20] Matveev A.G. *et al* 2012 *Vestnik of SibGAU* **2** 48–54

IOP Conf. Series: Materials Science and Engineering **516** (2019) 012008 doi:10.1088/1757-899X/516/1/012008

Development of a manipulator control system via the Internet

A V Pilipenko, A S Vetrov, A Y Pilipenko, N V Goncharova

OGU named after I.S. Turgenev, Oryol, Russia
E-mail: a@pilipenko.info

Abstract. Abstract. In this paper, the authors examine in detail the technology of developing an IoT project based on the AVR Atmega328 microcontroller, provide a detailed description of the system code and approbation of the results, as well as the authors touch on topical and important issues on creating a web server on Arduino and how it interacts with the client.

1. Introduction

In the world, information technologies are developing every day and the number of "connected" devices is growing, and with it the number of examples of using the Internet of Things (Internet of Things, IoT) in the economy, energy, industry, housing and utilities, agriculture, transport, health care and others [1]. In foreign practice, there are successful examples of IoT implementation at the initiative of both the government and business. For example, with the support of the state in the countries of the European Union, South Korea, China and India, smart city technologies are being introduced, which make it possible to increase the efficiency of energy management and traffic flows. Industrial IoT technologies underlie Industry 4.0: according to scientists, their implementation will increase the productivity of industrial enterprises by 30% until 2025. In this regard, the implementation of research on the concept of IoT and Industry 4.0 elements is necessary for more intensive development of technological processes.

In this paper, the authors solve the problem of controlling the servos of the Robot Arm Kit using the Ethernet interface, avr microcontroller and the IoT platform. To achieve the goal, the following tasks were performed: analysis of decisions in the field; development of the block diagram of the designed device; development of the electrical circuit of the device being developed; development of the algorithm in the form of a flowchart (block diagram); code development for the control system.

2. Concept

During the study of information on this topic, it was decided to use the Arduino UNO board in conjunction with the corresponding Ethernet Shield [2]. Arduino Uno controller is built on the ATmega328 [3]. The platform has 14 digital inputs / outputs (6 of which can be used as PWM outputs), 6 analog inputs, a 16 MHz crystal oscillator, a USB connector, a power connector, an ICSP connector and a reset button.

In the Robot Arm Kit manipulator used, mobility is provided by precision servo drives of two types: DF15MG and DF05BB [4]. These two types of servo drives have significant differences in geometrical dimensions, due to which the first servo drive has twice the torque and angle of rotation. Servomotors use for their performance voltage in the range from 4.8V to 5.7V, so the power will be connected to the servos directly from the board. Since it is necessary to interact with 6 servos, they will connect to 3, 4, 5, 6, 7 and 8 discrete outputs on the board, they will be controlled by the standard library Servo.h (Figure 1). While studying the possible implementations of the software component of this project, it

IOP Conf. Series: Materials Science and Engineering **516** (2019) 012008 doi:10.1088/1757-899X/516/1/012008

was decided to use Arduino, with the Atmel mega328 AVR microcontroller, as the "server" of the receiving and processing POST and GET requests, via the Ethernet2.h library. The answer to the GET request will be the html code of the page. POST requests will be processed and controlled by servos, depending on the information contained in them.

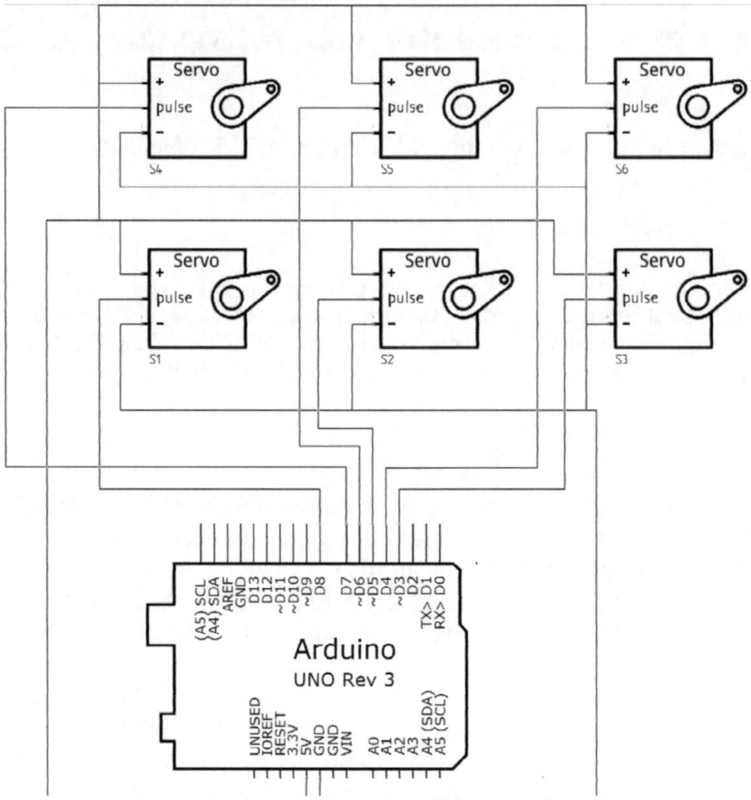

Figure 1. Electrical wiring diagram of servos.

The structural diagram of the project under consideration is presented in Figure 2. This structural diagram shows the interaction of the system elements with each other, namely: "Client" is the system element that sends requests via TCP / IP to the server (the device being designed); Ethernet Shield - an element of the system mediating data transfer between the client and the server; "Server" is an element of the system responsible for processing information and the subsequent management of the elements of the system; Robot Arm Kit is an element of the system that needs to be controlled depending on the information received from the "client" [5,6].

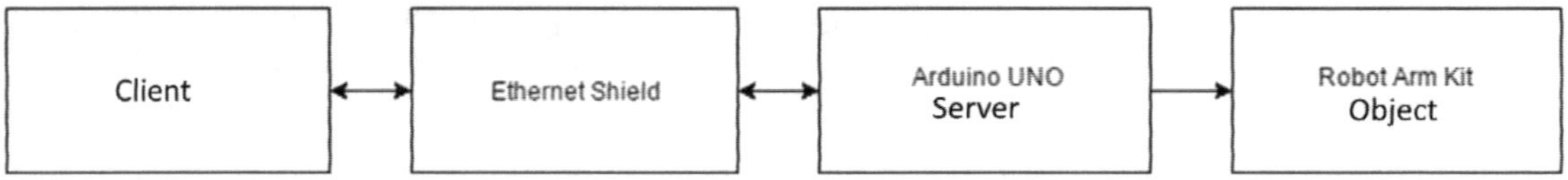

Figure 2. Block diagram.

Arduino is an open-source platform that consists of two main parts: the board itself (often called a microcontroller) and software (a special shell for programming the board) or IDE (Integrated Development Environment) [7]. The software runs on a personal computer and allows you to write the developed code to the board.

3. Software development

Requests for Arduino UNO will be processed through the library (Ethernet2.h). After connecting the device to the network, it will be ready to receive and process requests. To work correctly, you must open the browser and in the address. To work correctly, you need to open the browser and enter the device's ip address in the address bar, after which the browser (the client) will form a socket with a GET request and send it to the server (device), after which the html code of the page will be sent in response. two signed text fields and a button that will be located in the iframe, when you fill in the fields and press the button, a POST request will be sent to the server, which will contain data from the fields. When a request is received, the server will determine its type (POST or GET), and in accordance with this, it will either manage the servos, depending on the data received, or respond to the page's html code.

When receiving data on the need to control the servo drives, voltage will be supplied to the corresponding leg of the Arduino UNO via the standard library (Servo.h). The processing of the information obtained in the requests is performed by cycling through the elements, and sorting them in the same way.

First, libraries, variables, etc. are initialized. Library connection: #include <SPI.h> is a library that allows the use of serial synchronous data transfer protocol used by microcontrollers for data exchange. #include <Servo.h> is a library for managing servos. #include <Ethernet2.h> is a library for processing requests. Declare strings for intermediate storage of information from a query in the program memory: String tmp_string = String (); String query = String (); String query_get = String (); String query_post = String (); String query_method = String (). Servo Announcement: Servo servo1; Servo servo2; Servo servo3; Servo servo4; Servo servo5; Servo servo6. Declaration of variables for information processing: int j; int i = 0; int serv; int angle. Setting server settings: IPAddress ip (192, 168, 1, 177); byte subnet [] = {255, 255, 255, 0}; byte mac [] = {0xDE, 0xAD, 0xBE, 0xEF, 0xFE, 0xED}. Setting the server port 80: Ethernet Server (80).

The following is the contents of the Setup function, in which the server and servo drives are initialized once: Ethernet.begin (mac, ip, subnet); server.begin (); Serial.print ("server is at"); Serial.println (Ethernet.localIP ()); servo1.attach (2); servo2.attach (3); servo3.attach (4); servo4.attach (5); servo5.attach (6); servo6.attach (7).

The contents of the infinite loop will be divided into functional blocks. Getting information from the client: EthernetClient client = server.available (). Checking the relevance of the information: if (client). Preparing variables for storing information: j = 0; query = ""; query_get = ""; query_post = ""; query_method = ""; boolean CLB = true. The loop works as long as the client is connected: while (client.connected ()). The loop works as long as there is raw data: while (client.available ()). Gathering melon requests: char c = client.read (); query = query + c. Checking the end of the request: if (c = = '\ n' && CLB). If the request is not completed, then a cycle is started, working as long as there is POST data from the while (client.available ()) request, in which the POST data is saved: char b = client.read (); query_post = query_post + b. Next, using the for loop, the incoming data is sorted by the permutation method. After that, the end of the else if request (c = = '\ n') is re-checked, if the result is true, then the value in the variable CLB = true changes, if it is false, then an additional check is performed else if (c! = '\ R '), and when obtaining a logical unit, the reverse change of the value CLB = false is performed. The initial values of the variables responsible for the servo drive number and its rotation angle are set (serv = 0; angle = 0;). Then the code for checking the status of the request is written (if (query_post.indexOf ("serv")! = -1 && query_post.indexOf ("&")! = -1 && query_post.indexOf ("serv") <query_post.indexOf ("& "))) and if a logical unit is received, then information is obtained about which servo drive must be controlled: j = 0; tmp_string = ""; for (i = query_post.indexOf ("serv"); i <query_post.indexOf ("&"); i ++) {if (j == 1) {tmp_string = tmp_string + query_post [i];} if (query_post [i] == '=') {j = 1;} serv = tmp_string.toInt (). Now you need to get information about the angle of rotation of the servomotor, with the received index. The status of the request to change the angle if (query_post.indexOf ("angle")! = -1) is checked and, when a logical unit is received, the values of the angle are obtained: j = 0; tmp_string = ""; for (i = query_post.indexOf ("angle"); i

<query_post.length (); i ++) {if (j == 1) {tmp_string = tmp_string + query_post [i]; } if (query_post [i] == '=') {j = 1; }} angle = tmp_string.toInt ().

When the initial parameters are configured, the html page is created for the user interface and the progressive html code is sent to the client: client.println ("HTTP / 1.1 200 OK"); client.println ("Content-Type: text / html; charset = utf-8"); client.println ();

When the initial parameters are configured, the html page is created for the user interface and the progressive html code is sent to the client: client.println ("HTTP / 1.1 200 OK"); client.println ("Content-Type: text / html; charset = utf-8"); client.println (); client.println (); client.println ("<! DOCTYPE html>"); client.println ("<html lang = \" ru \ ">"); client.println ("<head>"); client.println ("<meta charset = \" UTF-8 \ ">"); client.println ("<title> Servo </ title>"); client.println ("</ head>"); client.println ("<body>"); client.println ("<form id = \" PostForm \ "method = \" post \ "ac-tion = \" 192.168.1.177 \ ">"); client.println ("<p> <b> Servo: </ b>
"); client.println ("<input name = \" serv \ "type = \" text \ "size = \" 20 \ ">"); client.println ("<p> <b> Angle: </ b>
"); client.println ("<input name = \" angle \ "type = \" text \ "size = \" 20 \ ">
"); client.println ("<input type = \" submit \ "value = \" Send \ "/>"); client.println ("</ form>"); client.println ("</ body>"); client.println ("</ html>").

It remains to consider the classic servo control algorithm, in which the state of variables with the servo drive number and rotation angle (if (serv! = 0 and angle! = 0)) is checked if the value has changed, if the value has changed, then the servo motor will be completely searched according to the following principle: if (serv == 1); else if (serv == 2); else if (serv == 3). If the state of the variable coincides with the index, for example, with the number 1, this engine is rotated to the specified angle: servo1.write (angle). After the transfer is completed, it is correct to pause the exchange of information with the client (client.stop ()).

The algorithm of the system in the form of a flowchart is presented in Figures 3-4 and describes the following functions: 1 - Start; 2 - check for power; 3 - The end of the program; 4 - getting information about the client connection; 5 - check for the presence of a client; 6 - resetting variables for storing and processing request information; 7 - check client connection; 8 - cyclic check of connection availability; 9 - character read request data; 10, 11, 12, 13, 14 - check for the end of the request header; 15 - check the availability of data of the request body; 16 - read the request body data; 17, 18, 20, 21, 22, 23, 24, 25 - parse the request header and get data about the type of request; 26 - the beginning of the formation of a response to the client; 27 - checking the presence in the request body of data on the servo number that needs to be controlled; 28, 29, 30, 31, 32, 33, 34 — retrieving data from the request body for the servo number; 35 - checking the presence in the request body of the angle data on which the servo needs to be turned; 35, 36, 37, 38, 39, 40, 41, 42 — retrieving data from the angle request body; 43 - completion of the formation of a response to the client; 44 - checking the availability of information about the servo number and the angle at which it must be rotated; 45, 46, 47, 48, 49, 50, 51, 52, 53, 54, 55, 56 — select the appropriate servo number and rotate it to the required angle; 57 - closing the connection with the client.

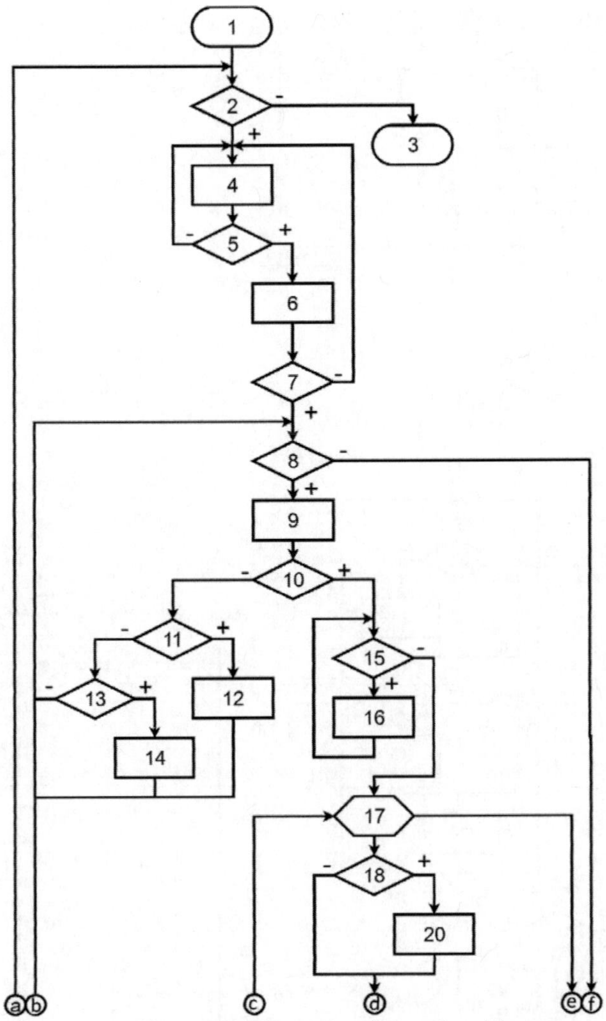

Figure 3. First part of the algorithm.

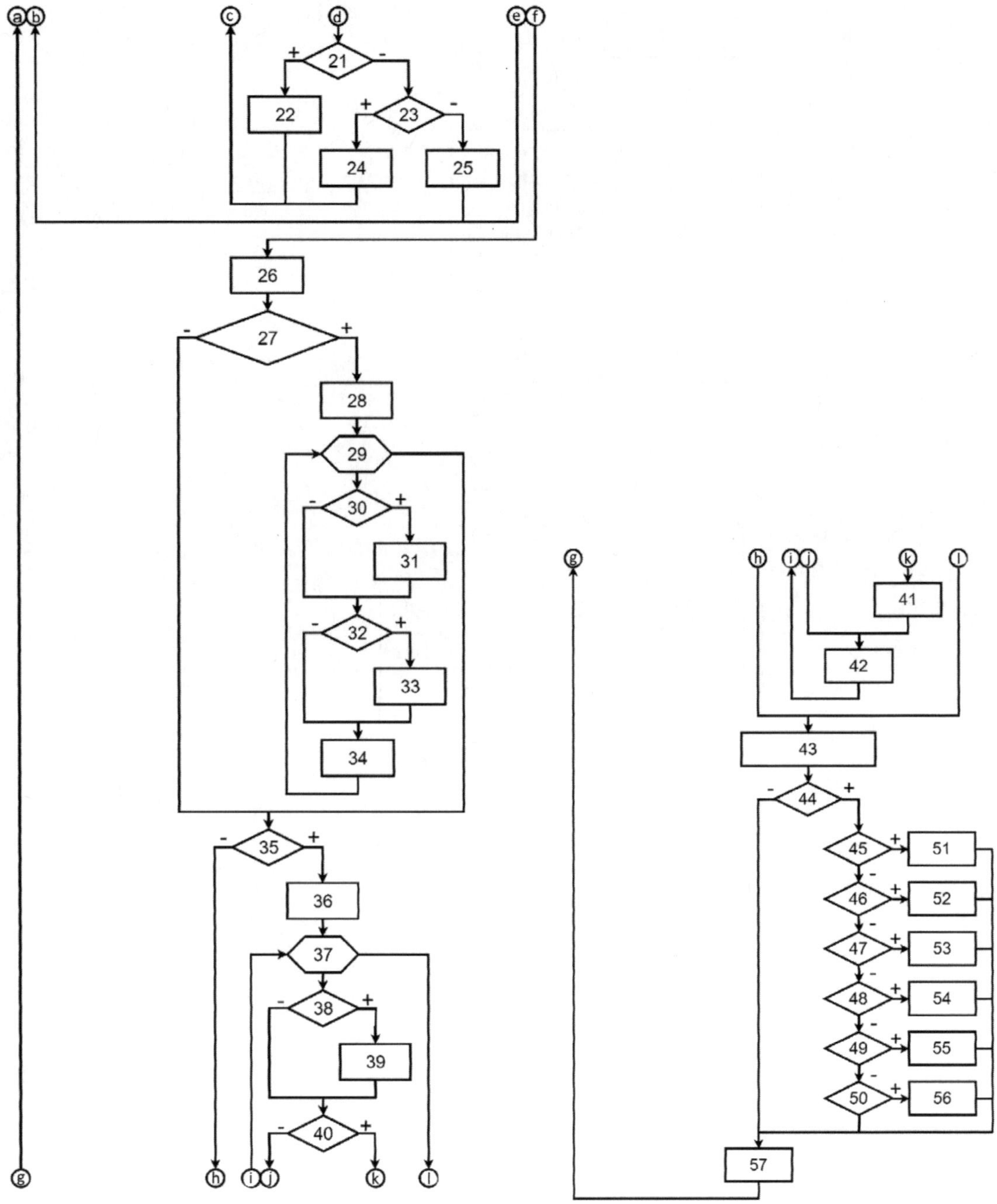

Figure 4. Second part of the algorithm.

4. Results

As a result of testing, the functionality of the device was confirmed through repeated testing and a "screenshot" of its work was made, which is presented below in Figure 5. Figure 6 shows an image of the assembled device.

IOP Conf. Series: Materials Science and Engineering **516** (2019) 012008 doi:10.1088/1757-899X/516/1/012008

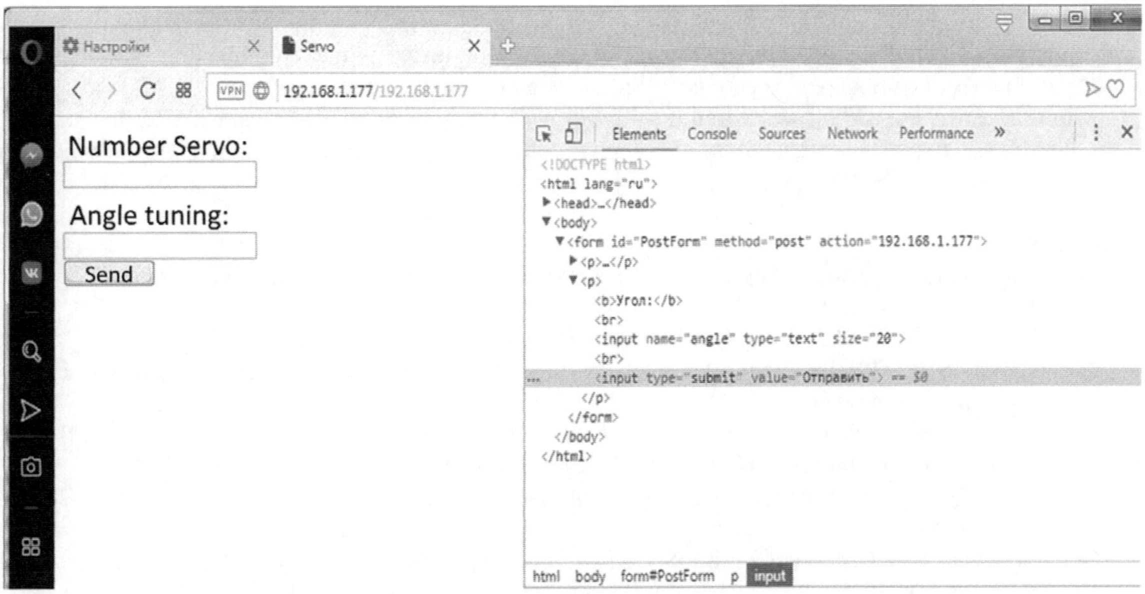

Figure 5. Screenshot of the program.

This "screenshot" of the program shows that when you send a GET request to the server on the Arduino UNO, the "client" (in this case, the browser) receives the html page code in response. On which you can see 2 fields to fill in and the button to send this data POST request to the server, with subsequent processing of this data [8].

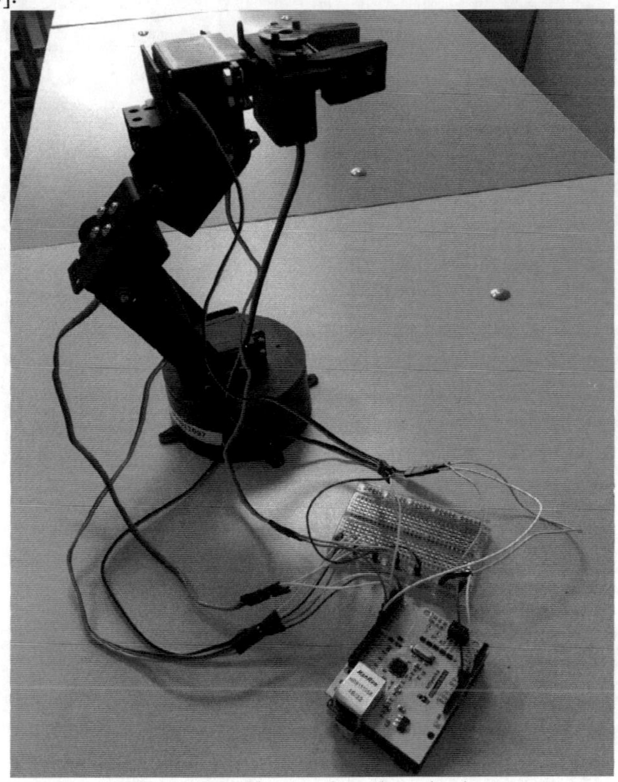

Figure 6. Photo of testing device.

During the creation of this project, there were problems displaying the html code page sent by the server, most likely they occurred because the device was tested without connecting the client to the "clean" Internet (without a proxy server). She managed to solve when using the browser "Opera".

In this paper, the code is described in full, but without using operator brackets. After the successful assembly, 30 requests were sent to the server with various servo control data, all of which were processed correctly by the system.

References

[1] Kupriyanovsky V.P., Namiot D.E., Sinyagov S.A. 2016 *International Journal of Open Information Technologies* **4 (2)**

[2] D'Ausilio A. 2012 *Behavior research methods* **44(2)** 305–313

[3] Barrett S. F. 2010 *Arduino Microcontroller Processing for Everyone* (Morgan and Claypool Publishers: San Rafael)

[4] Kazanzides P., Chen Z., Deguet A., Fischer G. S., Taylor R. H., DiMaio S. P. 2014 *IEEE International Conference on In Robotics and Automation (ICRA)* 6434–6439

[5] Tristo G. et al. 2015 *The International Journal of Advanced Manufacturing Technology* **78 (9-12)** 1511–1521

[6] Murdoch S. J., Stephen L. 2005 *Embedding covert channels into TCP/IP* (Springer: Berlin, Heidelberg)

[7] D'Ausilio A. 2012 *Behavior research methods* **44(2)** 305–313

[8] Ramadas M., Shawn O., Brett T. 2003 Detecting anomalous network traffic with self-organizing maps (Springer: Berlin, Heidelberg)

IOP Conf. Series: Materials Science and Engineering **516** (2019) 012009 doi:10.1088/1757-899X/516/1/012009

Amplitude control system of drive-mode oscillations of MEMS gyroscopes

T G Nesterenko, E S Barbin, P F Baranov, Lo Van Hao

National Research Tomsk Polytechnic University, Tomsk, Russia
E-mail: ntg@tpu.ru

Abstract. Two operating modes can be singled out for a gyroscope, namely drive mode and sense mode. During the drive mode, electrostatic forces of gyroscope enable drive-mode harmonic oscillations of a moving mass. With the angular velocity and translational acceleration of an object in the sense mode, a moving mass moves due to the Coriolis and inertial forces, thereby causing changes in capacitances of comb structures. Resultant capacitance changes are proportional to parameters to be measured, i.e. the rotational rate and linear acceleration. The aim of this work is to investigate and create a system of amplitude control of drive-mode oscillations. To improve the mems efficiency, drive-mode oscillations are excited at a resonant frequency.

1. Introduction

MEMS gyroscopes operates on the Coriolis principle. Two operating modes can be singled out for a gyroscope, namely drive mode and sense mode. During the drive mode, electrostatic forces of gyroscope enable harmonic oscillations of a moving mass. With the angular drift velocity and translational acceleration of an object in the sense mode, a moving mass moves due to the Coriolis and inertial forces, thereby causing changes in capacitances of comb structures. Resultant capacitance changes are proportional to parameters to be measured, *i.e.* the rotational rate and linear acceleration.

The aim of this work is to investigate and create a system of amplitude control of drive-mode oscillations. To improve the gyroscope efficiency, drive-mode oscillations are excited at a resonant frequency. The amplitude of drive-mode oscillations should be constant which is provided by the closed circuit system or a feedback circuit with the automatic level control (ALC) feedback [1–3]. The ALC feedback configuration is very important for the system stability based on a proportional-integral (PI) controller which is decisive for the amplitude control of drive-mode oscillations.

2. Mechanics of drive mode

The drive mode is provided by a mechanical resonator, whose simplified equivalent circuit model is given in Figure 1 [4].

The damping of drive-mode oscillations in the sensor is characterized by μ_{y1}, μ_{y2} and μ_y damping factors. Equaling in value and opposite in direction electrostatic forces $F_{el\,1}$ and F_{el2} are applied to moving masses. The motion equations for moving masses are as follows:

$$m_{y1} \cdot \ddot{y}_1 + (\mu_{y1} + \mu_y) \cdot \dot{y}_1 + \left(k_{y1} + k_y\right) \cdot y_1 - \mu_y \cdot \dot{y}_2 - k_y \cdot y_2 = F_{el1} \quad (1)$$

$$m_{y2} \cdot \ddot{y}_2 + (\mu_{y2} + \mu_y) \cdot \dot{y}_2 + \left(k_{y2} + k_y\right) \cdot y_2 - \mu_y \cdot \dot{y}_1 - k_y \cdot y_1 = F_{el2}\,, \quad (2)$$

where m_{y1}, m_{y2} are moving masses which perform drive-mode oscillations in opposite directions; k_y is the rigidity of spring elements between moving masses directed along Y-axis of drive-mode oscillations; k_{y1}, k_{y2} are joint rigidities of spring elements suspending the outer and decoupling frames.

For the ideal system we assume that $m_{y1}=m_{y2}=m_y$, $\mu_{y1}=\mu_{y2}=\mu_{y0}$, $k_{y1}=k_{y2}=k_{y0}$. Then Equal (1) can be rewritten as

$$m_{y1}\cdot\ddot{y}_1+(\mu_{y0}+2\cdot\mu_y)\cdot\dot{y}_1+\left(k_{y0}+2\cdot k_y\right)\cdot y_1=F_{el1} \tag{3}$$

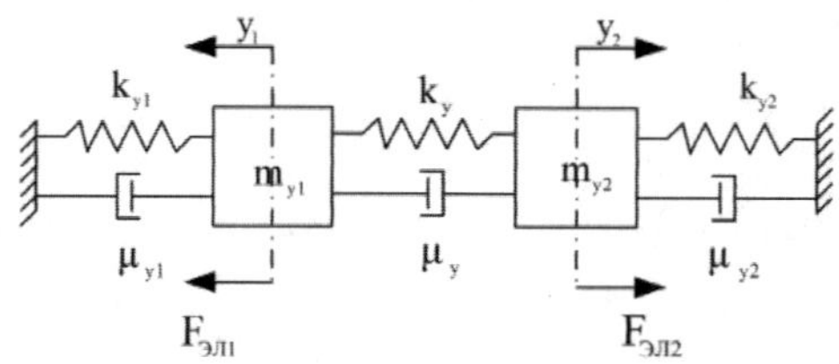

Figure 1. A simplified equivalent circuit model of mechanical resonator.

Harmonic oscillations usually occur in the drive mode, when an AC input signal $V_{in}=V\sin(\omega t)$ and DC voltage V_{dc} should be applied to the comb drive actuators *via* anchors and moving mass, as illustrated in Figure 2.

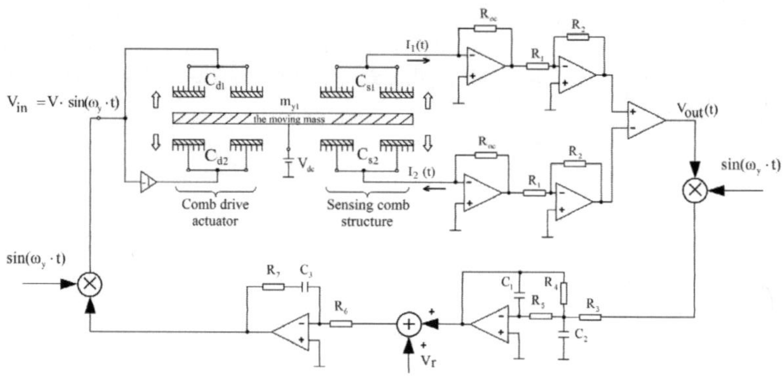

Figure 2. Interface electronics of the drive mode.

Electrostatic forces that act between two electrically conductive combs $C_{d1}$and C_{d2} are defined by the following relations [5–7]:

$$F_{el1}=2\cdot b_1\cdot V_{dc}\cdot V\cdot\sin(\omega_y\cdot t)\text{ and }F_{el2}=-2\cdot b_1\cdot V_{dc}\cdot V\cdot\sin(\omega_y\cdot t), \tag{4}$$

Where b_1 is the factor to approximate electrically conductive combs C_{d1} and C_{d2} [4].

The correlation between the mechanical motion $y_1(s)$ and applied force $F_{el1}(s)$ can be determined by

$$\frac{y_1(s)}{F_{el1}(s)}=\frac{1}{m_{y1}}\cdot\frac{1}{s^2+\frac{\mu_1}{m_{y1}}\cdot s+\frac{k_1}{m_{y1}}}, \tag{5}$$

where $\mu_1=\mu_{y0}+2\mu_y$ is the damping factor; $k_1=k_{y0}+2k_y$ is the rigidity of spring element.

The correlation between the motion and the input signal is obtained from

$$\frac{y_1(s)}{V_{in}(s)}=\frac{1}{m_{y1}}\cdot 2\cdot b_1\cdot V_{dc}\cdot\left[s^2+\left(\frac{\omega_y}{Q_y}\right)\cdot s+\left(\omega_y\right)^2\right]^{-1}, \tag{6}$$

where $\left(\omega_y\right)^2=k_1\cdot\left(m_{y1}\right)^{-1}$ and $Q_y=\left(\mu_1\right)^{-1}\cdot\sqrt{k_1\cdot m_{y1}}$.

Vibrations of the moving mass m_{y1} enable changes in the equivalent electrically conductive combs $C_{s1}=a_2+b_2\cdot y$ and $C_{s2}=a_2-b_2\cdot y$. These changes result in the generation of currents I_1 and I_2 passing between sensing electrodes, such that

$$I_1(t) = V_{dc} \cdot b_2 \cdot \dot{y} \text{ and } I_2(t) = -V_{dc} \cdot b_2 \cdot \dot{y} . \tag{7}$$

The voltage output V_{out} at the current to voltage converter can be written in the form

$$V_{out}(t) = 2 \cdot V_{dc} \cdot b_2 \cdot R_{oc} \cdot R_2 \cdot (R_1)^{-1} \cdot \dot{y} . \tag{8}$$

The correlation between the input and output signals is found from

$$\frac{V_{out}(s)}{V_{in}(s)} = \frac{1}{m_{y1}} \cdot \frac{R_2}{R_1} \cdot R_{oc} \cdot 4 \cdot b_1 \cdot b_2 \cdot V_{dc}^2 \cdot s \cdot \left[s^2 + \left(\frac{\omega_y}{Q_y} \right) \cdot s + (\omega_y)^2 \right]^{-1} . \tag{9}$$

3. Amplitude controls of drive-mode oscillations

A feedback circuit used for the amplitude control in the drive mode is a nonlinear system. It is difficult to analyze the behavior of nonlinear system, especially when it achieves the third order [8]. However, the only important driver is a change in the amplitude response of the system rather than its complete performance. The dynamic equilibrium of the feedback circuit is provided by the force driven by the negative feedback.

According to Equal 6 we obtain the dynamic equation for the drive mode:

$$\ddot{y} + \frac{\omega_y}{Q_y} \cdot \dot{y} + \omega_y^2 \cdot y = \frac{1}{m_{y1}} \cdot 2 \cdot b_1 \cdot V_{dc} \cdot V_{in} . \tag{10}$$

This motion and the input signal can be expressed by

$$y \approx y_m(t) \cdot \sin(\omega_y \cdot t) , \qquad V_{in} \approx V_{m,in}(t) \cdot \cos(\omega_y \cdot t) , \tag{11}$$

Where $y_m(t)$, $V_{m,in}(t)$ are amplitudes of motion and input signal, both being changed in time.

Now we can derive the amplitude ratio between the input and output signals

$$\frac{V_{m,out}(s)}{V_{m,in}(s)} = \frac{1}{2} \cdot \frac{A \cdot K_{A/V}}{s + \frac{1}{2} \cdot \frac{\omega_y}{Q_y}} , \tag{12}$$

where $A = \frac{4 \cdot V_{dc}^2 \cdot b_1 \cdot b_2}{m_{y1}}$, $K_{A/V} = R_{oc} \cdot \frac{R_2}{R_1}$.

Therefore, instead of a total model we can use a simplified model of the feedback circuit presented in Figure 3 [9].

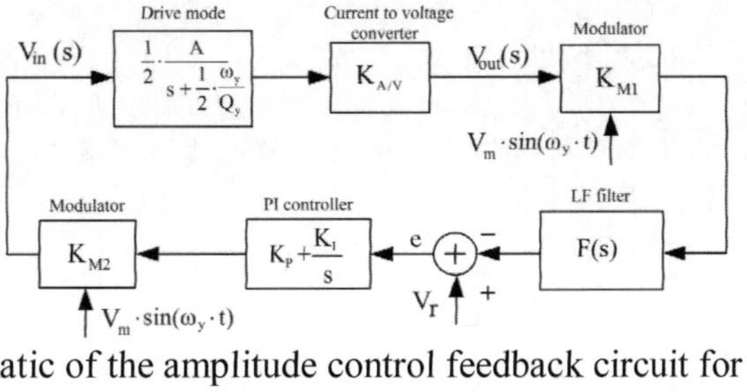

Figure 3. Schematic of the amplitude control feedback circuit for the drive mode.

As can be seen from Figure 3, the feedback circuit incorporates two signal modulators, the low-frequency (LF) filter and PI controller. The voltage output V_{out} at the current to voltage converter is compared in the summer to the reference voltage V_r. Mismatched signal *e* transfers to PI controller to apply a correction for the drive signal.

When the amplitude of the output signal is lower than that of the reference signal, the error signal of the control system becomes positive, thereby increasing the electrostatic force. When the amplitude of the output signal is higher than that of the reference signal, a reverse process occurs. Owing to this mechanism, the amplitude of the output signal is controlled by the reference value.

Let us accept two criteria for the PI controller design. The first is a short buildup time, and the second is a sufficient phase margin.

In order for the first criterion to be met, the pole cancellation [8] should be performed in the simplified model of the feedback circuit, *i.e.* the following condition must be satisfied:

$$s+\frac{1}{2}\cdot\frac{\omega_y}{Q_y}=s+\frac{K_I}{K_P}\rightarrow\frac{K_I}{K_P}=\frac{1}{2}\cdot\frac{\omega_y}{Q_y} \qquad (13)$$

Where K_P, K_I are gain factors for PI controller.

After the pole cancellation, the transmission gain in the feedback circuit is set to provide the system maximum bandwidth with the sufficient phase margin (approx.. 60 degrees). The phase margin has a direct effect on setting time and transient overshoot. Hence, the optimum phase margin is a necessary criterion for constructing the system for the amplitude control of drive-mode oscillations. After the pole cancellation in the simplified model of the feedback circuit, the system order depends on the order of the LF filter. The second criterion is met in conformity with the following equation [9]:

$$K_P=4\cdot\pi\cdot f\cdot\left(A\cdot K_{A/V}\cdot K_{M1}\cdot K_{M2}\cdot K_F\right)^{-1}, \qquad (14)$$

where F(s) is a transfer function of the LF filter; f is the frequency for 30 degree phase shift; $K_F=|F(s)|_f$ is the transmission gain of LF filter; K_{M1}, K_{M2} are transmission gains of modulators;

Figure 4 presents a total model of the amplitude control system of drive-mode oscillations. The modeling is performed using design factors of the simplified model of the feedback circuit. Using Eqns (10) and (11), we find the transmission gains of K_P= 2 and K_I= 360 for PI controller.

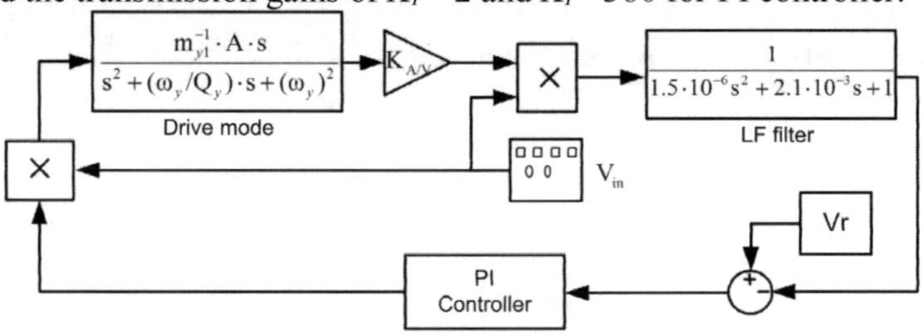

Figure 4. Numerical model of the amplitude control system in the drive mode.

The output signal of the low-frequency filter is depicted in Figure 5. We also ascertain setting times of 50 ms. This signal describes the amplitude value for drive-mode oscillations.

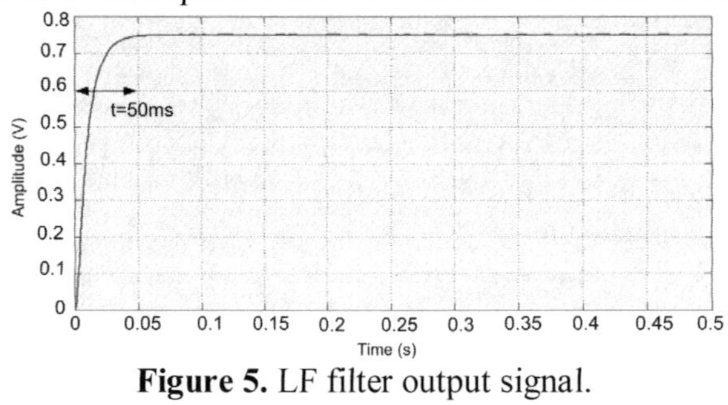

Figure 5. LF filter output signal.

Drive-mode oscillations with stabilized amplitude are shown in Figure 6. After the transient process, the output amplitude is constant and corresponds to the given reference signal.

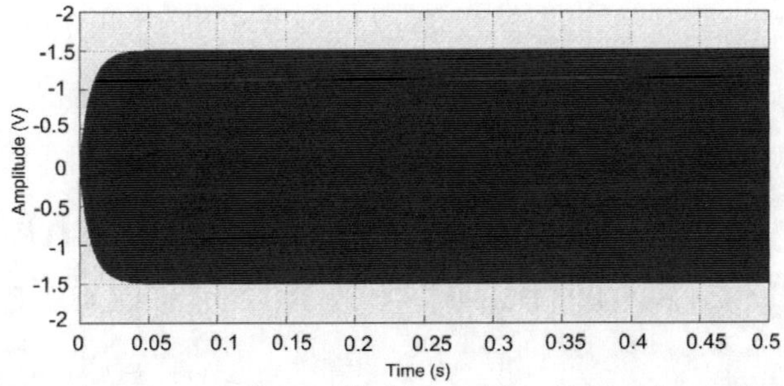

Figure 6. Drive mode oscillations with stabilized amplitude.

4. Conclusions

We proposed a theoretical analysis and a method of parameter determination for the amplitude control system in the drive mode. Modeling showed that the proposed control system stabilized the amplitude of drive-mode oscillations at a level corresponding to that of the reference signal.

Acknowledgments
This work was performed in National Research Tomsk Polytechnic University. The authors would like to acknowledge the financial support of the Ministry of Science and higher education of the Russian Federation (Agreement № 14.578.21.0232, unique identifier RFMEFI57817X0232).

References

[1] Cui J., Chi X. Z., Ding H. T., Lin L. T., Yang Z. C., Yan G. Z. 2009 *J. Micromech. Microeng.* **19**
[2] Huan-ming Wu, Hai-gang Yang, Tao Yin, Ji-wei Jiao 2013 *Sensors* **13** 12564
[3] Cui J., Guo Z. Y., Yang Z. C., Hao Y. L., Yan G. Z. 2011 *J. Micromech. Microeng.* **21**
[4] Barbin E.S. 2016 dis. cand. Tech. Science 111
[5] Hao L.V., Nesterenko T.G. 2018 *Izvestiya SFedU. Engineering Sciences* **2**
[6] Alexander A.T., Andrei M.S. 2007 *J. Micromech. Microeng.* **17** 1583
[7] Aaltonen L., Halonen K. A. I. 2009 *Analog Integrated Circuits and Signal Processing* **63** 465
[8] Bura kEminoglu. 2011 A thesis submitted to the Graduate school of natural and applied sciencesof Middle East technical university 146
[9] Sung W.-T., *et al* 2008 *J. Micromech. Microeng.* **18**

IOP Conf. Series: Materials Science and Engineering **516** (2019) 012010 doi:10.1088/1757-899X/516/1/012010

Adaptive fuzzy neural production network with MIMO-structure for the evaluation of technology efficiency

V I Syryamkin, S V Gorbachev, M V Shikhman

National Research Tomsk State University, Tomsk, Russia
E-mail: egs@sibmail.com

Abstract. The paper presents an example of modeling the algorithm operation. The paper analyses the modified Wang and Mendel MIMO-architecture of the adaptive fuzzy neural production network with a logical conclusion. It is distinguished by the automatic generation of a set of fuzzy rules based on a fuzzy decision tree and a hybrid algorithm for parameter adaptation by the neural network (centers, widths of membership functions, and conclusions) starting from the leaf nodes to the root nodes of the tree.

1. Introduction

When solving problems of evaluating the effectiveness of technologies, traditional statistical methods have some drawbacks, especially when two or more classes are connected by a chain of internally connected sampling objects, have a non-spherical form, are linearly inseparable sets, or when the density or volume of classes is different [1]. In the case of class crossing, contact or chaining, the classification model is blurred; therefore, fuzzy procedures are preferable when choosing mathematical methods for solving the problem of automatic classification [2]. The theory of fuzzy sets operates with such concepts as the weighted membership of the object under investigation to the set and offers a rather flexible apparatus for a formal description of similar situations. Therefore, the target setting moves from the question "Does the object belong to the class?" to determine the degree of membership of the object to a particular class. Summarizing, we can say that the fuzzy-multiple approach to the solution of classification problems in certain cases allows separating clusters of complex shape providing new opportunities for interpreting classification results.

Comparing the reliability of classification results obtained by different authors, we can identify the most effective classifiers, which include neural networks [1,3], fuzzy logic, and the methods of cluster analysis [4,5]. The best results were obtained using three-layer perceptron, probabilistic neural network, and Kohonen self-organizing maps. During testing, Takagi-Sugeno-Kang fuzzy network showed good results of complex functions approximation [6]. At the same time, studies that we have analyzed, did not consider the classification of the effectiveness of scientific and technological solutions and technologies, and did not use neural-fuzzy networks for technology classification.

2. The structure of Wang and Mendel fuzzy neural production network

From the functional point of view, Wang and Mendel fuzzy neural production network (L.-X. Wang and J.M. Mendel) is similar to the zero-order Takagi-Sugeno-Kang network (TSK) [7]. The components of the expressions that determine the prerequisites of the production rules are identical for both networks. Differences are observed in presenting conclusions. In the Takagi-Sugeno-Kang network, the result is polynomial:

$$r_{i0} + \sum_{j=1}^{m} r_{ij} x_j,$$

whereas in the Wang and Mendel network, the output variable in the rules is represented by the constant r_i, which can be considered as a zero-order polynomial that determines the center of the membership function of the conclusion of the corresponding rule.

Thus, Wang and Mendel fuzzy neural production network implements a fuzzy product model based on the following rules:

Rule *i*: **IF** x_1 is A_{i1} **AND** … **AND** x_j is A_{ij} **AND** … **AND** x_m is A_{im}, **THEN** y is r_i,
$i = 1, \ldots, n,$

where n is a number of rules;
m is a number of input variables.

Figure 1 shows an example of a Wang and Mendel network with a MISO- structure (Multi Inputs Single Output) containing two inputs, three rules, and one output.

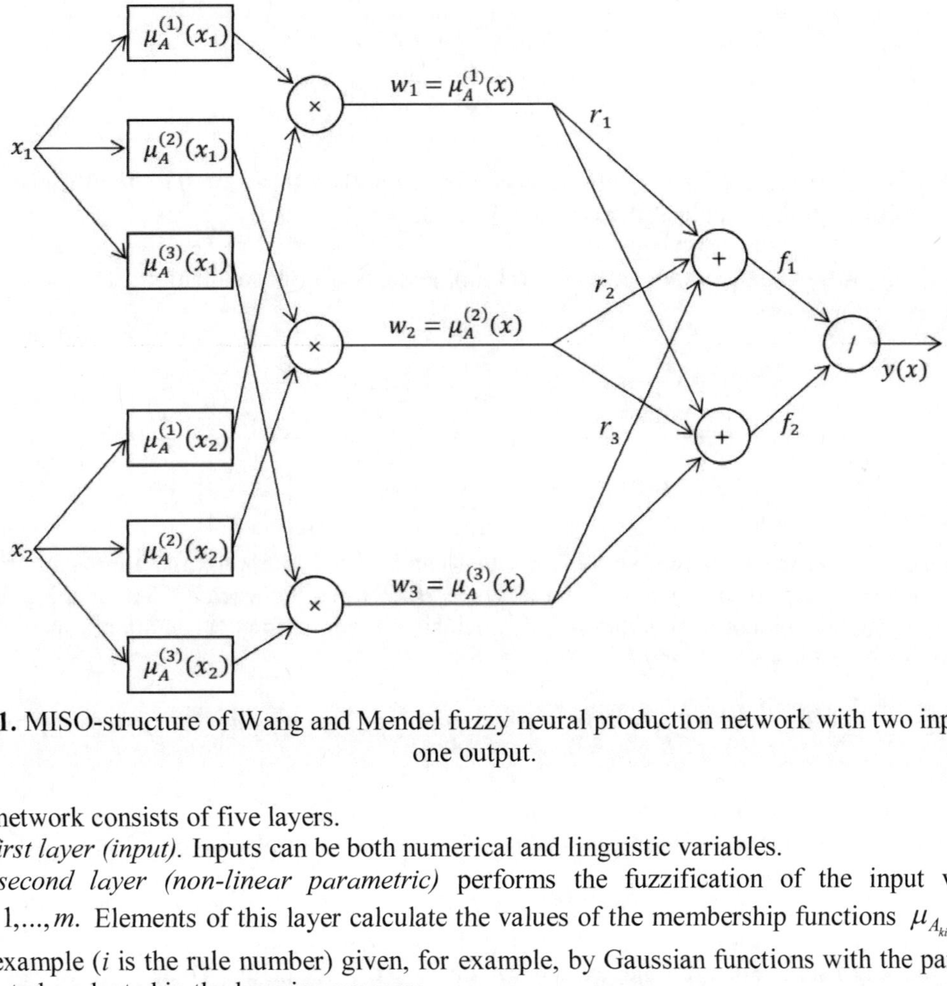

Figure 1. MISO-structure of Wang and Mendel fuzzy neural production network with two inputs and one output.

This network consists of five layers.

The first layer (input). Inputs can be both numerical and linguistic variables.

The second layer (non-linear parametric) performs the fuzzification of the input variables x_k, $k = 1,\ldots,m$. Elements of this layer calculate the values of the membership functions $\mu_{A_{ki}}(x_{kj})$ in the *j*-th example (*i* is the rule number) given, for example, by Gaussian functions with the parameters a_{ij} and b_{ij} to be adapted in the learning process:

$$\mu_{A_{ki}}(x_{kj}) = \exp\left[-\left(\frac{x_{kj} - a_{ki}}{b_{ki}}\right)^2\right].$$

This layer contains m×n×2 parameters of the Gaussian function.

The third layer (nonparametric) contains the number of elements equal to the number of rules in the database; it aggregates the degrees of truth of the conditions of the corresponding rules.

The fourth layer (linear parametric) consists of two neurons. The first neuron serves to activate the conclusions of the rules r_i in accordance with the values of the preconditions of rules aggregated in the previous layer. The second neuron generates a normalizing signal for subsequent defuzzification of the result. This is the parametric layer, in which linear weights r_i interpreted as centers of membership functions of the conclusion of the i-th fuzzy inference rule are the subject to adaptation.

The stage of accumulation of conclusions of fuzzy production rules in this algorithm is absent because of clear values of output variables.

The fifth layer (nonparametric) consists of one output neuron and performs defuzzification forming an output signal. The output variables are single-point fuzzy sets, therefore, their values are calculated using the modified center-of-gravity method (center-of-gravity for singletons), and another name for this method is fuzzy mean [7]. It allows clarifying the output variable without accumulating the activated conclusions of the individual rules as follows:

$$y_j(x) = \frac{\sum_{i=1}^{n} w_{ij} r_i}{\sum_{i=1}^{n} w_{ij}}, \tag{1}$$

where w_{ij} is an aggregated degree of truth for all conditions of the i-th rule in the j-th example;
r_i is a value of the output variable of the i-th rule ($y_i = r_i$);
n is a number of rules in the database.

Thus, for the j-th example, the neural network implements the approximation function, which can be written as follows:

$$y_j(x) = \frac{\sum_{i=1}^{n}\left(r_i \prod_{i=1}^{m} \mu_{ki}\left(x_{kj}\right)\right)}{\sum_{i=1}^{n}\prod_{i=1}^{m} \mu_{ki}\left(x_{kj}\right)} = \frac{\sum_{i=1}^{n}\left(r_i \prod_{i=1}^{m} \exp\left[-\left(\frac{x_{kj}-a_{ki}}{b_{ki}}\right)^2\right]\right)}{\sum_{i=1}^{n}\prod_{i=1}^{m} \exp\left[-\left(\frac{x_{kj}-a_{ki}}{b_{ki}}\right)^2\right]}. \tag{2}$$

Expression (2) defines a continuous function that can be used to approximate an arbitrarily given continuous function $g(x)$ from the many variables x_i, that form the vector x. With the appropriate selection of the parameters of the condition and conclusion, the function (2) can approximate the given function $g(x)$ with arbitrary accuracy ε.

3. Development of the modified MIMO-structure of Wang and Mendel fuzzy neural production network

3.1 Selection of the type of membership function

Fuzzification of the input space of attributes is associated with the formalization of input parameters in the form of a vector of interval values (fuzzy interval); the hit in each interval is characterized by a certain degree of uncertainty. We assume that based on primary data, experience, and intuition, experts usually do not have difficulties during quantifying the boundaries (intervals) of possible (admissible) values of the parameters. For example, experts can characterize the primary data with triangular fuzzy numbers with the following membership function $\mu_A(x)$ of the fuzzy set A:

IOP Conf. Series: Materials Science and Engineering **516** (2019) 012010 doi:10.1088/1757-899X/516/1/012010

$$\mu_A(x) = \begin{cases} 0, & x \le a \\ \dfrac{x-a}{b-a}, & a \le x \le b \\ \dfrac{c-x}{x-b}, & b \le x \le c \\ 0, & c \le a \end{cases}.$$

This expression models the following expert position: "parameter A is approximately equal to b and uniquely enters the segment [*a, c*]" (Figure 2).

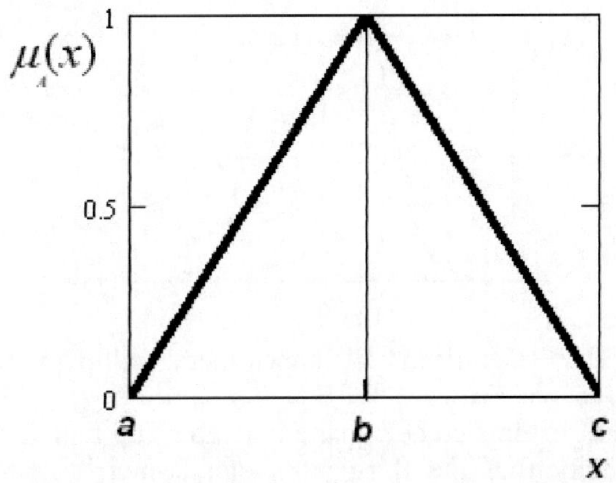

Figure 2. Triangular membership function.

One of the main advantages of triangular membership functions is their simplicity.

In practice, an analytical representation of the membership function is often used, for example, the Gaussian membership function has the form shown in Figure 3:

$$\mu(x) = \exp\left[-\left(\frac{x-c}{\sigma}\right)^2\right],$$

where *c* is a center,
σ is a width coefficient (variation).

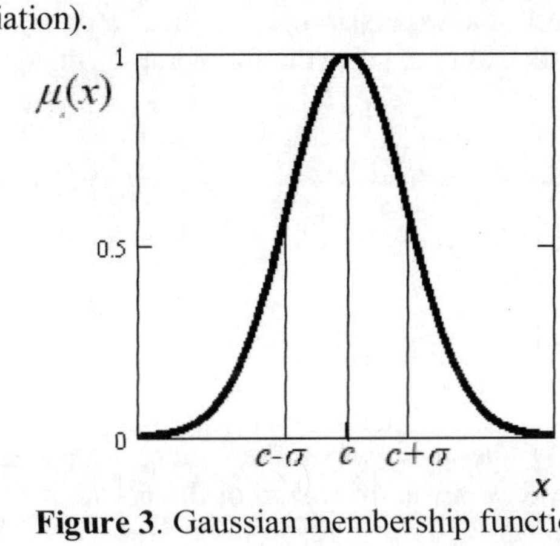

Figure 3. Gaussian membership function.

To develop the modified MIMO-structure (Multi Inputs Multi Output) of the Wang and Mendel fuzzy neural production network, we will use a symmetrical bell-shaped two-sided membership function controlled by three parameters and described by an analytical expression:

$$\mu(x) = \frac{1}{1 + \left| \frac{x - c}{\sigma} \right|^{2b}}$$

where c determines the location of the center (the coordinate of the displacement point of the curve symmetry center along the X axis);
σ is the parameter that determines the width of the upper part of the curve;
b is the coefficient of flatness of the left and right parts of the curve (Figure 4).

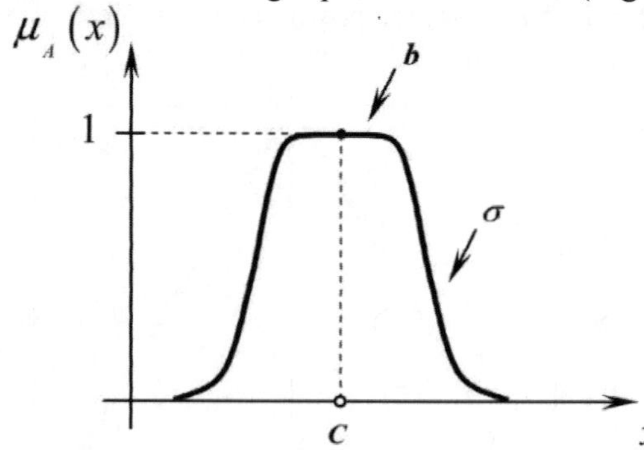

Figure 4. Generalized bell-shaped membership function.

The difference between the generalized bell-shaped membership function and the above-considered membership functions is addition of the third parameter, which allows implementing a smooth transition between fuzzy sets.

3.2 Generating a knowledge base using a fuzzy decision tree
To form the base of fuzzy rules, we propose to use a fuzzy decision tree [8, 9], which can be regarded as a hierarchical network with direct signal propagation. Therefore, in order to adapt its parameters (centers and width of membership functions) and the parameters of rule conclusions in tree leaves, we propose to use a hybrid algorithm for back propagation of the error.

To isolate a single layer in the network where the neurons correspond to the leaves of the tree, in the formula (1) we select the normalized coefficients $\overline{w_{ij}}$ of the truth degree of the conditions for each rule in the j-th example. In this case, the expression (1) for the output y_j of Wang and Mendel fuzzy neural production network in the j-th example takes the following form:

$$y_j = \frac{\sum_{i=1}^{n} w_{ij} r_i}{\sum_{i=1}^{n} w_{ij}} = \frac{w_{1j} r_1 + \ldots + w_{nj} r_n}{\sum_{i=1}^{n} w_{ij}} = \frac{w_{1j} r_1}{\sum_{i=1}^{n} w_{ij}} + \ldots + \frac{w_{nj} r_n}{\sum_{i=1}^{n} w_{ij}} = \overline{w_{1j}} r_1 + \ldots + \overline{w_{nj}} r_n = \sum_{i=1}^{n} \overline{w_{ij}} r_i ,$$

where $\overline{w_{ij}} = \frac{w_{ij}}{\sum_{i=1}^{n} w_{ij}}$ is the normalized degree of truth conditions on the i-th rule in the j-th example.

Thus, in addition to formula (1), the last layer in the Wang and Mendel fuzzy neural production network (Figure 1) functionally realizes a weighted sum of the normalized degrees of truth conditions for each rule with weight coefficients r_i. Therefore, in accordance with the expression obtained, we replace the *4*-th and *5*-th layers of the network in a modified structure isolating a separate layer, the neurons of which correspond to the leaves of the tree [10].

Layer 4 (nonparametric) performs normalization of the truth condition degrees for each of the rules:

$$\overline{w_{ij}} = \frac{w_{ij}}{\sum_{i=1}^{n} w_{ij}}.$$

Layer 5 (linear parametric) consists of the leaves of the tree, each of which activates the corresponding rule conclusions r_i in accordance with the values of the rule conditions normalized in the previous layer:

$$y_{ij} = \overline{w_{ij}} r_i.$$

Layer 6 (nonparametric) performs defuzzification summing up the rule conditions for each class that were activated in the previous layer:

$$y_j = \sum_{i=1}^{n} y_{ij}.$$

The number of neurons in this layer is also determined by the number of classes used in the task, at the same time the output signal of each neuron is the degree of truth to which the example agrees with the corresponding class.

4. Conclusion

Using the structural-functional diagram of the modified Wang and Mendel fuzzy neural production MIMO-network, we have implemented the adaptive neuro-fuzzy classifier with fuzzy output based on a combination of "hierarchical techniques (decision trees) + fuzzy logic + neural networks". This successful combination of intelligent techniques provides a synergistic effect allowing complex usage of the strengths of decision trees (simplicity, high speed, the use of any type of variables, and generation of rules in areas where experts have difficulties in formalizing their knowledge), fuzzy systems (interpretability of knowledge), and neural networks (the ability to learn on the data). The results of the study are listed below.

Acknowledgments

The paper was supported by the competitiveness improvement programme of National Research Tomsk State University (grant No 8.2.24.2018) and by the Russian Foundation for Basic Research (grant No 16-29-04388).

The authors are grateful to Tatiana B. Rumyantseva from Tomsk State University for English language editing.

References

[1] Udovenko S., Sorokin A. 2014 *Information processing systems* **10** 248-254

[2] Gorbachev S. V., Syryamkin V. I 2014 *Neuro-fuzzy methods in intelligent systems for processing and analyzing multidimensional information* (TSU Publishing House: Tomsk)

[3] Smirnov V. I. 2011 *Forecasting and classification of economic systems under conditions of uncertainty by methods of artificial neural networks* (Orenburg)

[4] Khaydukov D. S. 2009 *Philosophy of mathematics: actual problems*

[5] Boyarshinova V. V., Eltyshev D. K. 2015 *Automated control systems and information technologies: materials Vseros. Scientific-practical* 243–248

[6] Rutkovskaya D., Pilinsky M., Rutkovsky L. 2006 *Neural networks, genetic algorithms and fuzzy systems* (Hotline – Telecom: Moscow)

[7] Borisov V. V., Kruglov V. V., Fedulov A. S. 2016 *Fuzzy models and networks* (Hotline – Telecom: Moscow)

[8] Janikow 1993 *Proceedings of the Sixth International Symposium on AI* 360–367

[9] Janikow C. Z. 1998 *IEEE Transactions on Systems, Man, and Cybernetics, Part B: Cybernetics* **28(1)** 1–14

[10] Gorbachev S. V. 2014 *INNOVATIKA-2014* 102–107

IOP Conf. Series: Materials Science and Engineering **516** (2019) 012011 doi:10.1088/1757-899X/516/1/012011

Method of objective assessment of the quality of digital TV image transmitted over NGN network using GPON access technology

P Dunayev, Ye Sarsikeyev

Kazakh Agrotechnical University named after S. Seifullin, Astana, Republic of Kazakhstan
E-mail: dunayev.kz@mail.ru

Abstract. One of the most prospective technologies and one of the ways of data transmission are determined. The parameters of optical loss resource in the network are given. The influence of optical power in the line on quality of digital television image during transmission in NGN telecommunications network using GPON access technology is considered. Equipment NetUP IPTV Combine, Optical Line Terminal, Optical Network Terminal, optical switch, optical attenuator were used as the head station During the experiment. Step dependence of the image quality on optical power in the channel is revealed. Image quality was evaluated by subjective and objective method of evaluation. The least squares method, method of equation of straight line passing through two points were used. There are several figures attached to the text for the purpose of results explanation. Recommendations of the International Telecommunication Union were used in the calculations.

1. Introduction

One of the most developed and actively implemented modern solutions of subscriber broadband access to telecommunication networks is use of PON (Passive Optical Network) technology by laying a fiber-optic cable. Taking into account strategic advantage of the optical network, it will become economically feasible to have optical access to each subscriber's apartment, i.e. FTTH (Fiber to-the-Home) technology in the near future. In this case, it is possible to use any data transfer methods: SDH, ATM, Ethernet. However, preference is mainly given to Ethernet protocols [1, 2]. In currently deployed PON networks, downstream flow from Central node to subscribers goes at wavelength 1490 nm and has speed 622 Mbit/s using ATM Protocol, supporting QoS (Quality of Service) function. Upward data flows from subscribers go at wavelength 1310 nm using Protocol of multiple access with time division TDMA (Time Division Multiple Access) and have total speed from 155 to 622 Mbit/s. In some cases downward channel at wavelength 1555 nm is used for video transmission. Thus, WDM (Wavelength Division Multiplexing) technology is used in optical subscriber access networks [3, 4]. GPON (Gigabit PON) provides a scalable frame structure at transfer rates from 622 Mbit/s to 2.5 Gbit/s, and allows systems with the same rate of forward and reverse flow in PON tree, and with the various ones. GPON is based on ITU-t standard, G. 704.1 GFP (Generic Framing Protocol), providing encapsulation in synchronous transport Protocol of any type of service, including TDM. Services based on GPON have a wide range of applications in terms of functionality and consumer characteristics. This technology is stable and promising, user-friendly [4, 5].

IOP Conf. Series: Materials Science and Engineering **516** (2019) 012011 doi:10.1088/1757-899X/516/1/012011

2. Experimental studies using GPON access technology

Nowadays GPON access technology is not only the most promising, but also the only cable environment for high-quality transmission of various signals. However, there are also problems adversely affecting the quality of digital images during operation of such networks. Purpose of this experiment was to reveal the dependence of digital image quality on optical power in the channel.

2.1. Optical loss resource

Loss resource requirement for PON network, based on Recommendation of ITU G. 983.4, is 22 dB of total loss resources for Class B PON network and 27 dB for Class C PON network.Class B and Class C PON networks are distinguished by the used laser and, indirectly, quality of optical components. Such resource loss is quite tough, especially in case of use in the design of splitters with large number of ports. Splitters in PON network cause internal losses because input power is divided between several outputs. Splitter loss depends on branch ratio and is about 3 dB per 1×2 splitter, increasing by 3 dB each time the number of outputs is doubled. 1×32 splitter has splitter loss of at least 15 dB. Loss of optical power in the channel, in practice, can cause poor quality of optical patch cord and application of number of optical splitters greater than the norm, cable aging, increasing the number of plexus, failure of terminal or station equipment.

2.2. Experimentation

Network configuration: NetUP IPTV Combine (Figure 1) – optical switch – GPON OLT (Optical Line Terminal) – GPON ONT (Optical Network Terminal). Laser class B+.

Figure 1. NetUP IPTV Combine.

Value of optical power in the line was: - 27 dBm, using optical attenuator (Figure 2), signal attenuation was introduced to the limit: - 32 dBm (step: - 1 dB), in which there were changes in the quality of digital TV images.

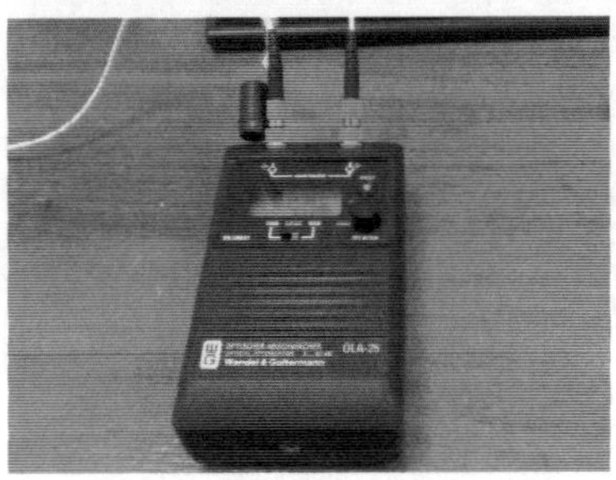

Figure 2. Optical attenuator.

Image quality was controlled at the output of the head end (Figure 4) and at the output of the subscriber line (Figure 4) [6].

Figure 3. Quality of digital TV images with the power of the input signal ONT -27 dBm.

Figure 4. Quality of digital TV images with the power of the input signal of -35 dBm ONT.

3. Subjective assessment of digital television image quality

First stage of analysis of the experimental data is calculation of an average assessment Q_{HDi} and confidence interval for each of the demonstration, according to recommendation ITU – R BT. 500-13 (Table 1) [7].

Table 1. Calculated average values of the estimates taking into account confidence intervals.

Number of a sequence	Q_{HDi}	Q_{HDimax}	Q_{HDimin}	d
1	4.8	5	4.6	- 27
2	4.6	4.9	4.3	- 29
3	4.8	5	4.6	- 31
4	1.1	1.3	0.9	- 32
5	1.4	1.75	1.05	- 33
6	1.1	1.3	0.9	- 35

Parameter d – value of optical power in the line, Q_{HDi}– number of average values, Q_{HDimax} – number of maximum values, Q_{HDimin} – number of minimum values. Based on the data (table 1, let's construct a graphical dependence of digital TV image quality assessment on the optical power in the line (Figure 5).

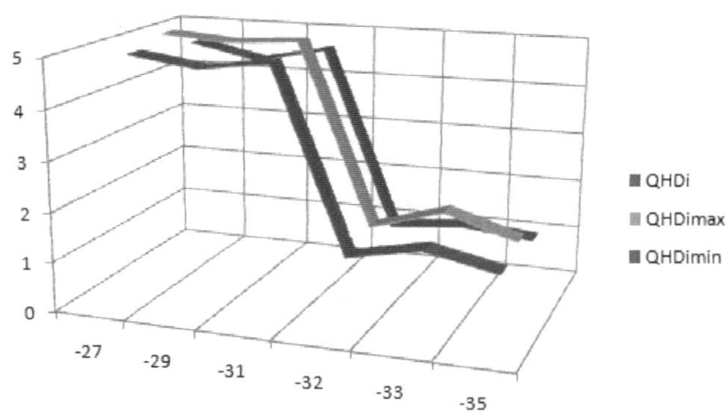

Figure 5. Graphic dependence of number average, maximum and minimum digital image quality on the value of optical power in the line.

To choose constructing method of necessary functional dependence, which can be used to objective assessment of quality of digital television images, it is necessary to determine law of statistical function of estimates distribution.

3.1. Definition of the law of statistical function of estimates distribution

To find out whether estimates distribution in case of test demonstration is normal or not, use test β_2 (calculation of function kurtosis coefficient, i.e. ratio of the fourth order moment to square of the second order moment). If β_2 is between 2 and 4, distribution can be considered as normal. This procedure is recommended in case of application of DSIS method, and DSCQS method and alternative methods [8]. Mean value of Q_{HDi}, standard deviation *Sjkr* and kurtosis coefficient β_2 are calculated for each test demonstration, where β_2 is given by the expression [8]:

$$\beta_{2jkr} = \frac{m_4}{(m_2)^2}, \tag{1}$$

where m_x is defined by the following formula:

$$m_x = \frac{\sum_{i=1}^{N} (Q_{HDijkr} - Q_{HDi})^x}{N}. \tag{2}$$

Moment of the fourth order $m_4$is 0.0832 in accordance with the expression (1), moment of the second order m_2 is 0.16 in accordance with the expression (1) for the first test sequence (table 1). Function coefficient of kurtosis β_2 was equal to 3.25 in accordance with the expression (2). Since the condition $2<\beta_2<4$ is satisfied, distribution for the first test sequence can be considered as normal [9, 10]. Condition $2<\beta_2<4$ is also satisfied for the remaining test sequences.

4. Subjective assessment of digital television image quality

Based on the obtained dependence of the subjective assessment of digital television image quality (Figure 6), it can be seen that dependence of $Q_{HDi}(d)$ is close to functional dependence close to the step function. Let's define mathematical model of the signal using dynamic representation [8], namely using Heaviside function [9–13]. Let's denote the areas (Figure 6).

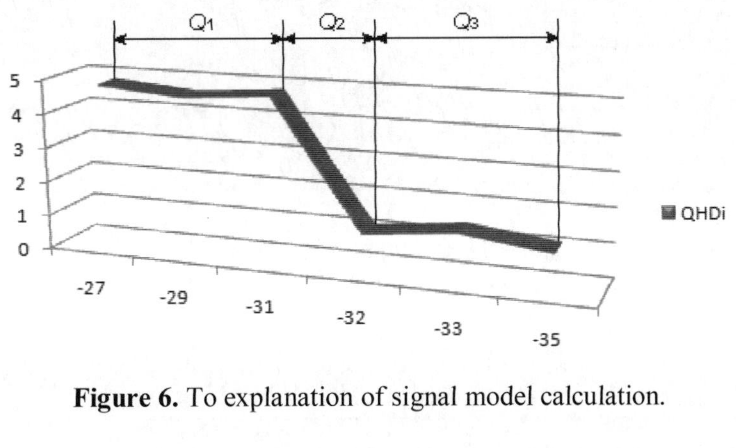

Figure 6. To explanation of signal model calculation.

Mathematical model of the signal (figure 6) is given by the system of equality:

$$Q_{HDi}(d)=\begin{cases} Q_1(d), d \geq 27 \\ Q_2(d), 31 \leq d \leq 32 \\ Q_3(d), d \geq 32 \end{cases}. \tag{3}$$

Let's express Q_1 and Q_3 through multiple regression equations. The least squares method is used to estimate the parameters of the multiple regression equation.

4.1. Estimation of regression equation for Q_1 and Q_3
Let's define vector of regression coefficients estimates. According to the least squares method, vector s is obtained from the expression:

$$s=\left(X^T X\right)^{-1} X^T Y. \tag{4}$$

Regression equation for area Q_1:

$$Q_1 = 4{,}73 + 0d. \tag{5}$$

Regression equation for area Q3:

$$Q_3 = 1{,}91 + 0{,}0214d. \tag{6}$$

4.2. Estimation of equation for Q_2
Estimation of equation for Q2 is calculated using equation of straight line passing through two points, using the expression:

$$\frac{Q_2 - Q_1}{Q_3 - Q_1} = \frac{d - d_1}{d_2 - d_1}, \tag{7}$$

Where:

$$Q_2 = -0{,}214d^2 + 2{,}16d + 92{,}15. \tag{8}$$

4.3. Mathematical expression for objective assessment of digital TV images quality
Let's determine values of Heaviside function for Q1, Q2, Q3.
For Q_1:

$$h(d-27)=\begin{cases} 1, & d \geq 27 \\ 0, & d < 27 \end{cases} \tag{9}$$

For Q_2:

$$h(d-31)=\begin{cases} 1, & 31 \leq d \leq 32 \\ 0, & d < 31 \end{cases} \tag{10}$$

For Q_3:

$$h(d-32)=\begin{cases} 1, & d \geq 32 \\ 0, & d < 32 \end{cases} \tag{11}$$

Let's make mathematical expression for objective assessment of digital TV images quality:

$$Q_{HDi\,ob}(d) = Q_1 h(d-27) + (Q_2 - Q_1)h(d-31) + (Q_3 - Q_2)h(d-32) \tag{12}$$

Let's make a table of number of subjective and objective assessments of digital television image quality (Table 2).

Table 2. Calculated average values of the estimates taking into account confidence intervals.

d	-27	-29	-31	-32	-33	-35
Subjective Q_{HDi}	4.8	4.6	4.8	1.4	1.1	1.4
Objective $Q_{HDi\ ob}$	4.73	4.73	4.73	1.27	1.25	1.21

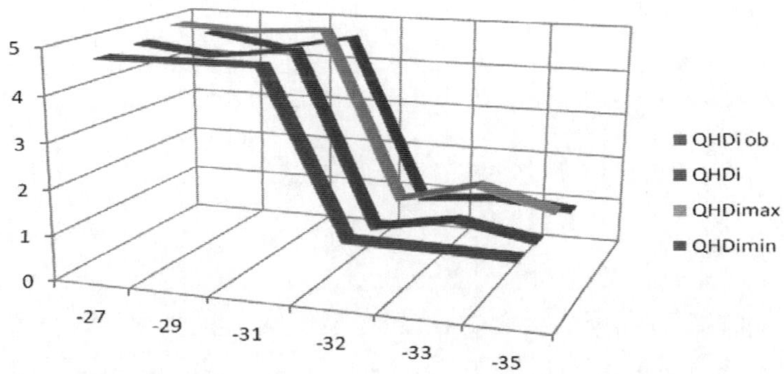

Figure 7. Graphic dependence of number of subjective and objective assessments of digital image quality on value of optical power in the line.

5. Conclusion

Study of GPON access technology revealed that change in optical power in the channel leads to distortions in digital TV image.

Studying dependence of digital TV image quality on optical power in channel, step dependence within the range from – 31 to – 32 dB is determined.

The law of statistical distribution in the subjective assessment of image quality is determined.

Mathematical model of objective assessment of quality of digital TV image was offered on the basis of system of equality of Heaviside function, corresponding to subjective assessment of image quality according to recommendation of ITU-R BT.500-13.

References

[1] Pfeiffer T. *et al* 2005 *Technology and means of communication* **4** 28–33
[2] Manchev G. V. 2009 *the Use in broadcasting of the Internet Protocol* (SibSUTIS: Novosibirsk)
[3] Leonov A., Konyshev V. 2007 *CONNECT-2007* **7** 110–114
[4] GOST R 52023-2003 *Networks of distribution systems of cable television. Main parameters. Specifications.* Measurement and test methods p. 61
[5] Onokopylov I. G. *et al* 2016 *IOP Conf. Ser.: Mater. Sci. Eng.* **189(1)** 012010 doi: 10.1088/1757-899X/189/1/012010
[6] Dcommendation ITU-R BT.500-13 *Methodology for subjective assessment of quality of television images* p. 46.
[7] Obukhov S.G., Sarsikeyev Ye.Zh., Sholokhova I.I. 2016 ICIEAM 2016 7911454
[8] Plotnikova I. V. *et al* 2018 *IOP Conf. Ser.: Mater. Sci. Eng.* **363 (1)** 012006 doi: 10.1088/1757-899X/363/1/012006
[9] Zhumagulov K., Sarsikeyev Y., Kapanova D. 2018 *MATEC Web of Conferences* **155** 01048

[10] Baskakov S. I. 2000 *Radio circuits and signals* (Higher education: Moscow)
[11] Akimzhanov T. B., Kharlov N. N., Borovikov V. S. and Ushakov V. Ya. 2014 *IFOST 2014* 6991138
[12] Akimzhanov T.B., Zhumazhanov S.K., Ivanova Y.V., Masabaev B.S. 2018 *IOP Conference Series: Materials Science and Engineering* **457(1)** 012019
[13] Plotnikova I. V. *et al* 2016 *IOP Conf. Ser.: Mater. Sci. Eng.* **457(1)** 012021 doi: 10.1088/1757-899X/457/1/012021

The research work of a submersible washing machine with fluid activation by air bubbling from the outside of the laundered objects

A V Mayorov, N V Yanukov, D A Mikheeva, N E Yaytseva, D V Lukina

Mari State University, Yoshkar-Ola, Russia
E-mail: ao_maiorov@mail.ru

Abstract. The research aimed at improving the process of washing filled cylindrical cans in the production of tin cans is very relevant and of great importance in the food industry. So, a new two-section washing machine design is proposed. Tests performed at the machine under consideration. The results of experimental studies of the quality of surface cleaning metal cans were presented. The optimum operating mode of the washing machine was determined. The minimum energy consumption of the machine is $E_{min} = 69 \cdot 10^{-3}$ W·h / pcs at the optimal driving wheel speed $n_{opt} = 31$ min^{-1}. For comparison - for analogs the minimum energy intensity is $E_{min} = 200 \cdot 10^{-3}$ W·h / pcs. As a result of experimental studies, it has been established that for the qualitative cleaning of the surface of filled cans, which corresponds to the requirements of the standard, the rotational speed of the drive wheel $n_{opt} = 31$ min^{-1}, temperature $t = 74 ... 85$ ° C and concentration of the washing solution $c = 1.7... 3$ g / l. When the capacity of the washing machine is increased to 100 pcs / min. it is required to increase $n_{opt} = 31$ min^{-1} to $n_{max} = 35$ min^{-1}, with $c = 1.7$ g / l and the temperature of the washing solution $t = 72 ... 85$ ° C. At the same time, energy intensity increases by only 2%. The design features of the developed washing machine allow to reduce the power consumption of the cleaning process, and the productivity can be increased without loss of quality of cleaning the cans in comparison with existing washing machines.

1. Introduction

Every year, in the Russian market of tin cans appear more and more new trademarks, new manufacturers, which aspire to improve the quality of produced product and attract a buyer [1–3]. In this situation, the packaging and label of the goods are powerful means of promoting the goods on the market. Labels stick to clean, dry cans with no traces of fat on the surface. To take this into account, technological process of canning production included operation for degreasing cans with a washing machine [3]. A large number of modern washing machines work in an irrational regime and are energy and metal consuming, so their use is unprofitable. Consequently, the development of new efficient machines and the improvement of existing washing machines in the canning industry is great reserve to reduce energy consumption, materials and cost of the whole production process of canned food [4].

Therefore, research aimed at improving the washing process filled cans in the production of canned food are very relevant and have great importance in the food industry.

2. Experimental

Taking into account the flow of canning production, the results of the analysis of the dependence of the quality of cleaning on the type of relative movement of the cleaning object and the designs of existing devices for washing cans, we proposed a two-stage washing machine with planetary movement of the cleaning object and activation of the liquid by air bubbling from the outside of the washed objects [5–8].

The machine (Figure 1) includes washing tanks 1 and 2 separated by a partition 3. Arched guides 4 and 5 is placed in them, driving wheels 7 and 11 with an elastic rim are symmetrically mounted on the shafts above each guide. Perforated pipelines (bubblers) are fixed on two lateral sides of guides 4 and 5.

The washing machine works as follows. Jars are fed into the guides 4 of the washing machine after canning. Cans roll forward and are completely immersed in the cleaning solution, fall into the gap between the elastic rim of the wheel 7 and the guide 4 and move due to the frictional force between the rubber rim of the wheel and the can in the direction of rotation of the wheel. Further, along the guide of the first bath, jars come out of the cleaning solution and fall on a convex arc-shaped section of the transition from one bath to the other. At the same time, the spent washing solution of the first bath drains from the cans and carries particles of contaminants with them. Then, the cleaning objects are fed to the guide of the second bath 2, arranged similarly to the first one. The technological process is repeated. The processed cans are sent to the autoclave basket or to another shop receiver.

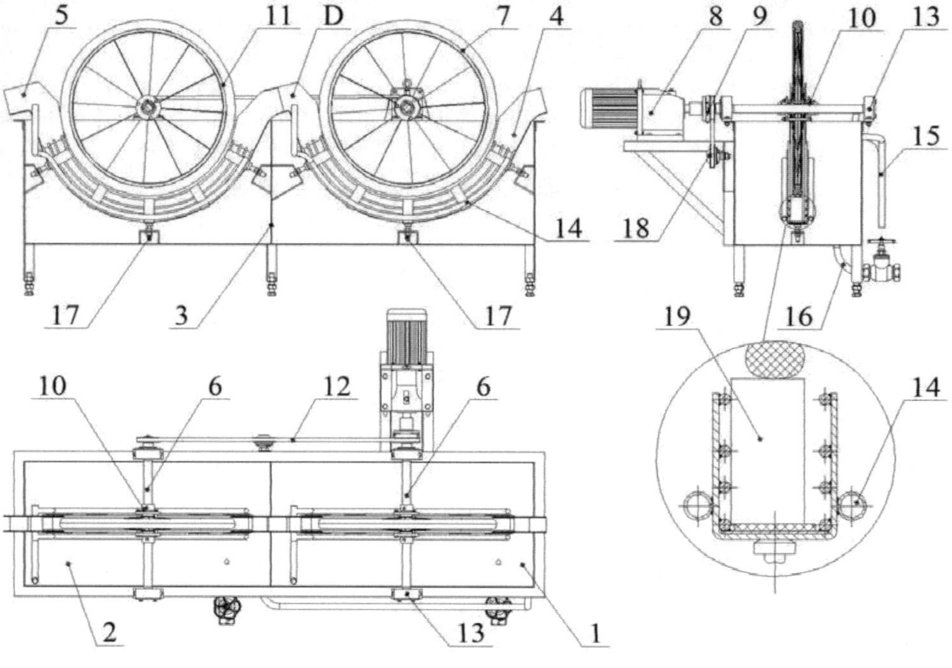

Figure 1. The scheme of the submersible washing machine with activation of the liquid by air bubbling from the outside of the washed objects, where 1, 2 are washing baths; 3 is partition; 4, 5 are guides; 6 is shaft; 7 is drive wheel; 8 is motor-reducer; 9 is coupling; 10 is the hub; 11 is driven wheel; 12 is a V-belt transmission; 13 is housing; 14 is bubbler; 15 is overflow branch pipe; 16 is outlet branch pipe; 17 is fixing and adjustment stud; 18 is tensioner; 19 are cans; D is section of the transition.

Serial measuring and recording equipment and devices were used to register process parameters during the washing machine work. In the study of the process of washing the surfaces of tin cans with a washing machine, the factors that are independent of each other and which most influence the purification process, such as the temperature and concentration of the washing solution, the speed of the driving wheel, aroused the greatest interest. These factors were investigated by us during the technological testing of the device. All other factors either depend on the three above, or have little effect on the washing process. The method of A.K. Koshcheeva with the use of cotton strips is used

IOP Conf. Series: Materials Science and Engineering **516** (2019) 012012 doi:10.1088/1757-899X/516/1/012012

as the feedback of the quality of purification. The essence of this method was to determine the area of the spot on the strips.

To determine the area of the spot on the indicator cotton strip, was used the program "AreaS", developed by the Samara State Agricultural Academy by Permyakov A.N. The error in determining the area does not exceed 0.001%.

The program of experimental studies included several stages and consisted of preliminary single-factor and complete three-factor experiments. Levels and intervals of variation of factors are given in the Table 1.

Table 1. Levels and intervals of factor variation

Coded factor Identification	The name of the factor, unit of measurement	The level of factor			The interval of variation
		-1	0	1	
x_1	Solution temperature, °C	65	75	85	10
x_2	Concentration of washing solution, g/l	1	2	3	1
x_3	Drive wheel rotation speed, min^{-1}	25	30	35	5

In order to study the influence of the speed of rotation of the driving wheel, the temperature and concentration of the solution individually on the quality of cleaning the outer surface of the cans, one-factor experiments were conducted in which the temperature and the concentration of the washing solution were changed at fixed frequencies of rotation of the driving wheel of the washing machine. Experimental studies were conducted with bubbling the washing solution. The pressure in the air supply system for bubbling was kept constant at 0.5 MPa. The diameter of the holes on the arcuate bubbler was 2.5 mm. The experiments were carried out three times.

3. Results and Considerations

Analysis of the results of single-factor experiments showed the following.

1. With the increase in the speed of rotation of the drive wheel of the washing machine at a constant temperature of the cleaning solution, the quality of cleaning meets the requirements of the standard at a lower concentration of the washing solution. At a frequency of n = 15 $min^{-1}$and a temperature of 85 ° C, the concentration of the washing solution must be at least 3.5 g / l. The same quality of cleaning is achieved at a concentration of 1 g / l of washing solution, its temperature is 85°C and the driving wheel speed n = 35 min^{-1};
2. As the speed of the drive wheel of the washer increases, the required quality of cleaning is achieved at a lower solution temperature. At n = 15 min^{-1}, the minimum temperature of the cleaning solution is 68 ° C at a concentration of 5 g / l. The same cleaning quality is achieved at a solution temperature of 48 ° C, a concentration of 5 g / l and n = 35 min^{-1}.
3. The quality of cleaning intensively increases with an increase in the concentration of the cleaning solution from 1 g / l to 3 g / l. Further increase in the concentration of the detergent solution improves the quality of the cleaning very slightly.

The orthogonal central composition plan of the three-factor experiment is realized in the study, which makes it possible to obtain an adequate dependence of the cleaning quality on the investigated factors: the temperature of the washing solution is t, ° C (x_1), the concentration of the washing solution is C, g / l (x_2) n, min^{-1} (x_3). For the feedback function, the quality of cleaning is accepted - K, % (Y).

After implementing the plan of experiment and processing the results, a regression equation was obtained that was tested for adequacy according to the $F_{0.95}$Fisher's criterion:

$$Y = 4.969+1.501\cdot x_1+14.378\cdot x_2-0.015\cdot x_1\cdot x_3-0.139\cdot x_2\cdot x_3-1.012\cdot x_2^{2}+0.038\cdot x_3^{2}.$$

To study the effect of factors on the optimization criterion Y, two-dimensional sections of the response surface were used (Figures 2–4).

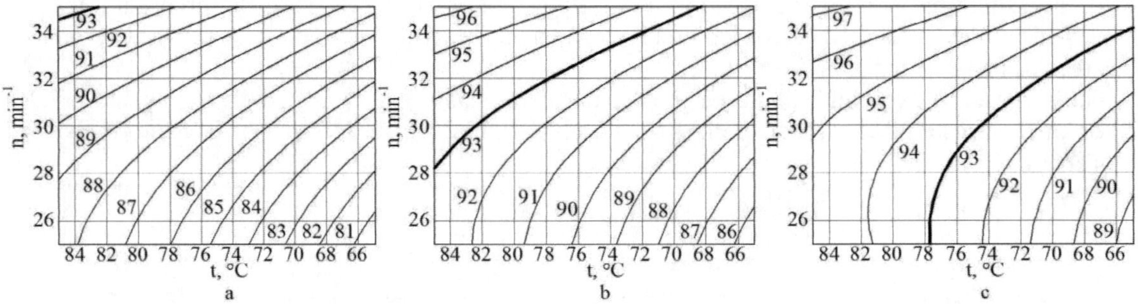

Figure 2. The two-dimensional section of the response surface characterizing the dependence of the cleaning quality (*Y*) on the factors x_1 (the temperature t of the washing solution) and x_3 (the frequency n of rotation of the driving wheel) for a fixed value of the factor x_2: a - at $x_2 = -1$ (C = 1 g / L); b - at $x_2 = 0$ (C = 2 g / l); c - at $x_2 = 1$ (C = 3 g / l).

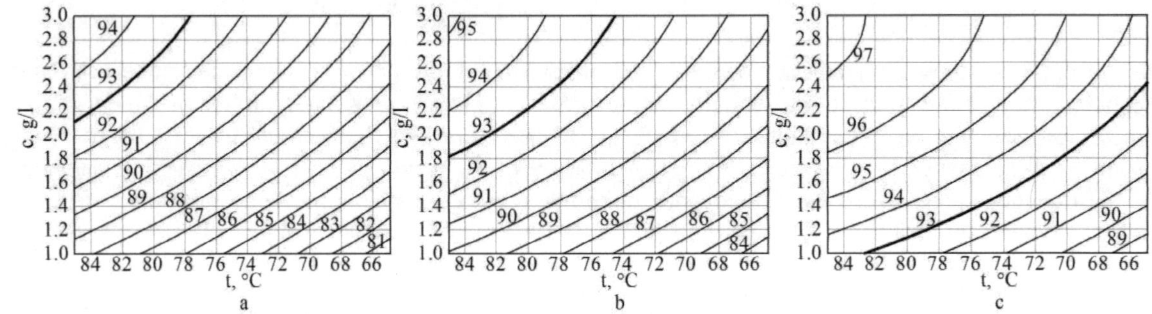

Figure 3. A two-dimensional section of the response surface characterizing the dependence of the purification quality (Y) on the factors x_1 (washing solution t temperature) and x_2 (concentration C of the washing solution) for a fixed value of x_3: a - at $x_3 = -1$ ($n = 25$ min^{-1}); b - at $x_3 = 0$ ($n = 30$ min^{-1}); c - at $x_3 = 1$ ($n = 35$ min^{-1}).

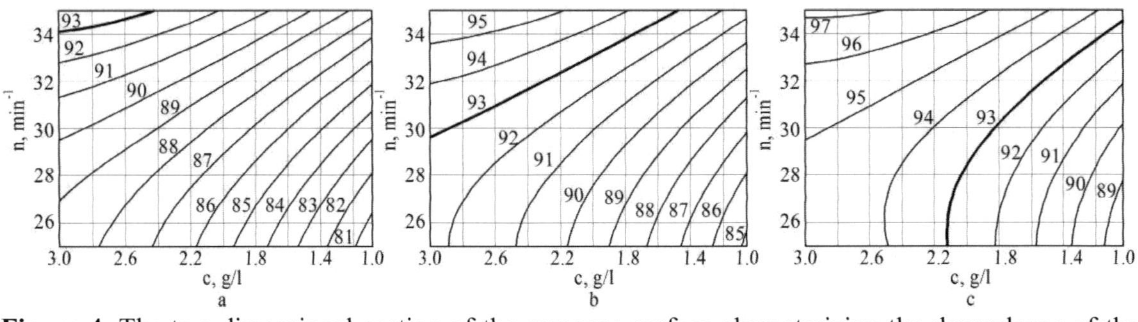

Figure 4. The two-dimensional section of the response surface characterizing the dependence of the cleaning quality (*Y*) on the factors x_2 (concentration C of the washing solution) and x_3 (the frequency n of rotation of the driving wheel) for a fixed value of the factor x_1: a - for $x_1 = -1$ (t = 65 °C) ; b – at $x_1 = 0$ ($t = 75$ ° C); c - at $x_1 = 1$ ($c = 85$ °C).

On the basis of the analysis of two-dimensional sections of the response surfaces [1], it can be noted that the washing machine provides the required quality of cleaning cans at speeds of rotation of the driving wheel 30 ... 35 min^{-1}, the temperature of the washing solution 75 ... 85 ° C and the concentration of the solution is 1 ... 2 g / l.

To obtain the appropriate quality standard for cleaning, in our case power from 361.5 W to 417 W corresponding to the rotation speed of 30 $min^{-1}$and 35 $min^{-1}$of the driving wheel is required. At the same time, theoretically and experimentally determined efficiency coefficients of power use η will be in the range 0.24 ... 0.26 [2]. For comparison, for submerged washing machines of vibration type η = 0.1 ... 0.2, and for jet washers η = 0.05 ... 0.10.

The minimum energy consumption of the machine is $E_{min} = 69 \cdot 10^{-3}$ W·h / pcs at the optimal driving wheel speed n_{opt} = 31 min^{-1} [2]. For comparison - for analogs the minimum energy intensity is $E_{min} = 200 \cdot 10^{-3}$ W·h / pcs.

As a result of experimental studies [1], it has been established that for the qualitative cleaning of the surface of filled cans, which corresponds to the requirements of the standard, the rotational speed of the drive wheel n_{opt} = 31 min^{-1}, temperature t = 74 ... 85 ° C and concentration of the washing solution c = 1.7 ... 3 g / l. When the capacity of the washing machine is increased to 100 pcs / min. it is required to increase n_{opt} = 31 min^{-1} to n_{max} = 35 min^{-1}, with c = 1.7 g / l and the temperature of the washing solution t = 72 ... 85 ° C. At the same time, energy intensity increases by only 2%.

The design features of the developed washing machine allow to reduce the power consumption of the cleaning process, and the productivity can be increased without loss of quality of cleaning the cans in comparison with existing washing machines.

4. Conclusion

As a result of statistical processing of the experimental data, a mathematical dependence of the quality of cleaning cans on the main independent factors of the washing process was obtained, which makes it possible to determine the optimum technological parameters of the washing machine.

The required quality of cleaning the surfaces of cans is achieved at a drive wheel speed of 31 min^{-1}, the temperature of the washing solution is 74 ... 85 °C and the concentration of the washing solution is 1.7 ... 3 g / l. At the same time, the energy intensity of the process does not exceed $69 \cdot 10^{-3}$ W·h per one can.

References

[1] Rauf S. *et al* 2016 *Procedia Comput Sci* **83** 902–906
[2] Abdel-Fatah M. A. *et al* 2017 *Ain Shams Engineering Journal* **8(3)** 305–310
[3] Galtseva O. V. *et al* 2016 *IOP Conf. Ser.: Mater. Sci. Eng.* **132** 012003 doi: 10.1088/1757-899X/132/1/012003
[4] Jain S. V., Patel R. N. 2014 *Renew Sust Energ Rev* **30** 841–868
[5] Corcoran G. D. *et al* 1994 *J Hosp Infect* **27(4)** 307–315
[6] Plappally A. K., Lienhard V. J. H. 2012 *Renew Sustain Energy Rev* **16(7)** 4818–4848
[7] Yurchenko A.V. *et al* 2017 *Journal of Physics: Conference Series* **88(1)** 012034
[8] Tadesse M. G. *et al* 2017 *Smart Mater Struct* **26** 065016

IOP Conf. Series: Materials Science and Engineering **516** (2019) 012013 doi:10.1088/1757-899X/516/1/012013

The analysis of the process of complaints' consideration using statistical methods of control

I V Plotnikova, L A Redko, N V Chicherina

National Research Tomsk Polytechnic University, Tomsk, Russia
E-mail: inna@tpu.ru

Abstract. The paper considers the task of reducing the time for processing claims using statistical methods for monitoring and analyzing data. The main significant factors of occurrence of inconsistencies are identified. Descriptive statistics were used to obtain primary information. Control charts of individual values were constructed and analyzed on the basis of the Pareto methodology. A statistically significant difference between groups of data was determined on the basis of analysis of variance and comparison of graphs. Based on the results of the analysis, the stage of the claims processing process was identified, which had the longest duration. Developed system solutions that reduce the length of the claims process.

1. Introduction

Formation of an effective control system of the organization is one of key factors of its success and stability [1]. Now the problem connected with low-quality products is the important reason of loss of financial resources by the state [2, 3]. For improvement of quality of products the organizations set the tasks of management of discrepancies and the reasons of defects of a product [4]. For the most effective management of low-quality products the statistical methods [5–7] are used which demonstrate a real situation in production and also simplify calculations for managers of the organization [8]. As the international practice shows, the effective mechanism of improving competitiveness is introduction of the process focused model of management of the organization. In this case activity of the organization is represented as the system of the interconnected and interacting processes [9–11]. Such model of management allows providing high extent of adaptation of the organization to the rapidly changing external environment [12]. For this purpose statement of the purposes on levels of management is regularly made [13], monitoring of achievement of the goals, assessment of effectiveness of activity [14] and actions for increase in efficiency and effectiveness of processes are developed.

2. Research description

The production company considered in this article is a modern industrial enterprise with a full production and design cycle on release of devices of measurement of pressure.

The product range includes more than two hundred names and also various modifications of products. The purpose of the enterprise is the set result of its activity.

The purposes of the enterprise can be corrected by results of the complex analysis of the external environment and internal opportunities of business [15]. In the organization the account and the analysis of the arriving information from consumers are kept. For a research data for the first and fourth quarters 2016 and for the first quarter 2017 were obtained.

For the initial analysis the method of descriptive statistics was used. The task of descriptive statistics is to reduce values of selection to several totals using mathematical tools which give a representation of selection [16].

Control cards are considered as one of the most important tools of statistical management of the quality. They represent rather simple graphic means of assessment of controllability (stability) of process of production by results of comparison of separate measurements with the set control borders.

The theory of control cards distinguishes two types of variability.

The first type is variability because of "accidental (of usual sizes), caused by an uncountable set of the various reasons which are present constantly and which hardly or cannot be revealed. Each of such reasons makes very small share of the general variability, and none of them is significant in itself. Nevertheless, the sum of all these reasons is measurable and is supposed that it is internally inherent in process. The exception or reduction of influence of the usual reasons requires management decisions and allocation of resources on improvement of process and a system.

The second type is connected with the real changes in process. They can be a consequence of some defined reasons not inherent in process internally, and can be eliminated. These established reasons are considered as the "nonrandom" or "special" reasons of change. Breakage of the tool, insufficient uniformity of material, the production or control equipment, qualification of personnel, failure to follow procedures, etc. can be referred to them. The main objective of control cards using is to find and separate the accidental deviations connected with the usual reasons from the deviations caused by action of the special reasons.

The considerable number of discrepancies and delays arises because of limited number of the reasons. For identification of the most significant and essential factors influencing emergence of discrepancies Pareto's chart is used. It gives the chance to establish a priority to the actions necessary for a solution. Besides, Pareto's chart and Pareto's rule allow to separate important factors from little significant and insignificant ones.

It is necessary to develop the form of the table in which enter for creation of the chart:

— types (signs) of cases, the facts (it is the better to have data in the decreasing order - at the beginning of the table the event type having the greatest number of repetitions, at the end of the table - the smallest one);

— number of emergence (repetitions) of each type;

— the saved-up sum of number of each type (with the accruing result: to number of the previous type the following increases);

— number percent on each sign in total amount;

— the saved-up percent (with the accruing result).

On the basis of creation of Pareto's charts, by rule 80/20, it is possible to reveal groups of devices on which repair it's spent more time.

For more detailed comparison of the maximum terms of inquiry consideration, having data for every quarter, we will divide the period for decades and we will construct schedules.

For comparison of the maximum number of days for considerations of the complaints in a quarter we will carry out the dispersive analysis. A task of the dispersive analysis is studying of influence of one or several factors on the considered sign.

3. Interpretation of results

On the basis of the conducted selective research and indicators of descriptive statistics calculated on this selection with the level of reliability of 95%, it is possible to draw a conclusion that the average value of time of consideration of the complaints counting on one device is twenty days in 2016; nine days in the first quarter of 2017. The minimum quantity of days of consideration of the complaints counting on device in 2016 is one day, in first quarter 2017 is one day. The maximum number of days of consideration of the complaints counting on one device in 2016 is eighty-seven days; in first quarter 2017 is twenty-eight days.

IOP Conf. Series: Materials Science and Engineering **516** (2019) 012013 doi:10.1088/1757-899X/516/1/012013

On the provided data the control cards of individual values and the sliding swings are constructed. The example of the control card is presented in Figure 1.

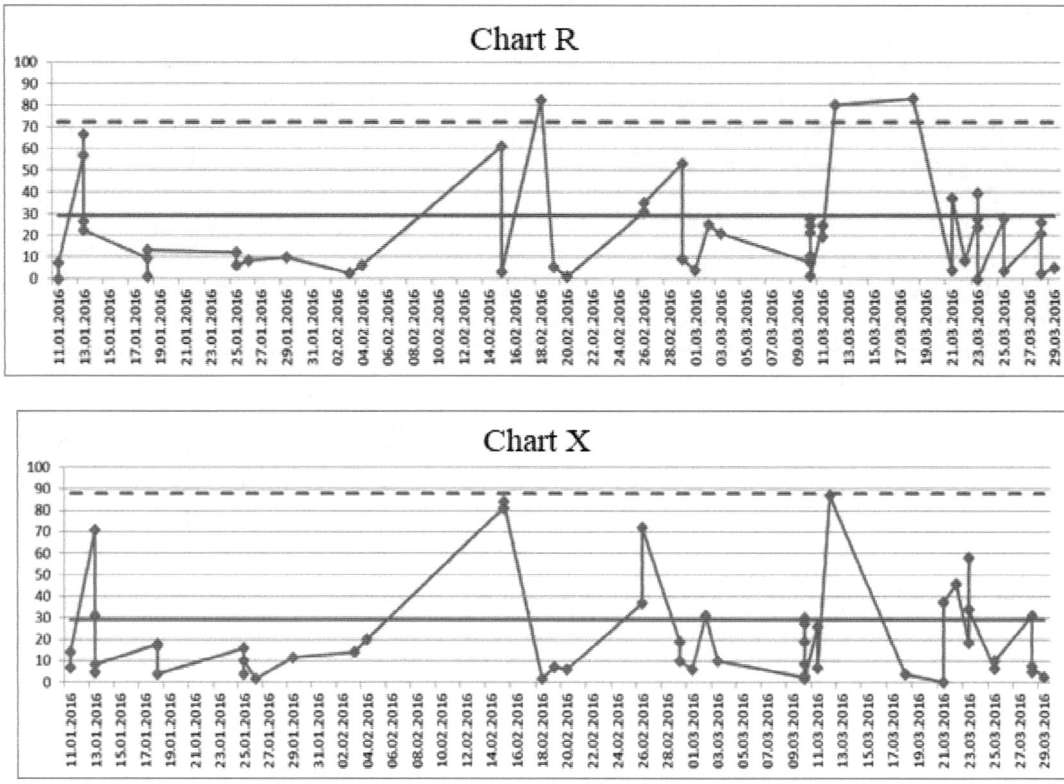

Figure 1. Control charts of individual values and sliding ranges.

At interpretation of control cards, it was revealed that the most stable indicators are indicators in the 1st quarter 2017. Control cards for 2016 show a significant amount of the points overstepping the upper control bounds.

On the basis of creation of Pareto's charts, using the principle 80/20, revealed groups of devices on which repair it's spent more time, the maximum number of days of consideration of the complaints counting on one device. Such groups on all available data in the first quarter 2016, for the fourth quarter 2016, in the first quarter 2017 were defined. The example of the Pareto's chart on types of devices is given in Figure 2.

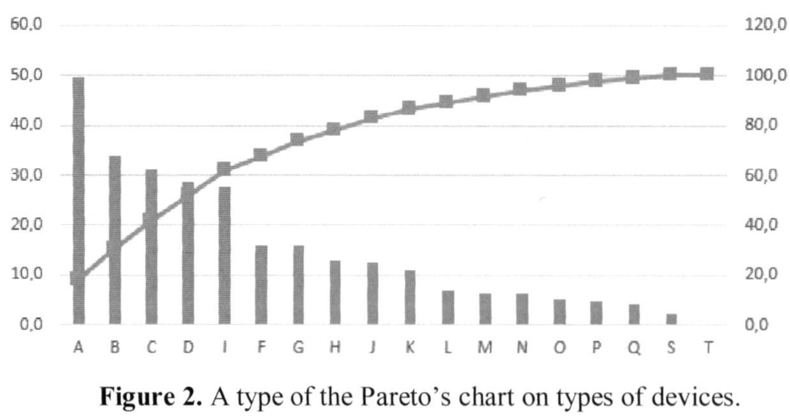

Figure 2. A type of the Pareto's chart on types of devices.

Or more detailed comparison of the maximum terms of inquiry consideration, having data for every quarter it is possible to construct the schedule. The example of comparison of the maximum terms of inquiry consideration between the first quarters 2016 and 2017g., is shown in Figure 3. The same way schedules of inquiries between other quarters 2016 and 2017.

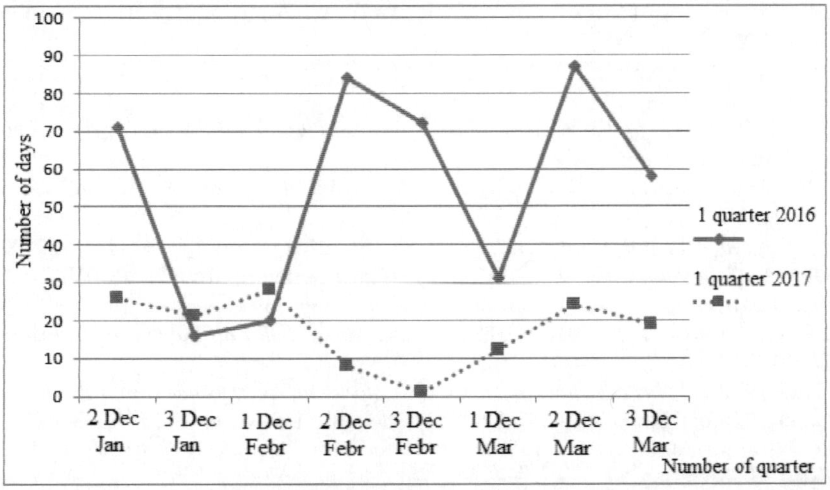

Figure 3. The schedule of comparison of the maximum terms of inquiry consideration between the first quarters 2016 and 2017.

In the course of analysis of variance, comparing the figures for 2016 and 2017, we found out that there is a big difference between the first quarter of 2016 and the first quarter of 2017. This is due to the fact that in 2017, a number of indicators have improved.

In spite of the fact that indicators of 2017 most answer the purpose of the enterprise, nevertheless there is a group of devices, time of consideration of complaints for which goes beyond average value. Having studied the register of receipt of complaints for devices, it is possible to divide the entire period of complaints consideration into three stages:

— Date of receipt of the complaints for the device.
— Date of drawing up of the act (control dismantling).
— Date of a release of the order.

Further time to the control dismantling and the general time of complaints consideration are analyzed. It is established that time of complaints consideration is defined by time of dismantling of the device. Then devices for which more working days is spent are revealed.

4. Conclusions

Being based on the statistical data it is possible to draw a conclusion that the reasons causing a delay of consideration complaints of devices in most cases are:

— a delay of terms of adoption of the response decision (letter) from the consumer.
— the increased terms of the procedure of consideration of the complaints, namely time to control dismantling.

Having adapted this method of process optimization, improvements of process of complaints consideration at a stage from receipt of the device on the claim to control dismantling were developed.

Thus, in order to achieve the target time for processing claims, the stage of the process was optimized, the loss of time on which was most significant. As measures, parallel execution of work and elimination of unnecessary coordination between those responsible for the stages of the process were used.

References

[1] Evans J. R. 2005 *Total Quality: Management, Organization and Strategy* (South-Western College Pub: US)

[2] Nizhegorodov A. I. *et al* 2018 *Bulletin of the Tomsk Polytechnic University, Geo Assets Engineering.* **329(5)** 67–75
[3] Nizhegorodov A. *et al* 2016 *Vibroengineering* **18** (6) 3734–3742. doi: 10.21595/jve.2016.16994
[4] Redko V. V. *et al* 2016 *J. Phys.: Conf. Series* **671(1)** 012049 doi: 10.1088/1742-6596/671/1/012049
[5] Neave H. R.1987 *Journal of the Royal Statistical Society Series D The Statistician* **36(5)** 561–570
[6] Plotnikova I. V. *et al* 2015 *IOP Conf. Ser.: Mater. Sci. Eng.* **132(1)** 012023 doi: 10.1088/1757-899X/132/1/012023
[7] Plotnikova I. V. *et al* 2016 *IOP Conf. Ser.: Mater. Sci. Eng.* **457(1)** 012021 doi: 10.1088/1757-899X/457/1/012021
[8] Deming W. E. 1986 *Out of the Crisis: Massachusetts Institute of Technology* (Center for Advanced Engineering Study: Cambridge, USA)
[9] Obukhov S. G. *et al* 2017 *IOP Conf. Ser.: Mater. Sci. Eng* **189(1)** 012008 doi: 10.1088/1757-899X/189/1/012008
[10] Obukhov S. G. *et al* 2017 *IOP Conf. Ser.: Mater. Sci. Eng* **289(1)** doi: 10.1088/1757-899X/289/1/012026
[11] Odnokopylov I. G. *et al* 2016 *IOP Conf. Ser.: Mater. Sci. Eng.* **189(1)** 012010 doi: 10.1088/1757-899X/189/1/012010
[12] Vavilova, G.V., Ryumkin, A.V. 2018 *IOP Conf. Ser.: Mater. Sci. Eng* **289**(1) 012017 doi: 10.1088/1757-899X/289/1/012017
[13] Gavrilin A *et al* 2014 *Proceedings of 2014 International Conference on Mechanical Engineering, Automation and Control Systems, MEACS* 6986947 doi: 10.1109/MEACS.2014.6986947
[14] Yurchenko V *et al* 2018 *IOP Conf. Ser.: Mater. Sci. Eng* **457(1)** 0120237 doi: 10.1088/1757-899X/457/1/012023
[15] Siegel A. 2011 Practical Business Statistics Academic Press
[16] Abishev N. K. *et al* 2016 *Rupkatha Journal on Interdisciplinary Studies in Humanities* **8(2)** 117–127 doi: 10.21659/rupkatha.v8n2.14

IOP Conf. Series: Materials Science and Engineering **516** (2019) 012014 doi:10.1088/1757-899X/516/1/012014

Non-contact detection of hidden objects under clothes

A I Eremeev, S E Shipilov, R N Satarov, I S Fedyanin, V L Khmelev, V P Yakubov

Tomsk State University, Tomsk, Russia
E-mail: aleksandr_eremeev_93@mail.ru

Abstract. The security portal is a set of technical devices united by a single software in an information environment with a common database. We design the portal for highly effective detection of the maximum number of threats: metal objects and electronic products hidden on the body and in clothes.

1. Introduction

The security portal allows to detect hidden radio-electronic devices (cell phones, bomb fuses, transmitters, miniature video cameras, listening devices). Figure 1 shows the appearance of portal.

Figure 1. Appearance of portal.

The portal is a combination of technical devices [1]. The metal detection subsystem provides real-time positioning of single or group objects under a person's clothing, which are made of metal or containing metal. The subsystem allows accurately control the location of a metal object on the human body. The subsystem splits the scanned space into 21 control zones. The porta controls seven zones in the vertical direction and three in the horizontal directions.

The subsystem of the radio-electronic location detects hidden radio-electronic devices such as cell phones, bomb fuses, transmitters, miniature video cameras, listening devices, metal elements on the human body, etc. [2]. In the same time, devices could be in the off state. The sub-system controls the location of electronic objects with high accuracy on the human body. The portal controls four zones in the vertical direction and two in the horizontal. The explosives detection subsystem uses supersensitive semiconductor sensors [3], which reacts to the content of NO2 vapors in air.

The portal could enhanced with the biometrics system by installing a video camera. Figure 2 shows software interface window.

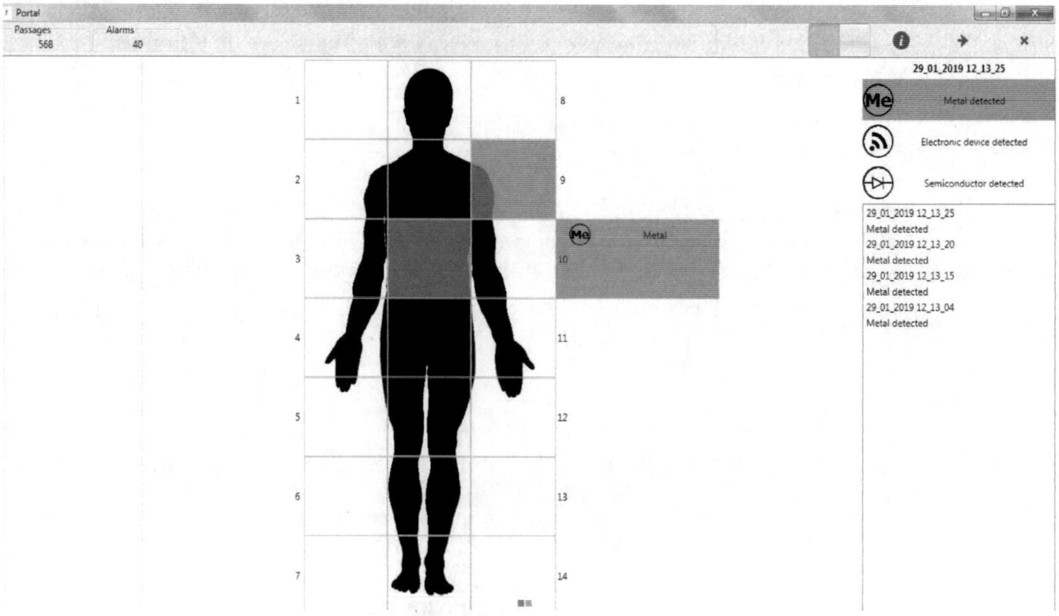

Figure 2. Interface window.

2. Device and operation

The design has two side panels. In each side panel there are seven receiving and one transmitting magnetic coils, passage sensors, and also a detector of semiconductor elements. Which is designed to search and detect electronic devices, both in active / disactive state.

We applied the eddy current method with a harmonic magnetizing field to detect metallic objects. The system of coils creates a uniform field in the control zone, which gives uniform sensitivity in the control zone. Digital processing of signals from receiving coils provides high product selectivity. The detector of semiconductor elements is based on the property of semiconductor elements to re-radiate the second and third harmonics when irradiated elements with microwaves. Maximum response from semiconductor elements of artificial origin at the second harmonic of the probing signal. Oxide films of natural origin and the accumulation of small metal fragments, which are often found in explosive elements during irradiation, give the maximum response at the third harmonic. Digital processing of the demodulated signal gives maximum sensitivity. The detector of semiconductor elements analyze responses from irradiated objects for the second and third harmonics of the probe signal, which makes it possible to identify electronic devices and natural oxide semiconductors in the same time. Also, the portal can simultaneously identify the levels of the second and third harmonics (Figure 3). The operator controls and changes the parameters of the portal with the help of a personal computer.

All subsystems combined on the software complex into the common information environment. Its designed for highly efficient detection of the maximum number of threats. The embedded software determines the type of prohibited item on the human body and shows to operators an approximate location on the human body.

Basic software features:

- Identify the type of hidden object (metal or electronic device).
- Determining the location of hidden objects on the human body.
- Sensitivity setting.
- Archiving events and photos on a remote server.
- Networking from multiple portals.
- Remote control.

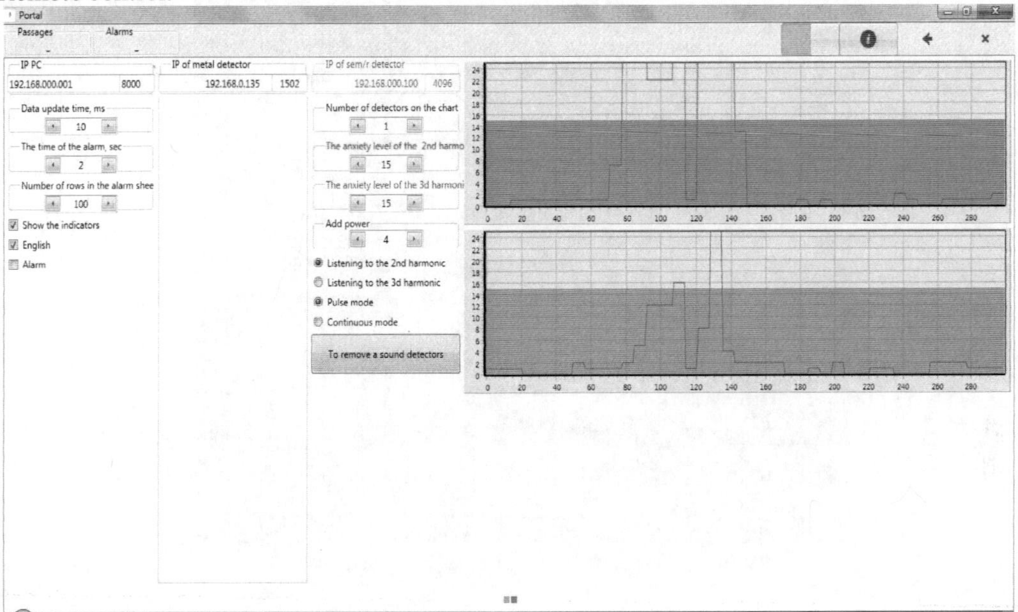

Figure 3. Nonlinear object detection.

3. Comparisons with analogues

Now, there are several approaches to detecting prohibited items under clothing:

Radio-wave scanners are designed for non-contact, safe and prompt inspection of citizens in order to prevent any substances, materials and products from 3–5 mm in size to the protected area [4, 5]. They are used at checkpoints organizations and enterprises

Advantages: high resolution, human safety

Disadvantages: requires a qualified operator (with a shift every 30 minutes), the high cost of the device (about $ 200,000), low throughput (six people per minute).

X-ray scanner.

Advantages: high resolution, detection of objects (drugs) hidden inside a person. *Disadvantages*: requires a qualified operator (with a shift every 30 minutes), the high cost of the device (about $ 100,000), low throughput (5 people per minute), unsafety for people in cases of frequent use.

The portal (Figure 4).

Advantages: low cost, the ability to detect a wide range of prohibited items, ease of maintenance does not require a specially trained operator, high throughput (up to 30 people per minute). Change a standard metal detector to the portal and starting to use. It is enough.

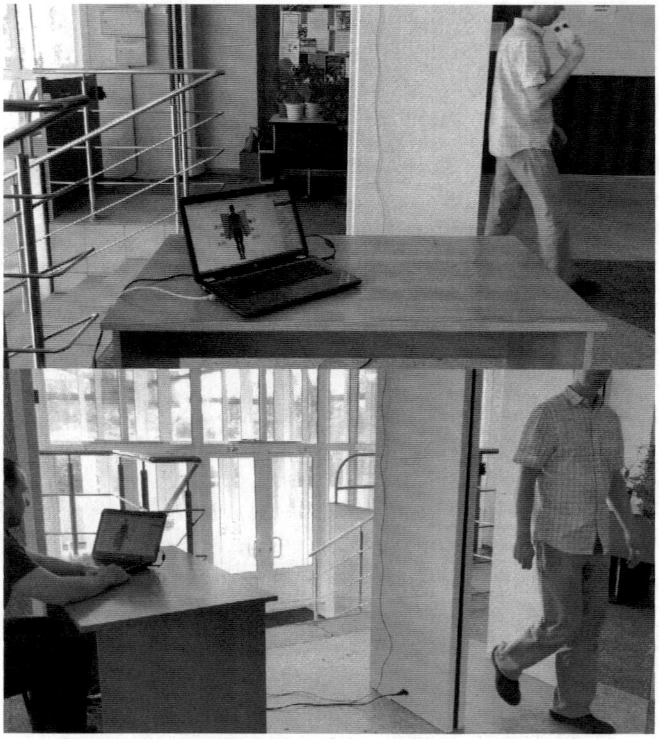

Figure 4. An example of the safety portal works.

4. Area of use

- Aviation, rail, water transport.
- Government agencies and organizations, corporate headquarters.
- Objects of increased danger, including military facilities, nuclear and hydroelectric power plants.
- Industrial enterprises, including non-ferrous metallurgy and jewelry enterprises.
- Warehouse terminals, archives, storage.
- Banks, data centers.
- Health care institutions, educational institutions, universities, museums.
- Embassies and diplomatic missions.

Acknowledgments
The research was supported by the Russian Ministry of Education and Science as a part of the state order No 3.2068.2017/4.6.

References

[1] Shipilov S. *et al* 2018 *High energy and special materials: demilitarization, antiterrorism and civil application «HEMs-2018»* (Tomsk : Publishing house of Tomsk State University)
[2] Antipov V. *et al* 2018 *MATEC Web of Conferences* **155** 01010 doi:10.1051/matecconf/201815501010
[3] Muksunov T. *et al* 2014 *Izvestiya vysshih uchebnyh zavedenij. Fizika* **57**(9) 123–128 doi: 10.17223/9785946215596/66
[4] Sukhanov D.Y. *et al* 2014 *Technical Physics* **59**1854–1858
[5] Sukhanov D.Y. *et al* 2015 *Technical Physics* **60** 1529–1534

IOP Conf. Series: Materials Science and Engineering **516** (2019) 012015 doi:10.1088/1757-899X/516/1/012015

Intelligent monitoring, modelling and regulation information traffic to specify the trajectories of the behaviour of organizational agents in the context of receipt of difficult-interpreted information

E L Loginov[1], V V Grigoriev[2], A A Shkuta[3], V Y Bortalevich[4], D D Sorokin[4]

[1] International Research Institute for Advanced Systems, Moscow, Russia
[2] Moscow State Institute of International Relations, Moscow, Russia
[3] Central Economics and Mathematics Institute of the Russian Academy of Sciences, Moscow, Russia
[4] Financial University under the Government of the Russian Federation, Moscow, Russia
E-mail: dima.dd.sor@mail.ru

Abstract. To prevent massive destructive actions committed by aggregated groups of organizational agents (people) it is necessary to establish the control over the intellectual dynamics of the behavioural activity of agents in relation to interrelated political, economic, social and other processes in the context of difficult to interpret information. Intellectual monitoring, modeling and regulation of information traffic creates the possibility of operating with the working parameters of the multiparameter monitoring system natural, technical and social processes. The regulation of the work of a set of elements of a controlled set of information and computing capacities in the framework of increasing the observability of information and telecommunication networks, providing agents with information about what is happening, allows the agent to receive information on the topics identified during the monitoring (informational irritants), change of communication services or rules of their provision, etc. to specify the trajectories of organizational agents.

1. Introduction

The management of the intellectual dynamics of the behavioural activity of organizational agents (people) in relation to interconnected political, economic, social and other processes is an important direction in preventing the catastrophic nature of massive destructive actions committed by aggregated groups of agents [1].

2. Integration of various methods of psychoengineering, distant influence, psychocorrection and psychosensing

If the agent constantly receives information about an object (country, political leader, corporation, etc.) or event (informational “stuffing” of a particular blogger, the next stage of reform, etc.), it is possible to manipulate the vector benevolent or irritable perception of information at the unconsciously-reflexive level (“transfer of learning”). Impact tool: when the received information package is accompanied by an information marker, implemented as an additional informational message of the media, or a temporary dotted change in the rhythm of information transfer (the number of resolution points on the

IOP Conf. Series: Materials Science and Engineering **516** (2019) 012015 doi:10.1088/1757-899X/516/1/012015

TV or computer screen, the temporary appearance of unauthorized but audible noise transmissions, etc., or a partial change in the frequency of the broadcasting pulses of television or radio signals).
That is, further integration of various methods of psycho-engineering, distant influence, psycho-correction and psycho-sounding with synchronization of psychophysiological and psycho-semantic influences during the convergence of the process of perceiving informational stimuli, where the manipulative concentration of agents and the interpretation of information are “combined” into a single composition within the framework of human interpretation - pictures of the surrounding real and fictional world with flowing lines of behaviour [2].

It is necessary to develop a set of monitoring (testing) of agent behaviour in information and telecommunication systems on a wide range of profiles (news, education, entertainment, participation in network communities, etc.), which will allow to highlight its interests, motivation, patterns of interpretation of what is happening, serving as the basis quasi-independent choice of actions based on the analysis of the retrospective, the current situation and modelling the development of the situation [3; 4]. Such testing, as an independent computer program, can automatically identify an individual agent (person), adapt to his addictions and, in permanent mode, specify the details of his psycho-semantic subjectivity. A full-fledged psycho-semantic portrait of an agent can be built even without his direct connection with openly destructive elements (people, sites, territories, demonstrations, etc.), but only on the basis of indirect data (lack of interest in official news programs, switching off the gadget or switching channel when speaking or transmitting news about a particular political leader, watching only foreign films or avoiding watching films on patriotic-historical topics, etc.) [5].

3. The influence of electromagnetic fields and the corresponding frequency ranges considering the biorhythms of a specific biological agent

Electromagnetic fields of natural or man-made origin can provoke a desynchronization of brain processes, manifested in the dominance of rapid fluctuations (beta, gamma), acting as a source of brain activity disorders.

Low-frequency electromagnetic fields penetrate with low attenuation almost everywhere - into the soil, into the water column, into confined spaces.

At the same time, the strongest effects appear within the framework of frequency acquisition and phase matching (synchronization) of electromagnetic effects and oscillatory processes of biological systems with the possibility of resonance. Obviously, the magnitude of the response of living systems (amplitude modulation) and the reorganization of the bio-rhythm of physiological processes (frequency modulation) can also change.

When exposed to low-frequency electromagnetic fields or if the microwave modulation frequency falls into the spectrum of its own rhythmic activity, for example, the brain, the amplitudes of individual rhythms of electroencephalography may increase.

The result may be deregulatory disorders, accompanied by a violation of the general functional state and pathological abnormalities in the work of various body systems.

The increase in the intensity of electromagnetic fields is characterized by changes in spectral power in all frequency ranges of the brain: for delta, theta and be-ta-frequencies most pronounced in the frontal areas, for the alpha range in characteristic alpha activity of the occipital and parietal areas brain.

These changes lead to a functional reorganization of the activity of the cortex of the big hemispheres and subcortical centres in accordance with the processes of self-regulation of the functional state of the brain.

At the same time, within each of the frequency bands of the main rhythms of electroencephalography there is a narrow frequency sub-band, which is most sensitive to changes in electromagnetic fields. Unidirectional synchronous oscillations in different subjects were detected in these frequency subbands.

In fact, the influence of intense electromagnetic fields and the corresponding frequency ranges, considering the biorhythmic of a specific biological object, introduces many people into an altered state of consciousness that is close to trance. Such a state blocks the majority of human conscious reactions (including self-control, caution in perception, etc.) in case of discrepancies in the received, but outwardly

IOP Conf. Series: Materials Science and Engineering **516** (2019) 012015 doi:10.1088/1757-899X/516/1/012015

convincing and seemingly reliable information and the surrounding reality. That is, it facilitates the impact on it, including the use of neuro-linguistic programming elements for this, with correction or even a complete change of the reflexive matrix (the matrix of key reflective reactions) with the corresponding change and fixation of the new model of interpretation of the events.

The influence of electromagnetic fields and the corresponding frequency ranges, taking into account the biorhythms of a specific biological agent object (detected indirectly through monitoring and analysing the dynamics of using electronic communication services, for example, analysing the millisecond reaction of pressing the keys of the gadget and its correlation with the intensity of electromagnetic fields) can enhance the effects of imprinting the cognitive-reflexive model and the corresponding reflexive matrices through an automated adjustment - dynamic adaptation intensity ICT exposure to the agent selected during the monitor-ring clip (information stimuli).

In this case it is necessary to operate the working parameters of multiparametric monitoring of the system "bioelectrical activity of the brain – psycho-semantical subjectivity of the person (agent) - imprinted reflexive matrices - informational stimuli" on various types of technical systems that receive data on natural, technical and social processes. With appropriate computing power, the received information is processed from information systems for general and individual use, communication systems, technological equipment containing intelligent devices (smart grids, sensors, chips, programmable controllers, etc., which allow them to receive data remotely), on processes that were previously inaccessible for analysis, but which are involved in the configuration of metastable states of personalities as a driver of interrelated political, economic by social, social and other processes.

3.1 Increasing the observability of the analyzed psycho-semantic subjectivity of the person (agent)

Considering ongoing processes of convergence of information, telecommunication and computing services in global information networks and information systems, practically any user access to electronic content regardless of equipment, communication channel and communication method can be considered a particular case of accessing a quasi-integrated unified distributed database of electronic data. In this database, if necessary, the appeal of a specific person, even hiding his identity, will soon be possible to identify in terms close to real time. In addition, traces of access to electronic content now remain there for a long time, allowing you to apply the accumulated methods of monitoring analysis, user identification, identifying its relationship with other users, belonging to explicit and implicit groups, etc.

A modern computer, smartphone or TV remote with intelligent functions allows not only to accurately identify a person by the manner and speed of keystrokes, but also to determine his positive / negative attitude to the current electronic content present on the screen and even identify the degree of irritation and aggression in relation to this content. These same devices can fix not only the territorial point of a person's location, but also link it with the intensity indices of the electromagnetic fields at this point, change the parameters of the equipment located near this point, affecting the electromagnetic fields. Soon, gadgets will be able to record the temperature, blood pressure, heart rhythms, and so on. Biorhythms of a person.

In our case, the results of monitoring (testing) should provide the possibility of increasing the observability of the analyzed psychosemantic subjectivity of the person (agent) in the context of ambiguity or lack of information about it for a reasonable prediction of transitions between the quasi-stationary states of the intellectual dynamics of behavioral activity emergency conditions.

Combining the selected data from all possible forms of electronic content into a package of information about the psycho-semantic subjectivity of an individual (agent), combining structured, complex-structured and conditionally structured data, makes it possible to detect its behavioral activity and hidden qualities that characterize the sociopathic tendencies of the individual (agent), explicit or hidden political or religious affiliation, phobias, ways of informal self-realization in the social sphere, network and hierarchical relationships, role structure in the family and in the community, by operating the volume-proxy resources, etc.

The starting point of the development of the destructive activity of specific individuals is receiving information from the media (the dynamics of informational stimuli).

It is intended to develop a subject-adapted configuration of the basic characteristics of a complex of monitoring systems and the management of the formation of individual and group cognitive-reflexive models for identifying and interpreting the pro-source, serving as the source of the actions of individuals and their groups. Here it is necessary to identify system-parametric relationships (relationships), incl. the magnitude of the flows of information affecting an individual (from electronic media, from colleagues in the network community, etc.) and its perception, penetration into the level of unconsciously-it-named ideological stereotypes and use for identification and interpretation of what is happening.

Configuring ways of communicating information to a consumer will create a basis for forming based on domestic cognitive semantic protocols, a framework for imprinting agents of semantic transcription models for agents to interpret "imprints of reality" [6, 7].

4. Results

The regulation of the work of a set of elements of a controlled set of information and computing capacities within the framework of increasing the observability of information and telecommunication networks that provide agents with information about the current situation allows the agent to manage the process of receiving information from electronic media by the agent during the monitoring), change of communication services or the rules of their provision, the possibility of the presence of incognito agents on the Internet, etc.

The mechanisms of optimization monitoring and processing of data on the sociopathic reactions of agents arising in one of the iterations in the implementation of higher brain functions, such as sensory perception, motor commands, spatial thinking, conscious thinking and language should be aimed at creating software interfaces that allow regulate the state of the aggregated groups of agents that make up the segment of society.

Thus, programming of cognitive-behavioural stereo-types will be implemented as a basis for event chains of agents' actions in order to properly function social control systems based on people's interpretations of the surrounding real and virtual world with flowing lines of behaviour.

Acknowledgments

Theses were prepared with the financial support of the Russian Foundation for Basic Research (Project No. 19-07-01066 "Creating an artificial intelligence system as a component of a digital platform for monitoring the behavioral activity of large groups of people based on the use of methods for analyzing large weakly structured data, building thematic models with cognitive and multiparameter semantic interpretation, exploratory search and collaborative filtering with convergent control "

References

[1] Ageev A.I., Loginov E.L., Shkuta A.A. 2018 *Economic strategies* **20** 70–87
[2] Tumanyan A.G., Bartsev S.I. 2016 *Computer Research and Modeling* **8** 941–950
[3] Raikov A.N. 2016 *Economic strategies* **18** 172–179
[4] Golubev V.I. 2016 *Information technologies* **22** 233–239
[5] Loginov E.L., Eriashvili N.D., Bortalevich V.Y., Loginova V.E. 2017 *Vestnik Moskovskogo universiteta MVD Rossii.* 5 250–256
[6] Loginov E.L., Shkuta A.A. 2017 State service **19** 24–29
[7] Zakharova A., Vekhter E., Shklyar A., Zavyalov D. 2017 *Communications in Computer and Information Science* **754** 215–230

Design principles for electronic measuring path of ultrasonic tomography robotic system

I O Bolotina, G V Garanin, A M Lider, D A Sednev

Tomsk Polytechnic University, Tomsk, Russian
E-mail:bolotina@tpu.ru

Abstract. The paper discusses the requirements introduced to ultrasonic testing for industrial applications including robotic inspection. The specifics of ultrasonic tomography systems development are discussed in relation to the measuring path design. Requirements to robotic scanner are considered. The block diagram of analog component of single channel system of ultrasonic testing is provided.

1. Introduction

Development of advanced ultrasonic tomography robotic system represents complex engineering task engaging solutions from various engineering fields, starting from precision mechanical scanner development, application of advanced hardware components in amplification paths, ultrasonic signal filtering and analog-to-digital conversion, and concluding with implementation of ultrasonic scanning data processing and imaging algorithms based on parallel processing [1].

At present ultrasonic testing is performed in manual and automated testing modes [2]. During manual ultrasonic testing the ultrasonic transducer is moved manually while during automated ultrasonic testing the ultrasonic transducer performs automated movements within the specified program with application of specific mechanical system that is scanner. Significant progress in speed and quality of ultrasonic tomography system performance can be achieved with development of robot data measurement systems enabling to perform in real time mode the primary ultrasonic signals processing including those in ultrasonic multichannel tomography systems.

Equipment designed on the principles of ultrasonic tomography systems as of the promising method of nondestructive testing comprises the following functional units:

- mechanical system enabling movement of the ultrasonic multielement transducer along the specified trajectory near the surface of the object under control;
- electronic multichannel system for amplification, filtering and digital conversion of ultrasonic signals;
- computation system enabling processing, interpretation and imaging of defects spacing in the volume of the object under control.

Analysis of the available on the market final products that enable realization of the functions of ultrasonic tomography systems elements allows to use the major share of the existing finished components; thus lessening the period and reducing the costings for inspection systems development [3–6]. The most significant and quite specific point for the system development is software design. In this way, the process of ultrasonic tomography system development presupposes systemic integration of the above mentioned functional units which represents complex engineering task.

2. Requirements to ultrasonic testing for industrial applications

The equipment and instruments developed in compliance with corresponding state standards, unified operational procedures and maintenance manuals are to be used for inspections.

Calibration of testing setups and equipment is performed periodically following the schedule set in compliance with the guidelines for these setups and equipment technical maintenance, also stated in equipment certificates and maintenance manuals. The data about periodical calibration and parameters under control is put into the designated sections of the equipment certificates (service list).

Metrological guarantee is realized in compliance with the requirements of the state standards of the Russian Federation: for setups and equipment defined as measuring instrumentation, the pattern approval certificate is issued, while for setups and equipment not defined as measuring instrumentation, the certificate of conformity is issued.

The largest scale enterprises that use nondestructive testing systems such as Gazprom, Russian Railways, Rosneft, Rosatom developed their own industry specific requirements for nondestructive testing procedures. The requirements are set as branch standards determining the scope, inspection intervals and nondestructive testing methods and techniques assigned for equipment and products. Also, the branch standards determine the types of transducers and equipment for ultrasonic nondestructive testing that are recommended for the use in inspection of the given objects under control.

2.1 Requirements to ultrasonic robotic testing

Manual nondestructive testing is known to possess a number of limitations that is low productivity and reliability of inspection, not sufficient results recording, and considerable human factor effect.

Application of instruments and systems of robotic testing partially or completely eliminates these limitations. Mechanical scanning systems widely available on the market of nondestructive testing equipment can be divided into two categories: systems of manual and automated positioning. At that, for both categories there are single-axis scanners as well as two-axis scanners available. Both categories are used for single channel systems as well as for multichannel systems with the transducers based on phased arrays.

Single-axis scanners are used for linear movement of the transducer as well as for the transducer positioning on the circumference. Such scanners are conventionally used for quality control of pipelines welded joints. Scanners to perform manual testing are equipped with encoders of linear and rolling type with high resolution capacity for ultrasonic testing data binding to the measurement point. Two-axis scanners are more often used for radial surfaces inspection, for example, for pipelines corrosion testing.

In case of the automated scanner positioning, the law of ultrasonic transducer motion is set by the computing unit which processes data acquisition during testing. In all the cases, application of single- and two-axis automated positioning systems is limited by the surface configuration of the object under control. For the objects of complex geometry there is the need in individual adjustment of the positioning system or development of custom made solution which together with upgraded software is responsible for ultrasonic data binding to the inspection point and imaging. Further, it is also possible to use robotic manipulators for scanning systems with greater number of degrees of freedom enabling to perform inspection in different dimensions.

2.2 Main requirements to ultrasonic testing instrumentation

The ultrasonic testing systems are characterized by the following parameters:
- absolute sensitivity;
- relative sensitivity;
- ultimate relative sensitivity;
- sensitivity factor;
- testing range;
- assessment of the dead zone;
- nonuniformity of sensitivity compensation within the testing range;
- relative resolution at the pulse edge;

- relative resolution at the occurrence depth;
- the function of the object under control parameters` effect (ultrasound speed, surface curvature and roughness) on deflection of relative sensitivity and error of the depth meter.

Measurements of the mentioned above parameters are performed in compliance with the procedures stated in GOST-R 55809, where standard calibration samples are used as objects under control. Generally these parameters are not specified or partly stated in specifications for flaw detectors and ultrasonic testing systems available on the market. At that, detailed specifications are provided for the parameters of the electronic module (measuring path) of ultrasonic flaw detector for the systems available:

- the excitation pulse parameters – amplitude, pulse leading edge duration, pulse duration;
- maximum sensitivity of the reception path;
- pass band of the reception path;
- amplitude characteristic of the reception path;
- parameters of time adjustment of the reception path sensitivity – maximum/minimum duration of time adjustment of sensitivity, maximum/minimum delay time of time adjustment of sensitivity;
- the threshold indicator parameters – tuning error of the threshold indicator, operation speed of the threshold indicator.

2.3 Specifics of ultrasonic tomography systems development

Ultrasonic industrial tomography is based on pulse-echo method and differs from conventional ultrasonic tomography in two main principles: inspection of the volume within the object under control and imaging of the testing results. While in conventional instruments, solely mechanical scanning of internal structure of the object under control by narrow-beam ultrasound is implemented; in ultrasonic tomography three types of scanning are used:

- mechanical scanning – the ultrasonic transducer is moving along the surface of the object with the input of the data about coordinates of the current location of the transducer;
- electron scanning is performed by physically formed beam of ultrasound phased array;
- virtual scanning is realized by computational method using the set of signals received during probing the object under control by elements of ultrasound phased array.

Different combinations of these types for inspection of the volume within the object can be applied.

Another difference of tomographic instruments from conventional ones is the testing results provision as images of internal structure of material of the object under control. The two dimensional images of plane sections of the object under control (tomograms) are used most often; however, there is also three dimensional (volumetric) imaging of the internal structure of the object. At that, in addition to convenience of the operator`s reading the testing results, images are quite adequate in relation to boundaries and material discontinuity actual location, in many cases imaging allows to measure actual dimensions of material discontinuity and not relative ones.

To receive sufficient volume of the testing data and build reconstruction of the internal structure of the object, single element transducers as well as phased arrays can be used as they allow to receive controlled acoustic field pulse edge.

In spite of all advantages of ultrasonic multichannel testing systems based on multielement transducers, using these transducers is not always feasible or economically viable. Using multielement transducers is efficient and viable to perform inspection of shallow depth which is related to considerably lower level of the excited ultrasonic signal due to small size of individual piezoelements of phased array, thus leading to signal-to-noise ratio decrease. Further, treatment of surface of the object under control is required for using multielement transducers to ensure specified degree of the surface roughness and planeness. This, for example, does not allow inspection of large-dimensional products casted from metals and alloys.

Due to this, using single element transducer to perform inspection of the object under control is more viable. Hence, mechanical scanning is the type of scanning to perform ultrasonic tomography.

IOP Conf. Series: Materials Science and Engineering **516** (2019) 012016 doi:10.1088/1757-899X/516/1/012016

In this case transmission and receiving of ultrasonic signals is performed periodically with different positioning of the transducer on the surface of the object under control. In order to perform transmission and receiving signals under various angles it is suggested to use robot manipulator which enables to perform scanning for the whole geometrically complex surface reconstruction.

During simultaneous space-time processing of all signals received by the transducer from all its positions on the surface of the object under control, synthesis of ultrasonic aperture of large wavelength takes place; images of the object internal structure are being reconstructed. At that, the synthesized aperture is virtually (that is after signals` processing) focused in each point of the visualized zone [7]. This method of volume inspection and obtaining its images is called the synthetic aperture focusing technique – SAFT.

3. Requirements to measuring path of ultrasonic tomography system

General equipment specification is applied to measuring path (electronic module) of ultrasonic tomography system. The specification is applied to electronic and automatic equipment, it comprises:
- General provisions.
- Requirements to the system performance specification.
- Requirements to the system technical specification.
- Requirements to electromagnetic compatibility.
- Requirements to reliability.
- Requirements to electric power supply.
- Requirements to design and others.

Measuring path of up-to-date conventional testing system contains not only the hardware component responsible for receiving and measurement preprocessing of signals from the transducers (analog path) but also the digital hardware component which operates measurements. At that, the digital hardware component is practically an integral part of the embedded or downloaded control software realized via control microprocessor or programmable logical circuit. This leads to specification of the requirements not only for hardware but also for software components of the measurement system for the ultrasonic tomography of the intelligence combined robotic system of digital mechanical processing production.

The block diagram for one of the channels of the ultrasonic multichannel nondestructive testing system (Figure 1) comprises blocks which enable the functions of attenuation, ultrasonic signal amplification, filtering and analog-to-digital conversion. This block diagram is partially or completely followed for each channel during the ultrasonic multichannel nondestructive testing system development.

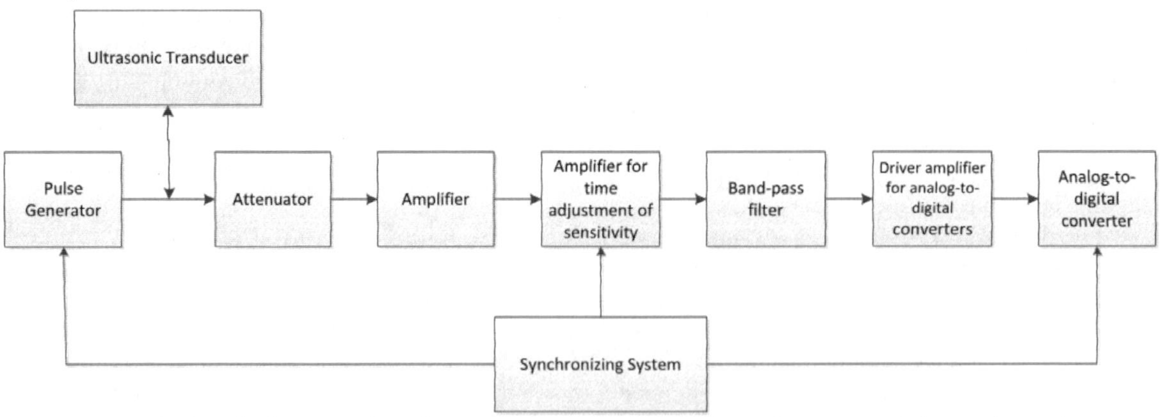

Figure 1. The block diagram of analog component of single channel system of ultrasonic testing.

The pulse generator is intended to generate the probe pulse, and is controlled by the synchronizing system which enables accurate time registration of the probing signal in relation to the scanning scale of A-scan. The input attenuator and amplifier are intended for attenuation or amplification of the echo-signal depending on its amplitude. The amplifier for time adjustment of sensitivity is used to compensate for ultrasonic signal attenuation in the object under control, the amplification coefficient is programmed in relation to strokes generated by the synchronizing system. One of the main requirements to the amplifier for time adjustment of sensitivity is control of the amplification coefficient within the wide range from 40 to 80 dB. The band-pass filter enables the desired signal filtering and noise suppression outside the service band. In case the ultrasonic testing system is designed for ultrasonic transducers with various service bands, it is required that the band-pass filter enables regulation of top and lower frequency of pass band of the band-pass filter. The driver amplifier for analog-to-digital converter (ADC) is actually the input analog-to-digital converter and matches the output resistance of previous stages with the input resistance of the analog-to-digital converter.

A wide range of low-noise high frequency integrated amplifiers and variable gain amplifiers (VGA) allows relatively easy and compactly realize the functions of the preamplifier and the amplifier for time adjustment of sensitivity. The options of designing the block of analog path – the mode convertible band-pass filter come into detailed consideration. The band-pass filter can be made up of passive components or as the active filter based on the operational amplifiers. The passive band-pass filter is easy to realize, however it possesses limitations: signal attenuation, narrow filter order. Active band-pass filters are more complex for realization; however they enable to design the filters of high filter order and, at that, with the specified amplification coefficient. Taking into consideration that during ultrasonic testing systems development ADC number of bits tends to increase, thus, the requirements introduced to the analog path noise characteristics are toughened. The active band-pass filters application seems reasonable, in spite of their sufficient complexity. In both cases when designing the mode convertible band-pass filter the circuit design task of changing the top and lower frequency of pass band arises. There is the option for mode conversion of the passive frequency setting elements or commutation of the entire channels of the active filter.

Conventionally used for these purposes analog switches based on metal oxide semi-conductor circuits operate well within the frequency range up to 1 mHz. Much higher frequencies require application of special high frequency analog codes. The performed analysis of the available on the market components reveals that the most optimal option to develop the mode convertible band-pass filter is using special wide-band amplifiers – multiplexers, which pass band reaches 500 mHz under low distortion factor while the amplification coefficient is normally equal to 2, as they change individual filter channels. The costings and complexity of the circuit design based on amplifiers – multiplexers will be practically the same as for the circuit design based on high frequency analog codes with much higher characteristics of analog path. At that, separate regulation of lower and top frequency of pass band of the band-pass filter requires relatively low costings.

In comparison to analog systems, the main advantage of digital systems of data acquisition and processing is their capacity to change the processing algorithm rapidly, append new functional capabilities when upgrading software responsible for signals processing and imaging. However, as long as ultrasonic signal is analog quantity evaluated as voltage signal which changes in time in the measuring transducer – the key component for digital systems of signals processing is the analog-to-digital converter. When it comes to ultrasonic signals recording, it is essential that the analog-to-digital converter possesses high sensitivity and wide dynamic range for simultaneous recording of significantly different in amplitude signals. In the latest years the world leading producers offered the available now 16-bit analog-to-digital converters with the conversion frequency up to 250 mHz within their products range. This enables new options for development of ultrasonic inspection systems which earlier were equipped with the transducers with the comparable operating speed; however they possessed lower digital capacity which was, as a rule, not higher than 12 bit.

The conventional products range from such leading producers of electronic components such as Texas Instruments, Analog Devices, Linear Technology, Maxim Integrated Products and others contains, as a rule, analog-to-digital converters families based on custom made architecture enabling the required conversion rate and being distinctive in its digital capacity, as, for example, the family AD92xx – AD94xx – AD96xx which is based on pipeline architecture and offers 12 to 16 bit ADCs range with

IOP Conf. Series: Materials Science and Engineering **516** (2019) 012016 doi:10.1088/1757-899X/516/1/012016

the conversion rate from 125 to 250 million conversions per second. This allows to use once developed design of the printed-circuit board of electronic path of ultrasonic system to program configurations optimized by the set of transducer characteristics as well as in relation to the costings.

Availability of choice from ADC models depending on their operating speed parameters and digital capability is vital; it is also important when considering the real signal-to-noise ratio reached in analog path of the measuring channel. In some cases the effect of the system self-noises reduces to zero the advantages of wide dynamic range and makes 14 bit and higher bit ADCs application senseless. Detailed studies of the electronic equipment self-noises effect lead to toughening the requirements introduced to input circuits design, analog components choice and tracing quality assurance and the printed-circuit board production. At that, in all cases, the higher ADC digital capacity is, the greater capacity loss will take place due to external noises and self-noises effect.

4. Requirements to robotic scanner

Up-to-date production processes and operating practices presuppose design, production and quality control of products using their digital representation. Starting from engineering development of a product in its digital representation, based on which the programme is developed for a range of machines with numeral program control and concluding with end products storage – at any production stage a range of digital models is being designed. The digital models describe the product dimensional components, technological and service properties. In such case, with relatively low costs it is possible to develop a program to control nondestructive testing system which is especially advantageous for geometrically complex objects' inspection. With the digital representation of the object under control available it is possible to use robotic inspection centers characterized by greater number of degrees of freedom. Conventional industrial robots that are in mass use in production processes possess 5–6 degrees of freedom with robot extension up to 1.5 m, accuracy of positioning repeatability is up to 0.1 mm which by the parameters meets the nondestructive testing requirements and considerably increases the capabilities of the systems in case they are used as the basis for automated positioning control device of the transducer. Solutions available on the market from the leading producers of the robotic systems such as FANUC, ABB, KUKA, SCHUNK, Reis, KAWASAKI, KC ROBOTICS, KAMAN CORPORATION and others provide control of their products based on recorded interfaces which enables fast integration of robotic equipment into inspection systems.

5. Design requirements to ultrasonic tomography instruments

The ultrasonic testing robotic system comprises the following individually produced and supplied electrical measuring and control units and modules:
- the unit for measuring acoustic path;
- the scanner control unit (two-axis mechanical or robotic);
- the programmable controller of utility system;
- the control computer of the measuring data acquisition system;
- the server for the measuring data processing and storage;
- uninterruptable power supply;
- network commutator and other modules.

Assorted product range of electronic equipment used in ultrasonic testing system development requires the unified mechanical module construction enabling integration of ultrasonic testing system elements based on unified mechanical structure which will allow to perform maintenance, introduce changes and complements avoiding mechanical processing of individual structural components.

At present the most widely used is the construction of so-called the 19 inch standard rack. Practically all producers of electronic equipment offer their products for assembling in the 19 inch standard rack cabinets and equipment racks. In case of nonstandard equipment the market offers wide range of cases and assembly adjusting accessories compatible with the 19 inch standard rack construction with the option to assemble this equipment with minimum efforts.

The cabinet construction comprises the following main elements: the frame, side frames, doors, cover, guidance elements with holes for equipment assembly. The cabinets provide equipment protection against moisture, dust and dirt, physical damage. The cabinets also serve for protection against electromagnetic radiation. The degrees of equipment protection against dust and moisture are stated as IP code, the cabinets with the degree of protection from IP20 to IP68 are available on the market.

6. Conclusion

Design requirements to ultrasonic testing system are introduced in relation to all its components: measuring path, software, robotic scanning system. At the present stage of the project realization, grounding our opinion on analysis of the characteristics and versions of flaw detectors available on the market, taking into consideration the objective to perform inspection of mold castings with complex geometry it is possible to formulate specific requirements to the system technical specifications and parameters.

The relevance of development of the custom made intelligence combined ultrasonic robotic testing system for digital mechanical processing production is conditioned by the lack of ready-made solutions in the flaw detectors market which would meet the requirements to interfaces open to exchange the measuring data, their transmission capacity. It is also characteristic for closed systems to present problems related to their integration with the automated and mechanization systems provided by the third party producers.

When developing the ultrasonic testing system it is vital to ensure module construction, compatibility with the conventional in-service equipment provided by the third party providers. Thus, it seems reasonable to introduce the requirements to mechanical performance of ultrasonic testing system individual elements using the 19 inch standard rack construction during design and production of measuring and control systems.

When developing the intelligence combined ultrasonic robotic testing system for digital mechanical processing production it is also vital to make provision for increasing the number of ultrasonic testing channels by parallel adding of the measuring path modules. At that, the system expansion is to be sustained by hardware construction as well as by software. This solution application will enable to increase testing speed due to parallel processing, when necessary.

Hence, the developed for ultrasonic testing system software is to meet the requirements conditioned by up-to-date tendencies of information processing technologies development: specification and development based on open software interfaces; client-server architecture; implementation of programming technologies for parallel code processing, multithread processing; implementation of programming technologies for graphic processing unit to perform computation.

Based on the analysis of the available solutions to develop mechanical scanning system for the objects of ultrasonic testing it is possible to conclude that for the objects under control with simple surface geometry it is sufficient to apply conventional two-axis scanners while for the objects under control with complex geometry it is relevant to apply the transducers positioning systems with regard to the object under control. The systems are industrial robotic platforms based.

Taking into consideration the objects under control dimensions, specifics of their production and mechanical processing technologies, it is expedient to choose single channel ultrasonic testing system with robotic positioning of the transducer. This configuration will enable to achieve sufficient productivity of nondestructive testing process as well as increase the testing data imaging efficiency; in case high speed system for measuring data acquisition and processing is used based on single operating station with the GPU accelerators card. This will considerably simplify software development and adjustment as well as it will reduce hardware costings.

Acknowledgements

The study is financially supported by the Ministry of Science and Higher Education of the Russian Federation within the Federal Target Program entitled "R&D in Priority Areas of Development of

IOP Conf. Series: Materials Science and Engineering **516** (2019) 012016 doi:10.1088/1757-899X/516/1/012016

Russia's Scientific and Technological Complex for 2014–2020" (unique project ID: RFMEFI57817X0251).

References

[1] Kvasnikov K. G., Soldatov A. I., Bolotina I. O., Krening M. K., Potapenko A. A. 2013 *Russian Journal of Nondestructive Testing* **49(11)** 625–630
[2] Samokrutov A. A. and Shevaldyukin V. G. 2007 *Plant Laboratory. Materials Diagnostics* **73(1)** 50–59
[3] Nondestructive Testing Equipment Available at URL: http://armada-ndt.ru/
[4] Up-to-date Market of Ultrasonic Testing Equipment Available at URL: http://www.acsys.ru/article/rynok-sredstv-ultrazvukovogo-kontrolya-sovremennoe-sostoyanie/
[5] Nondestructive Testing Equipment for Ultrasonic Mechanical Inspection Available at URL: http://diagnost.ru/novosti/credstva-nerazrushayushhego-kontrolya-ispolzumye-pri-provedenii-mehanizirovannogo-ultrazvukovogo-kontrolya.html
[6] Measuring, Monitoring and Testing Equipment Available at URL: http://www.geo-ndt.ru/catalog-18-yltrazvykovie-defektoskopi.htm
[7] Badalyan V G et al 2008 *Holographic Techniques for Ultrasonic Flaw Detection in Metals* (Mashinostroenie: Moscow)

Application of ellipsometry, spr-technic and raman-spectroscopy into diagnosis of colorectal cancer

M V Kruchinina[a,b,e], A A Gromov[a], Y I Prudnikova [a], V M Generalov[c,e], V N Kruchinin[d,e], V A Volodin[d], G V Shuvalov[e], O V Minin[f], S V Shehovtsov[a], S E Peltec[a] , I V Minin[e,f]

[a] Siberian Branch of the Russian Academy of Sciences, Novosibirsk, Russia
[b] Novosibirsk State Medical University, Novosibirsk, Russia
[c] Science State Research Center of Virology and Biotechnology "Vector", Novosibirsk Region, Koltsovo, Russia
[d] Rzhanov Institute of Semiconductor Physics, Novosibirsk, Russia
[e] Siberian State Order of the Red Banner of Labor Institute of Metrology, Novosibirsk, Russia
[f] Tomsk State University, Tomsk, Russia
E-mail: kruchmargo@yandex.ru

Abstract. Experiments with application of the "*Ellips-SPEK*" spectral ellipsometric SPR (Surface Plasmon Resonance) analyzer, validating the results on the basis of *ProteOn XPR36*, revealed the differences in the values of velocity constants of specific interactions of monoclonal highly specific antibodies with tumor M2-pyruvate kinase on the surface of blood serum biochips in patients with various stages colorectal carcinoma (CRC) and location of metastases (AUC 0.89).Scanning ellipsometry enabled to reveal a significant increase of biomolecule surface effective thickness resulting from the specific "antigen-antibody" interaction in the group of patients with extrahepatic and hepatic metastases, as compared to the control group with healthy patients ($p<0.0001$– 0.05).Raman spectroscopy of blood serum demonstrated the differences in peak intensity within the range of 1005–1520 cm^{-1} in the same groups of patients ($p<0.0001$–0.05) with a 90% predictive accuracy of the method applied at early stages of the disease. The results obtained can be used for development of methods for early diagnosis of CRC, localization of different metastases, relapses and control of the quality of administered therapy.

1. Introduction

Colorectal cancer is known to be the fourth and most frequently diagnosed tumor in the world [1]. Currently the radiographic examination, comprising computer tomography, has become the most common method of preoperative evaluation of stages of the disease in patients with CRC [2]. Histological analysis of surgical samples often results in change of the initially diagnosed stage. In particular, surgery more accurately defines depth of tumorous invasion and involvement of lymphatic nodes. However, modern methods, used to determine a pathology stage, sometimes lead to omission of occult metastases. In postoperative period patients undergo thorough dynamic observation with the purpose of revealing the local or remote relapses as early detection leads to earlier treatment before the tumor expands [3]. Guidelines of the American Society of Clinical Oncology [4] recommend application of annual computer tomography in patients with the administered surgical alternatives as well as evaluation of a carcino-embryonic antigen (CEA) in blood serum which is to be carried out every three months in patients with II and III stages of the disease for no less than 3 years in case such patients are to undergo surgery or chemotherapy due to metastatic disease [5]. This intense postoperative monitoring aims at detection of resectable metastases. For instance, local metastases into the liver, in case of absence of the extrahepatic ones, can be resected [6, 7].

Thus, biomarkers facilitating determination of occult metastasis prior to or after surgery will improve determination of the CRC stage, which potentially influences upon the therapeutic approach.

The aim of this study is to determine the possibilities of detection of blood serum antigens in patients with CRC of various location of metastases to highly specific monoclonal antibodies, i.e. the tumor M2-PK by means of spectral ellipsometry close to observation of surface plasmon resonance in order to improve sensitivity rate of determination of biomolecules in ultra-low concentrations; to reveal the differences in Raman-spectra of blood serum of the above groups of patients.

2. Materials and methods

Groups of patients. 49 patients (mean age - 52±9 years; 22 females, 27 males) with CRC of various localization i.e. transverse colon (*n*=3), descending parts of large intestine (*n*=3), sigmoid colon (*n*=9), recto-sigmoid part (*n*=6), rectum (*n*=22), numerous primary tumors of large intestine (*n*=6) have been included into the study. Histological studies revealed adenocarcinoma of various stage of differentiation in all patients under observation. Diagnosis was verified at the oncological hospital based on the complex of clinical and instrumental studies (including ultrasonography, KT, multi-spiral computer tomography, colonoscopy, fibrogastroduodenoscopy, histologic examination).

The patients were divided into three groups depending on localization of metastases, i.e. Group 1 comprised 15 patients with local regional CRC (7 subjects – stage II and 8 patients – stage III of the disease); Group 2 consisted of 18 patients with hepatic metastases only (12 patients had singular metastases, the rest – the multiple ones with the diameter of the nodes ranging from 16 to 92 mm); Group 3 comprised 16 patients with extrahepatic metastases (including the ones into supraclavicular lymphatic nodes, lungs, bones and brain). Thus, patients of Group 1 and Group 2 had the fourth stage of CRC. 3 patients (20%) from Group 1, 6 subjects (33%) from Group 2 and 11 (69%) from Group 3 have undergone the adjuvant polychemotherapy for 3 months.

Control group comprised 19 relatively healthy subjects (mean age - 53±8 years) without oncologic pathology and pathology of inner organs.

Samples preparation. Blood serum of patients under study, obtained by means of centrifugation (2000 rpm for 20 min.) of whole blood taken at fasting stage, has been examined.

The method of detection of biological macromolecules of small concentration by means of spectral ellipsometry under the conditions of observing the surface plasmon resonance has been developed and tested for application in *spectral ellipsometric SPR-analyzer «Ellips-SPEK»* [8]. Within the course of study *we implemented optimization of ways of surface treatment* in order to increase selectiveness to binding with certain macromolecules and surface adsorption ability.

A one-hour activation of silicon plate surface with carbonyl diimidazole was performed, following the stage of hydroxylation with sulfochromic mixture or «piranha» (i.e. the mixture of sulfuric acid with peroxide at ratio of 3:1). After that the plate was rinsed with H_2O (bidistilled water; 5×10 ml), ethyl alcohol (3×10 ml), acetonitrile (1×10 ml). 200 mg of weighed sample of carbonyl diimidazole was dissolved in 20ml of acetonitrile. The silicon plate was submerged into the obtained solution, stirred slowly and stored for 12 hours at room temperature. The reaction being over, the plate was sequentially rinsed with acetonitrile (3×30 ml), acetone (1×10 ml) and dried under vacuum conditions.

Immobilization of protein probes. 0.3 mcl dots of buffer solution of protein molecules in concentration of 0.01–100 µg/ml (ranged from 0.3 pg to 360 ng) were applied to the activated surface, parent antibodies were diluted 10-fold, incubated for 12 hours at room temperature under the conditions of saturation with water vapors. Later on, the unconjugated biologic material was repeatedly rinsed from the surface with a 1x phosphate buffer (137 mM of NaCl, 2.7 mM of KCl, 10 mM of Na_2HPO_4, 1.76 mM of KH_2PO_4; pH 7.4), containing 0.1% of Tween-20.

SPR-technics and Ellipsometry. *The Anti-PKM2 antibody [EPR10138(B)], (Abcam RabMab, USA)* highly specific monoclonal rabbit antibodies were immobilized onto the surface for further study by means of spectral ellipsometry under the conditions of observing the surface plasmon resonance.

Surface response of silicon plates (initially covered with monoclonal antibodies) to *Tumor M2-PK* in reaction of interaction of blood serum of patients with CRC and healthy subjects, was studied applying the *«Microscan»* (ISP SB RAS) *scanning HD ellipsometer* [9]. Ellipsometric angles Ψ and Δ were read in central areas of the plates (76 mm in diameter) of 15×15 mm^2 in size, X and Y step size of 0.1 mm. The tilt angle of the light to the sample equaled 60°.

The-ProteOn XPR36 (BioRad) surface plasmon resonance-based device was used for parallel detection of antigens to monoclonal antibodies to tumor

PKM2 in blood serum taken from healthy patients and patients suffering from CRC. Antibodies to human *Tumor M2-PK* were covalently bound with the surface of the *GLC* sensor chip. 0.2 mcg of mouse monoclonal antibodies to human *Tumor M2-PK* protein was applied to 230 mcl of acetate buffer (pH 5,0). Detection of *Tumor M2-PK* protein was made in *PBST* buffer (137 mM of NaCl, 2.7 mM of KCl, 10 mM of Na_2HPO_4, 2 mM of KH_2PO_4, 0.1% Tween-20, pH 7.4).

Initially, binding of serum, diluted 10, 50, 250, 1250 and 6250-fold in *PBST*, was tested using several samples taken from the healthy subjects and the patients with CRC. Concentrations of 1/50 and 1/250 were selected for screening, which would secure the clearest difference between the samples under evaluation. 200 mcl of diluted serum was run through the canal with conjugated antibody for 400 sec at 30 mcl/min. Dissociation of complex antigen-antibody was carried out in 10 mM buffer of glycine-HCl (pH 2.0). The curves obtained demonstrate that protein concentration to *Tumor M2-PK* in serum is proportional to the binding level.

Raman spectroscopy. Spectra of combinational scattering of light of liquid blood serum of patients were recorded by means of a spectrometer with the triple *T64000* (*Horiba Jobin Yvon)* monochromator at room temperature. Ar^+ laser line with a 514.5 nm wave length was used for excitation; spectral resolution equaled 1.5 cm^{-1}. A silicon photo sensor array, cooled down by liquid nitrogen, served as a detector. A supplement for microscopic studies of Raman scattering was used. The power of laser beam, reaching the sample, comprised 2–3 megawatt (a spot size equaled 5-6 mcm).

3. Results and discussion

Tumor markers are known to be the substances, which at certain concentration in the organism signal about the presence of neoplasm [10]. Pyruvate kinase is one of such substances influencing upon the phosphoenolpyruvate metabolism. In normal cells it is present in several forms, i.e. L, R, M1, M2 that are located in kidneys and liver, erythrocytes, brain and muscles, lungs, correspondingly. They are highly specific. While tumor is growing the number of standard isoenzymes decreases and *Tumor M2-PK* appears. Primary isoenzyme, having four subunits in its structure, transforms into a low activity dimer, specific for tumorous cells [11].

In clinical practice, they make feces analysis aimed to reveal the presence of tumorous M2-pyruvate kinase, determined by enzyme (sensitivity 10^{-9}) in the following cases [12]:

- diagnosis of adenocarcinoma, recurrent tumors in postoperative period and dysplastic polyps;
- evaluate of the stage of adenocarcinoma;
- localization of bleeding tumors;
- routine examination of patients of risk group.

This study involved blood serum of the patients under study, as according to [13], studies of tumor pyruvate kinase of feces are not diagnostically efficient enough. Figure 1 demonstrates kinetics of interaction between highly specific monoclonal antibodies and *Tumor M2-PK* from blood serums of the patients from study groups defined by spectral ellipsometry in proximity to observation of plasmon resonance.

IOP Conf. Series: Materials Science and Engineering **516** (2019) 012017 doi:10.1088/1757-899X/516/1/012017

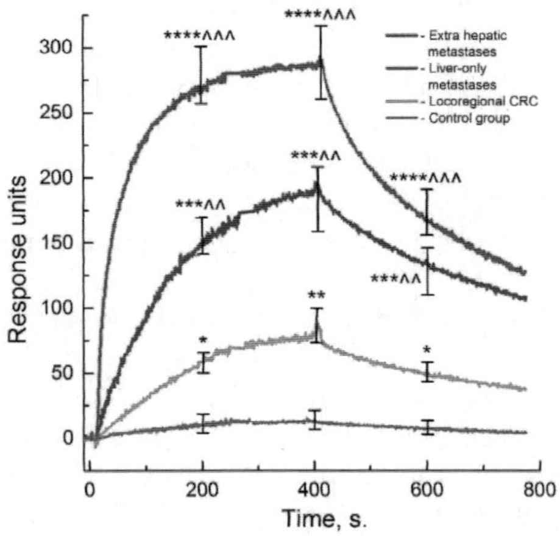

Figure 1. Sensograms of binding and decomposition of complexes upon interaction of antigens of blood serum and highly specific monoclonal antibodies to *Tumor M2-PK* in patients with CRC and control patients. Blood serum dilution rate equaled 1:250.

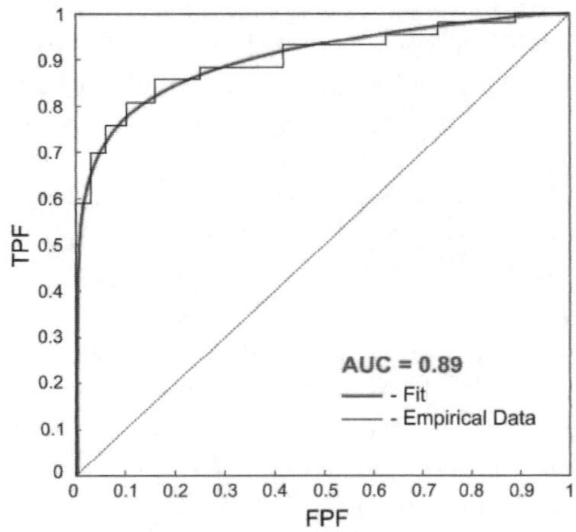

Figure 2. ROC-curve of serum tumor M2-PK levels in differentiating the patients suffering from CRC with metastases.

The mean level of serum *Tu M2-PK,* found in control, equaled 11.6 $\pm$ 4.5 RU/mL with no relation to sex (p = 0.51) and age (p = 0.61), which correlates with the findings of a number of studies [14, 15]. The mean values of serum level of *Tu M2-PK* (RU/mL) in patients with local regional CRC (Group 1) comprised 77.4$\pm$5.8; the level of 192.5$\pm$12.8 was registered in patients with CRC and liver metastases (Group 2) and 284.2$\pm$1466 – in the group of patients with extrahepatic metastases (Group 3) ($p<0.0001$-0.05), being approximately 6-23 times different as compared to the similar one in the controls. Tonus C., Neupert G. (2006) also revealed correlations of fecal Tu M2-PK level with the stage of the disease in compliance with the TNM classification and Duke's staging [16].

The defined low levels of serum *Tu M2-PK* in the control group and no apparent variations of readings in relation to sex and age can be explained by the fact that this enzyme is bound to tumor metabolism [11]. The obtained associations of serum level of Tu M2-PK with TMN classification of malignant tumors and Duke's staging of CRC have been proven by clinical studies by Mazurek S. et al.(2005), Christofk H.R. et al.(2008) [17, 18].

Diagnostic accuracy of detection of serum level of *Tu M2-PK* for determination of the difference between the cases of CRC and control comprised 100% which is consistent with the findings by Meng W. Et al. (2012) [15] and is considerably higher than in case of applying colonoscopy, examination of feces for occult blood and detection of fecal tumor M2 pyruvate kinase [19, 20]. In case when the cut-off level of serum *Tu M2-PK* comprised 11.6 $\pm$ 4.5 RU/mL, none of the CRC cases were missed and 43.3% of colonoscopic examination could be avoided. Diagnostic sensitivity for all cases of colorectal lesions increased upon the reduction of cut-off level of serum *Tu M2-PK*. The results obtained speak in favor of viability of the proposed early stage of CRC diagnosis.

At the same time, the diagnostic value of the method was recognized in detection of various localizations of metastases. Carrying out the ROC-analysis for detection of metastases in patients with CRC, the value of AUC for serum *Tu M2-PK* comprised 0.89 (0.84, 0.94) (95% confidence interval) (Figure 2). The findings, obtained by applying a spectral ellipsometric SPR analyzer «Ellips-SPEK», correlated with those obtained using the *ProteOn XPR36 (BioRad, USA)* and demonstrated the same sensitivity and accuracy rate as the one achieved using similar foreign equipment [8]. Thus, the diagnostic method of a specific «antigen-antibody» interaction, based on the surface plasmon resonance phenomenon, enabled to increase the accuracy of detection of biomolecules up to $\sim 10^{-11}$–10^{-12} M/ml.

The scanning ellipsometry technique demonstrated a significant increase of bimolecular surface effective thickness resulting from a specific interaction «antigen-antibody», ranging from control (Figure 3*a*) to the group of patients with hepatic metastases (Figure 3*d*) ($p<0.0001$–0.05). The peak areas (1005–1520 cm^{-1}) of Raman spectra appeared to be significantly lower in patients with CRC as compared to the controls ($p<0.0001$–0.05) (Table 1, Figure 4), correlating with the process stage ($r=-0.68$, $p<0.001$) and the presence of metastases ($r=-0.57$, $p<0.003$). The accuracy rate of the above technique comprised 85%, 90%, 94%, 96% of serum of control and the patients with local regional CRC as well as CRC with only hepatic metastases and CRC with extrahepatic metastasis, correspondingly.

Various levels of peak intensity of Raman scattering spectra, to a certain extent, reflects the existing differences in integral metabolomic profiles of blood serum in patients with CRC [21].

The results, confirming the fact that changes in metabolomic profile tend to depend upon the tumor localization, seem to be astonishing. The question is, whether the changes in circulating metabolites will reflect the differences in biology of tumor or changes in the host reaction to tumor or a combination of both mechanisms. The host reaction to tumor can change along with the spread of metastases [22], as a metastatic disease is biologically deferent from cancer, not expanding outside the original tissues, and many other aggressive tumors can provoke a more (or less) abrupt reaction of the host [23]. The host reaction can also differ because of the local effect of the tumor. For instance, a tumor can cause numerous paracrine effects in a tumor microenvironment as well as metabolic or inflammatory reaction of surrounding normal tissues can differ in case of localization of metastases in intestine, liver and other locations [24].

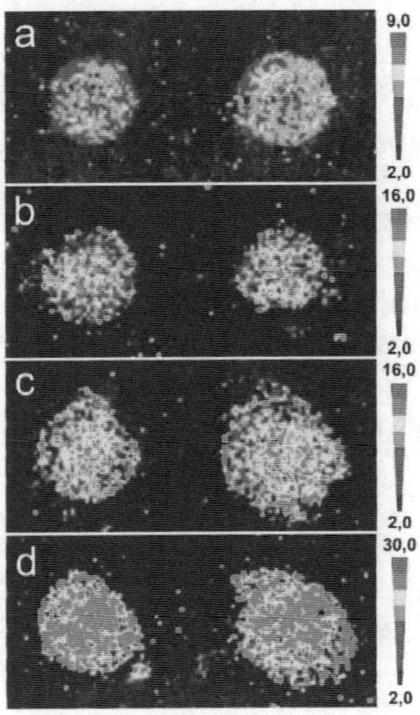

Figure 3. Scanograms of film surface obtained with scanning HD ellipsometer «Microscan» resulting from specific interaction of blood serum antigens - *Tu M2-PK* with highly specific monoclonal anti-PKM2 antibodies: a – in control; b – in patients with coloregional CRC; c – in patients with CRC and liver metastases; d – in patients with CRC and extrahepatic metastases. In right parts of diagram – distribution of effective thickness (in nm) of films from blue to red.

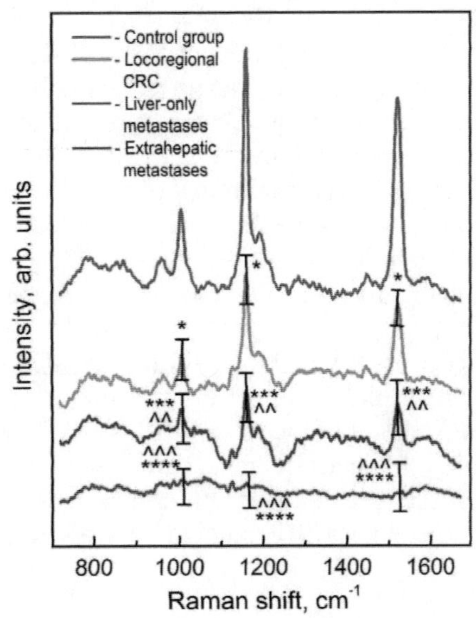

Figure 4. Raman spectra of blood serum of patients with CRC and the controls.

Table 1. Peak areas of Raman scattering in patients suffering from CRC with various localization of metastases and control (M±m).

Peak location, cm^{-1}	**Peak area**, arb. units.			
	Control (*n*=19)	Group 1 with coloregional CRC (T2-3) (*n*=15)	Group with CRC and hepatic metastases (T4) (*n*=18)	Group with CRC and extrahepatic metastases (T4) (*n*=16)
1005	990± 84	557± 103*	387± 90**^	129 ± 78***^^^$$
1157	2570± 301	2004± 264*	908± 115**^^	712 ± 132***^^^$$
1520	3258 ± 190	2404 ± 162**	714 ± 135***^^^	474 ± 108***^^^$$

Notes: M – mean value, m – standard error of a mean value,
– statistical significance (p) of difference as compared to control (- p<0.05, ** - p<0.02, *** - p<0.0001);
^ – statistical significance (p) of difference from Group 1 (^ - p<0.05, ^^ - p<0.02, ^^^ - p<0.0001);
$ – statistical significance (p) of difference from Group 2 ($ - p<0.05, $$ - p<0.01).

4. Conclusion

Studies with application of the «Ellips-SPEK» (ISP SB RAS, Russia) spectral ellipsometric SPR (Surface Plasmon Resonance) analyzer and validation of the results with *ProteOn XPR36 (BioRad, USA)* revealed distinction in velocity constants of specific interactions of monoclonal highly specific antibodies to the tumor M2-pyruvate kinase with blood serum tumor *M2-PK* on the surface of biochips in patients with colorectal carcinoma of various stages and location of metastases (AUC 0.89).

Application of scanning ellipsometry enabled to define a significant increase of biomolecule surface effective thickness, resulting from the specific «antigen-antibody» interaction in the group of patients with extrahepatic and hepatic metastases as compared to the control group of healthy patients (p<0.0001–0.05).

Raman spectroscopy of blood serum revealed the difference in peak intensity within the range of 1005-1520 cm^{-1} in the same groups of patients (p<0.0001-0.05) with 90% predictive accuracy of the method at early stages of the disease.

The results obtained are perspective for development of methods for early stage diagnosis of colorectal cancer, detection of differently located metastases, relapses and control of administered therapy. Another alternative method for registering colorectal cancer based on dielectrophoresis is discussed in [25].

Acknowledgments

The work was performed under the theme "Epidemiological monitoring of the public health and the study of molecular genetic and molecular biological manifestations of the development of common therapeutic diseases in Siberia to improve the approaches to their diagnosis, prevention and treatment" GZ № 0324-2018-0001, Reg. No. AAAA-A17-117112850280-2, development work code "Erythrocytes".

References

[1] Jemal A. *et al* 2009 *CA Cancer J. Clin.* **59**(4) 225–249 doi: 10.3322/caac.20006.

[2] Cai S.R. *et al* 2011 *Cancer Prev. Res. (Phila)* **4** 1572-1579 doi: 10.1158/1940-6207.CAPR-10-0377

[3] Pawlik T.M. *et al* 2005 *Ann. Surg.* **241** 715–724 doi: 10.1097/01.sla.0000160703.75808.7d

[4] Meyerhardt J.A. *et al* 2013 *J. Clin. Oncol.* **31**(35) 4465-–4470 doi: 10.1200/JCO.2013.50.7442
[5] Bathe O.F. *et al* 2009 *BMC Cancer* **9** 156–163 doi: 10.1186/1471-2407-9-156
[6] Pawlik T.M. 2008 *Oncologist* **13** 51–64 doi: 10.1634/theoncologist.2007-0142
[7] Shah S.A. *et al* 2007 *J. Am. Coll. Surg.* **205** 676–683 doi: 10.1016/j.jamcollsurg.2007.06.283
[8] Rykhlitskiy S.V. *et al* 2010 *Pribory i tekhnika eksperimenta* **2** 1–2
[9] Spesivtsev E.V. *et al* 1997 *Avtometria* **1** 100–105
[10] Wild N. *et al* 2010 *Clin. Cancer Res.* **16** 6111–6121 doi: 10.1158/1078-0432.CCR-10-0119
[11] Kaura B. *et al* 2004 *J. Obstet. Gynaecol. Res.* **30** 193–196 doi: 10.1111/j.1447-0756.2004.00187.x
[12] Ewald N. *et al* 2007 *Anticancer Res.* **27** 1949–1952
[13] Shastri Y.M., Stein J.M. 2008. *Br. J. Cancer* **99** 1366–1374 doi: 10.1038/sj.bjc.6604679
[14] Demir A.S. *et al* 2013 *Turk. J. Gastroenterol.* **24**(1) 36–42 doi: 10.4318/tjg.2013.0607
[15] Meng W. *et al* 2012 *World J. Gastrointest. Oncol.* **4**(6) 145–151 doi: 10.4251/wjgo.v4.i6.145.
[16] Tonus C. *et al* 2006 *World J. Gastroenterol.* **12**(43) 7007–7011 doi: 10.3748/wjg.v12.i43.7007
[17] Mazurek S. *et al* 2005 *Semin. Cancer Biol.* **15** 300–308 doi: 10.1016/j.semcancer.2005.04.009
[18] Christofk H.R. *et al* 2008 *Nature* **452** 230–233 doi: 10.1038/nature06734
[19] Walkowiak J. *et al* 2005 *Scand. J. Gastroenterol.* **40** 1493–1494 doi: 10.1080/00365520500319112
[20] Helm J. *et al* 2003 *Cancer Control* **10** 193–204
[21] Feng S. *et al* 2015 *Biomedical Optics Express* **6**(9) 3494-3502 doi: 10.1364/BOE.6.003494
[22] Zhou Q. *et al* 2010 *J. Transl. Med.* **8** 13–19 doi: 10.1186/1479-5876-8-13
[23] Giusca S. E. *et al* 2010 *Rom J. Morphol. Embryol.* **51** 73–79
[24] Meyerhardt J.A. 2015 *J. Clin. Oncol.* **33**(16) 1717–1720 doi: 10.1200/JCO.2015.60.8661
[25] Kruchinina M. V. et al 2019 *Proc. SPIE*, in press.

IOP Conf. Series: Materials Science and Engineering **516** (2019) 012018 doi:10.1088/1757-899X/516/1/012018

Velodyne LIDAR method for sensor data decoding

M V Okunsky, N V Nesterova

National Research Tomsk State University, Tomsk, Russia
E-mail: iamleftbrain@gmail.com

Abstract. This article discusses the main technical features of the LIDAR technology and work with the output data of the LIDAR sensor. We considered the main technical characteristics of the sensor company Velodyne. The main area of consideration was the output data packet of the VLP-16 sensor. Such a data packet contains a cloud of points, their coordinates and characteristics and is transmitted via the UDP protocol. The main goal of the work was to develop a decoding algorithm for raw data for use in other software.

1. Introduction

Lidar (now used as an acronym of Light Identification Detection and Ranging) is a technology for obtaining and processing information about distant objects by means of active optical systems that use the phenomena of absorption and scattering of light in optically transparent media. To a first approximation, lidar as a device is an active rangefinder of the optical range. In machine vision systems, scanning lidars form a two-dimensional or three-dimensional picture of the surrounding space, which can also be used for mapping of a given space.

Lidar uses a light transmitter that sends a short pulse of a laser beam in one direction. The light wave passes through space until it hits the target, reflects and returns. The receiver records the flight time and light intensity. After that, the flight time is calculated. The relationship between time, speed and the distance to the target can be expressed by the following equation:

$$d = \frac{t-t'}{2} * c, \qquad (1)$$

where d is the distance to the target, t is the flight time, t' is the start time of the beam, c is the speed of light propagation in the medium.

In expression (1), the total time is divided by 2 because the light takes the same path twice.

This article discusses LIDAR VLP-16 by Velodyne.

Velodyne VLP-16 is an intelligent sensor that has its own internal processor and network interface. It provides data via the Ethernet interface over UDP protocol.

The VLP-16 has 16 vertically arranged channels with a transmitter / receiver pair on each channel. Lasers rotate through 360° around the vertical axis. The range of the vertical field of view is 30°, which means that the distance between each pair of rays is 2°. The maximum measurement distance is 130 m (found experimentally); this corresponds to obtaining 300,000 scanned points per second.

Table 1 presents the main technical specifications of the VLP-16.

Table 1. Technical Specifications of the VLP-16.

Title of the product	**Number of working channels**	**Measurement range**	**Measurement accuracy**	**Angle of view**		**Rotary speed**	**Angular resolution**	
				Vertically	**Horizontally**		**Vertically**	**Horizontally**
VLP-16	16	Up to 100 m	±3 cm	30°	360°	5–20 hz	2°	0.1°

Each laser has its own angle of shift relative to the vertical axis. This arrangement of angles prevents interference of light and other interference [1–3]. Figure 1 and Table 2 present the values of degrees with the corresponding laser numbers.

Table 2. Numbering of laser beams with the corresponding direction.

Number of laser	**Angle of direction shift, °**	**Number of laser**	**Angle of direction shift, °**
0	-15	8	-7
1	1	9	9
2	-13	10	-5
3	3	11	11
4	-11	12	-3
5	5	13	13
6	-9	14	-1
7	7	15	15

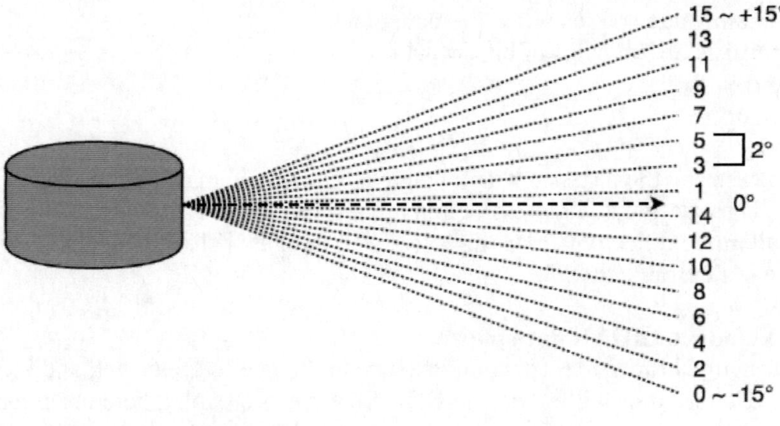

Figure 1. Vertical range of VLP-16.

2. Sensor data packeges

Data packages contain the following components:

1) 3D-data measured by the sensor;
2) the calibrated reflectivity of the surface from which the light pulse returned;
3) a set of azimuths;
4) 4 byte timestamp
5) 2 factory bytes identifying the sensor model and the mode of operation

The sensor transmits information about the distance relative to its position in spherical coordinates (r is the radius, ω is the height, α is the azimuth). Conversion from spherical to Cartesian coordinates is performed using the following formulas:

$$X = R * \cos(\omega) * \sin(\alpha)$$
$$Y = R * \cos(\omega) * \cos(\alpha) \quad (2)$$

$$Z = R * \sin(\omega).$$

Figure 2 show the relationship of the spherical coordinate system and the Cartesian coordinate system.

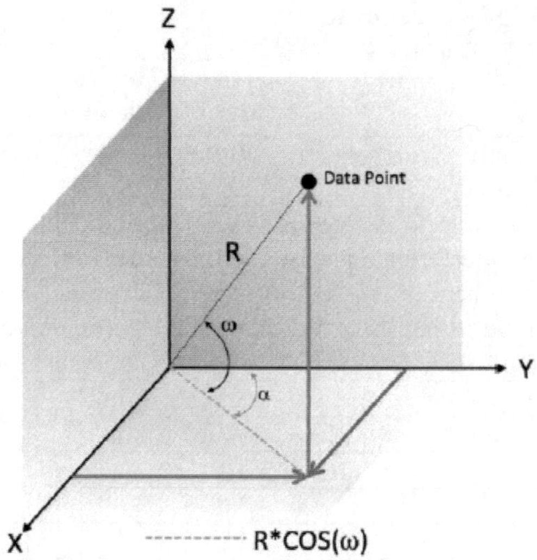

Figure 2. Spherical coordinates of the point.

Azimuth is 2-byte value; it is an angle in hundredths of a degree. The valid azimuth value is from 0 to 35999. One azimuth value is transmitted per data block.

A data block consists of 100 bytes of binary data:

- two-byte flag (0xFFEE);
- two-byte azimuth;
- 32 data points.

The result of measuring the distance with a single laser channel may be called as a data point out of a cloud of points. The data point is represented in a packet in three bytes — two bytes of distance and one byte of the calibrated reflectivity. The distance is an integer without a sign. Calibrated reflectivity is reported using a scale from 0 to 255.

3. Decoding the Velodyne LIDAR data packet

The data packet contains header files, data on distance and angle to the target, and information on the reflectivity; these data are transmitted via the UDP Ethernet protocol. There are three modes of data transfer: strongest, last, and dual [4, 5]. To implement the decoding method, we select the dual mode because in this case the data packets will receive information from two modes of operation at the same time. The data packet will consist of 1248 bytes.

Packet decoding operations can be divided into three parts: azimuth calculation (horizontal angle), distance calculation, and conversion to Cartesian coordinate systems.

After checking the flag file of the block (xFFEE), we can proceed to the calculation of the azimuth. First, the system receives two sequences of hexadecimal values. Second, we swap the bytes of the two hexadecimal values and combine them together into a two-byte value. Third, we convert the hexadecimal value to decimal and divide by 100, thereby obtaining the current value of the azimuth (horizontal angle) [6].

The first steps to calculate the distance are similar to the azimuth calculations. After receiving two sequences of hexadecimal values, we swap them and combine them into a two-byte value and convert from hexadecimal to decimal values. The final step is to multiply the decimal value by 2.0 mm. It is necessary to decode the distance value for each channel in order to build a cloud of points [7].

VLP-16 uses a spherical coordinate system; therefore, it is necessary to convert the values into the Cartesian system. To do this, we use expressions (2). The angle value ω (vertical angle) is a fixed value. Table 2 shows the angle vales according to the channel number.

The value of the angle α corresponds to the found value of the azimuth angle. The value of R also corresponds to the already found value of the distance. Thus, it is possible to decode data packets transmitted via the UDP protocol [8,9].

4. Results

Let us consider an example of data decoding.

Using the VLP-16 sensor, we scanned the room and saved the data in pcap format. Figures 3 and 4 show point cloud and data packets.

Figure 3. Point cloud received using VLP-16.

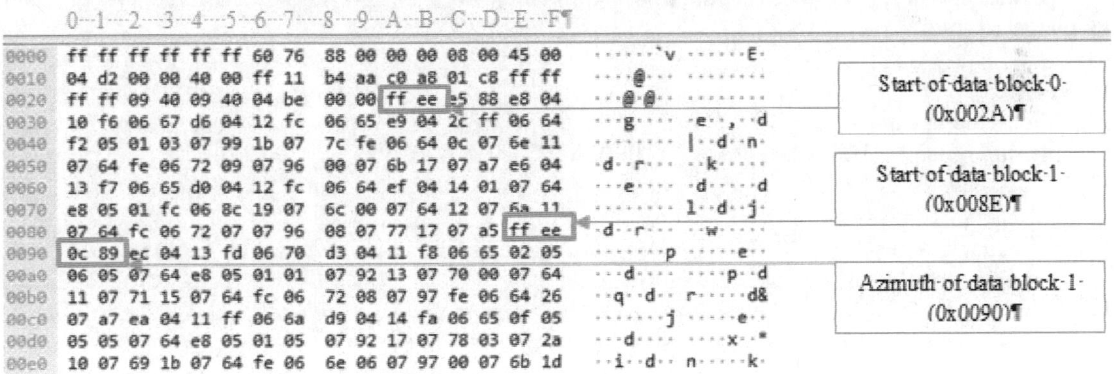

Figure 4. Data saved in pcap file.

Azimuth calculation:
-find azimuth values: 0×0C and 0×89 after flag-bytes 0×FF and 0×EE;
-swap azimuth values: 0×89 and 0×0C;
-unite bytes in one value: 0×890C
-convert the value to decimal: 35084;
-divide the resulting value by 100;
As a result, we get the value: 350.84°.

Calculation of the distance to the point:
-find the distance data: 0×EC and 0×04;
-swap the values found: 0×04 and 0×EC;
-unite data in one value: 0×04EC;
-convert the value to decimal: 1260;
-multiply the resulting value by 2 mm;
As a result, we get the value: 2520 mm or 2.52 m;
Reflectivity is equal to 0×13.
Conversion to Cartesian coordinate system:
-horizontal angle is found: $\alpha = 350.84°$;
-distance to the point is found: $R = 2.52$;
-using Table 2, we find the vertical angle corresponding to the 0th channel of the sensor: $\omega = -15°$;
-using formulas (2), we calculate values of coordinates XYZ:

$$X = R * \sin(\propto) * \cos(\omega) = -0.382$$
$$Y = R * \cos(\propto) * \cos(\omega) = 2.403$$
$$Z = R * \sin(\omega) = -0.652$$

5. Summary

As a result of the work, the space was scanned using the VLP-16 sensor and data were obtained in the form of a cloud of points and hexadecimal values containing information about this cloud of points. Using the developed algorithm, one can obtain the coordinates of each point at any time in the Cartesian coordinate system. In the results section, step-by-step calculation of the parameters of points for obtaining the final coordinates is shown. In further work, the device will be designed for decoding and converting LIDAR data in real time.

Acknowledgments
The research was carried out in Tomsk State University with the financial support of the Ministry of Education and Science of Russia, a unique project identifier: RFMEFI57817X0241.

The authors are grateful to Tatiana B. Rumyantseva fromTomsk State University for English language editing.

References

[1] Yu-Cheng Fan *et al.* 2018 *IEEE International Instrumentation and Measurement Technology Conference (I2MTC)* 1–4 doi: 10.1109/I2MTC.2018.8409759
[2] Shashev D. *et al.* 2015 *IOP Conference Series: Materials Science and Engineering* **81** doi: 10.1088/1757-899X/81/1/012101
[3] Li-Juan Zheng *et al.* 2017 *IEEE International Conference on Consumer Electronics* 17062572 doi:10.1109/ICCE-China.2017.7990982
[4] Shikhman M., Shidlovskiy S. 2018 *IOP Conference Series: Materials Science and Engineering* **363** (1) doi: 10.1088/1757-899X/363/1/012019
[5] Velodyne LIDAR VLP-16 user manual. [Electronic resource]. URL: https://velodynelidar.com/downloads.html.
[6] Shashev D.V., Shidlovskiy S.V. 2017 *Journal of Physics: Conference Series* **881** doi:10.1088/1742-6596/881/1/012029
[7] Shidlovskiy S. 2016 *MATEC Web of Conferences* **79** N01014. doi: 10.1051/matecconf/20167901014
[8] Shidlovskii S. 2006 *Journal of Computer and Systems Sciences International* **45** (2) 282–286 doi: 10.1134/S1064230706020122
[9] Sukhanov D., Zavyalova K. 2016 *MATEC Web of Conferences* **79** 01040

Social and humanistic aspects for the training of technical professionals

E Yu Liventsova, T B Rumyantseva, E G Syryramkina

National Research Tomsk State University, Tomsk, Russia
E-mail: evg.liv@mail.ru

Abstract. Development of soft skills of students is an urgent educational task of modern classical universities. The aim of the article is to demonstrate the experience of approbation of the educational technology for developing soft skills of students in a modern university. The article presents a new approach to the problem of developing soft skills of students at the National Research Tomsk State University (TSU). The authors analyze the essence of the educational technology 'Accelerator' as well as results gained during approbation of this technology based on the Park of Social and Humanistic Technologies of TSU (the structural subdivision of the University). Prospects for introduction of the described technology in classical universities are related to demand for the enhanced quality of soft skills of students in conditions of transition to a new technological paradigm dedicated to social and humanistic technologies. Approbation of the described technology in the modern university allows authors to draw conclusions about its effectiveness and to present results of the experience to a wider audience.

1. Introduction

Modern trends in the development of education and society as a whole not only strengthen the role of modern universities in transformation of the regional development, but also require a new approach to developing skills and over-professional competencies of university students, especially students of technical specialties, who will live in conditions of a new technological paradigm. In this connection, classical universities have a special role because social and cultural development of the society is their original mission. The methodological basis of the work is the idea of an activity and competence approach. The authors have used such methods as analysis of scientific literature, analysis of documents, desktop research, interviewing and questioning of participants in the educational process, educational experiment and description of practical experience of implementing a program of accelerating students' projects in a modern university (case study). Based on the study of foreign information resources and the author's experience, the article provides detailed structure of the educational technology 'Accelerator' for developing students' soft skills; this program includes the following components: training, mentoring, providing conditions for presenting the results of project activities in the community

In the field of innovative activities, the modern society has great expectations for the younger generation because young people are more sensitive to changes in the external environment and adapt to new conditions of professional activity flexibly and quickly.

At the same time, modern trends in the training of specialists show that the educational paradigm on assignment and building up knowledge by an individual has been replaced by a competence-based approach that requires managing the process of cognition itself and the ways of creating, generating, and assigning knowledge. It is the so-called 'broad education' (soft power), which is an intellectual power and cultural imperative. Education becomes a reflection and a way of creative and modified use of this knowledge in changing and complicating practice.

Therefore, the so-called 'hard' skills (which every professional should a priori have today), are replaced by 'soft' skills that allow a person to be successful regardless of the specificity of activity. These skills include effective communication and management, ability to work in a team, etc.

The fact that specialists with developed soft skills will be in demand in the future labor market is confirmed by the *Atlas of New Professions* developed by the Skolkovo School of Management and the Agency for strategic initiatives of the Russian Federation. According to the Atlas, future professions the in the IT field, robotics and engineering require transfessional skills and abilities such as customer focus, interdisciplinary communication, multilingual and multicultural communication, project management, ability to work with people, ability to work in conditions of uncertainty, and artistic skills. Thus, educational establishments need to develop new forms and programs aimed at formatting soft skills and developing transfessional competencies that are not provided by the basic educational programs of universities.

The article describes the experience of Tomsk State University on the use of social and humanistic technologies for training students of technical specialties. In particular, it summarizes the experience of implementing the educational technology of social accelerator during the period of 2015 - 2017, which, in our opinion, responds to two main educational challenges:

-intensification of education while maintaining its quality;

-training through activity and communication.

2. Educational technologies used for developing soft skills

The authors used the following methods: analysis of scientific literature and documents, desktop research, interviewing and questioning participants of the educational process, an educational experiment, and a case study (description of practical experience of the implementation of an acceleration program for socially oriented business projects in a modern university).

The methodological basis of the work was formed by the ideas of the activity and competence approaches. Competence is supposed to be an independent realizable ability to perform activities and solve life problems based on acquired life experience and its values. For our study, the definition of competence includes everything that can be used for an effective action: cognitive components (knowledge and skills) and non-cognitive (motivation, ethical attitudes, value orientations, and behavioral components) [1].

In modern psychological and educational studies, the problem of the development of values, social attitudes, and communication practices is discussed in the context of developing the theory and practice of humanistic technologies and providing psychological and educational support. Psychological and educational support is the process of creating conditions under which students acquire the experience of transforming their personal potential and the potential of the social environment into their own education resources [2].

The special role of humanistic technologies for developing these soft skills is crucial because these technologies focus on the design of educational environments that allow people to build life-trajectory and realize life resources as much as possible. Thus, the relevance of humanistic technologies is justified by the priority of education in teaching students new ways of thinking and new forms of activity [3].

Researchers emphasize the role of high humanistic technologies, which determine the personal presence of a student in his education process with the ability to influence it, provide training for innovation and form the competences of effective group interaction, communication and joint activities necessary for a person's life in an open educational space [4].

Taking into account social and humanistic aspects of training professionals, educational techniques for developing soft skills mean the form of organizing activities, in which students interact with the social sphere, learn it more deeply, master its values, actualize their own potential for performing socially beneficial activities, and acquire successful experience in solving social problems while developing social and communicative competences.

Nowadays, there is a wide variety of educational technologies used for developing soft skills. They include the following:

- acceleration technologies, that is intensive educational programs including weekly workshops, trainings, and practical meetings, which allow working out a business model of a project in a short time in order to present it to investors and the community;
- club technologies, that is informal meetings with specialists, networking, and presentation sessions for designing changes in various fields;
- *service learning* technologies (learning through action / compassion / service), that is the formation of project and entrepreneurial competencies through socially beneficial activities (social practices, internships in non-profit organizations, etc.).

The so-called *accelerators* are becoming increasingly popular among existing educational technologies aimed at developing transfessional competences in the university environment. Currently, the accelerator is a technology that includes educational activities, consultations, mentoring support, access to equipment, and other resources for accelerating projects development. The technology has the following features: a clearly defined period of conduct, group training, training with the participation of mentors and well-known specialists/entrepreneurs, and training that ends with the presentation of projects to investors (demo days) [5, 6].

Generally, the operation cycle of accelerators consists of five main phases: acquaintance, selection of applications, educational program, demo day, and further support.

The acquaintance phase consists in posting information about the launch of the accelerator on the Internet, social networks, and through the business community. It is also possible to conduct special informational events.

The phase of selection of applications consists in placing the registration form (usually in an electronic form) and the subsequent selection of candidates among those registered based on the specifics and themes of the project team and growth potential.

The phase of the educational program includes a series of lectures, seminars, trainings, workshops, and business games aimed at acquiring knowledge and gaining the necessary skills by the team members. Duration of training ranges from six weeks to several months. A mentor who is a professional in a particular area of knowledge and/or activity usually works with each team [7].

The phase of the educational program ends with demo day when participants get the opportunity to present their projects to investors. This day is important both for investors and for project teams because presentations of projects may result in receiving investments or establishing useful contacts, which further will subsequently lead to investments. There are various formats for demo days including public presentations, personal presentations, elevator pitch, and others.

Mainly, accelerators are focused on high technology, in particular on information technology field. At the same time, we can notice a general tendency to the emergence of social accelerators, which are aimed not only at developing entrepreneurship in the region, but also at promoting social projects for the benefit of certain categories of people and/or aimed at solving certain acute social problems in society. Examples of existing social accelerators are StartUp Chile (Chile), HSE Start Up (Russia), Social Entrepreneurship Accelerator at Duke (USA), Rybakov Foundation (Russia).

As for the form of organization of activities, social accelerators can function either as separate structures/programs or as substructures within the framework of a classical accelerator.

3. Experience of Tomsk State University

National Research Tomsk State University has a structural subdivision - the Park of Social and Humanistic Technologies (the TSU Park). Its activities are focused on providing consulting and information support for students' social projects, which have features of long-term sustainability and potential for growth into a social entrepreneurship regime.

Based on the study of foreign literature and authors' experience, we proposed a new approach to the development of the technical students' competence. The developed technology 'Social Accelerator' includes such components as training, mentoring support, and providing conditions for presenting the results of project activities in the community; the technology was tested on the basis of the Park of Social and Humanistic Technologies, which is a structural unit of National Research Tomsk State University.

Our interviews and questionnaire survey of students, mentors, and consultants (120 people in total) and summarizing accumulated practical experience show the effectiveness of the accelator as a technology used for the development of students' entrepreneurial competence.

The educational part contained consecutive topics related to launching a social startup: creating a roadmap for a project, marketing, finance, project team, sales, fundraising and working with investors, PR of a social business project, effective presentations and others. Selected experts competent in various areas of developing and promoting startups shared their knowledge and experience with students; and mentors accompanied the process of developing projects' business models and helped to determine effective project implementation plans. It is important to point out that this was learning by action, i.e. after completing the study of each topic, students had the possibility to perform a specific action to implement their project, for example, to launch a group on a social network, to organize and conduct the first fundraising event, etc.

We emphasize that within the framework of the *StartupSocial* accelerator, special attention was paid to mentoring support of students' projects. Acting entrepreneurs played the role of mentors and led from one to three projects. In our opinion, the applied 'competence model', in which the mentor connects training with practical application of knowledge, and the 'development model', which is based on the mentor's position as an assistant in achieving personal and professional growth through reflection, are the most acceptable for teaching social entrepreneurship to students. They enable students to learn from the mentor entrepreneurial experience in solving social problems and increase the confidence of young people in their own abilities as well as in making decisions. In turn, mentors perform not only the functions of motivators and consultants, but also help students, if necessary, to get in touch with the right partners, investors, experts, clients, because they have already established business contacts. The close interaction of students with mentors, joint consideration of a specific project allows moving from idea to action faster and more efficiently.

All student startups of the TSU Park were implemented by the teams, in which students with leadership qualities came together. Entrepreneurs are leaders because they need to be able to rally people, to inspire them to achieve the goal, and to convince them of the ability to achieve such a level of task fulfillment that previously seemed inaccessible to them. Only a person who can create optimistic plans, calculate risks, explain the essence of innovation, and set up a team has the power to do that. According to specialists in the field of psychology, a leader is a person who realizes the public interest by satisfying his own egoism. Leaders distribute wealth and provide work for hundreds of people by developing their own activities. At the same time, leaders stimulate progress in society and bring a revival to the economy, which gives impetus to the evolution of the society.

The project activity within the framework of the *StartupSocial* accelerator is also aimed at developing students' skills in communication with local and wider communities. Our experience shows, that projects were successful if they were based on building strong relationships with representatives of the local community, to whom student teams were able to demonstrate their discoveries and ideas. The entire period of students life in the university should become a school of civil and professional development, active participation in the organization of the life of the university, city, region, and country. By means of active participation in the scientific and social life of the university, city, and region students acquire skills of organizers, develop their personal qualities necessary for specialists, leaders, and public figures [8].

4. Results

The university *StartupSocial* accelerator as an educational program allowed ninety university students not only to dive into the basics of entrepreneurship in general and social entrepreneurship in particular but also to test their professional knowledge and skills and to assess their entrepreneurial potential in building social business. Ninety students went through the *StartupSocial* accelerator during the period of 2015–2017; some of them came to the Park of Social and Humanistic Technologies only with an idea of a social project, while others came with working social projects that needed commercialization.

During three years, 36 student teams have become participants in the educational program *StartupSocial* accelerator. According to the results of the survey and interviewing graduates of 2015 and 2016, 92% of respondents noted that participation in *StartupSocial* accelerator enabled them to acquire basic knowledge and skills for starting projects, 89% pointed out the importance of professional communication with members of the business community and mentoring.

Today, 16 projects out of 24 projects proposed by *StartupSocial* graduates are active. Currently, 13 teams are developing project activities. These projects are aimed at solving urgent social problems of the region, such as provision of social services to people with disabilities; improvement of social and educational spheres of the region; protection of environment in the region; rehabilitation and involvement of people to a healthy lifestyle and others.

Three years of experience in implementing the technology of *StartupSocial* accelerator allows us to evaluate its educational potential. First, it consists in the emergence of students' sustainable motivation for further activities as a whole and in the desire to continue studying in city, regional, All-Russian and international schools and programs. In addition, students develop a new look on social development and a willingness to present their ideas and projects to a wide audience, to present results of their activities to the local community and to demonstrate participation of the modern university in social and economic development. Performance of a social function is one of the key indicators of universities' development and their place in the society.

Of course, it is also important that the project activity of students within the framework of the accelerator contributes to expanding their horizons and to developing transfessional competencies and personal qualities, as well as increases the cost of graduates in the labor market.

5. Summary

Opportunities for the implementation of educational programs aimed at developing soft skills of students of technical specialties are also available at classical universities, as evidenced by the experience of Tomsk State University. The proven technology of *StartupSocial* accelerator is an innovative tool in achieving the task of higher education institutions to generate a new wave of specialists with transfessional skills. First, this technology allows recognizing students capable of entrepreneurial activity in order to provide subsequent assistance to them in developing relevant aptitudes. Second, this technology contributes to creative atmosphere in educational institutions, which favors development of students' independence, initiative, activity, and the desire to identify problems that are of great practical importance, as well as intensifies business communication between students and qualified specialists. Third, the technology contributes to a wide demonstration of various forms and types of entrepreneurship, stories and practices of existing successful entrepreneurs.

Moreover, the activity of the TSU Park in developing soft skills of students was reflected in the work on independently established educational standards of TSU. Such skills as systemic and critical thinking, development and implementation of projects, teamwork and leadership, and intercultural interaction were included into the group of universal competencies. Thus, the described technology of *StartupSocial* accelerator is replicable and promising for implementing in modern universities because of the increased demand for soft skills in the transition period from the fifth to the sixth technological order.

Acknowledgments
The paper is prepared within the framework of the competitiveness improvement programme of National Research Tomsk State University and grant of the Russian Foundation for Basic Research No 16-29-04388.

References

[1] Abakumova N. 2007 *Competence Approach in Education: Organization and Diagnostics* (Publishing House of Tomsk State University: Tomsk)
[2] Krasnorydtseva O. 2007 *Tomsk State University Journal* (305) 165–168
[3] Miller T., Wesley C., Williams D. 2012 *Academy of Management Learning and Education* **11**(3) 349–370
[4] Galazhinsky E. 2010 *Phycology of Education* (12**)** 5–21
[5] Kickul J., Lyons T. 2014 *Understanding Social Entrepreneurship: The Relentless Pursuit of Mission in an Ever Changing World* (Alpina Publisher: Moscow)
[6] Sokolova E. 2013 *Continuing education: the 21st century* **4** 1–11
[7] White Eric R. 2015 *The Journal of General Education* **64(**4) 263–277
[8] Belenky V., Paulman V. 2012 *Sociological research* (10) 11–20

IOP Conf. Series: Materials Science and Engineering **516** (2019) 012020 doi:10.1088/1757-899X/516/1/012020

Methane content of coal seams of Karaganda basin

A A Kenetayeva[1], Zh K Kenetayeva [2], Zh T Tokusheva [1], M Rabatuly [1]

[1]Karaganda State Technical University, Karaganda, Kazakhstan
[2] Translator, LLP "JV "Alaigyr", Karaganda, Kazakhstan
E-mail: aigul_tate@bk.ru

Abstract. Methods for studying the natural methane content of coal seams during geological exploration are adequately covered, so the main focus when considering them is landing on those features that are characteristic of the Karaganda basin. The method of studying the qualitative composition of gas in the basin establishes a change in the concentration of methane depending on the geological conditions of the bedding, while it is necessary to find the upper boundary of the methane zone, which is defined as the surface, and the concentration of methane is 80%.

1. Introduction

Safe and economical coal mining in the Karaganda basin, as in many mines in the country, depends on the content and distribution of methane in coal seams and rocks, as well as on the quality and degree of accuracy of the prediction of methane-bearing coal seams and mine methane abundance. Therefore, studies of natural methane-bearing and geological factors determining the formation, accumulation and preservation of methane in the coal layer are priority tasks when exploring new mine fields and developing deeper horizons of existing mining enterprises. The need for such studies is due to the high release of methane in the mine workings, which is associated with the use of high-performance equipment, increasing the depth of development and higher methane-bearing coal beds of the Karaganda basin, unlike other basins in the country.

The qualitative characteristic of natural methane content, which prevailed before the adoption of regulatory documents, gave way to a quantitative assessment with a given accuracy of determination. However, the further development of the work is hampered by insufficient knowledge of the factors responsible for the natural methane content of coal seams and rocks, as well as the weak feasibility of some of the adopted methodological provisions.

2. Methods for studying

Methods for studying the natural methane content of coal seams during geological explorations are adequately covered, so the main focus when considering them is landing on those features that are characteristic of the Karaganda basin [1].

During exploration in the basin, the method of study is the qualitative composition of the gas, as well as direct and indirect methods for determining the methane content of coal seams.

The method of studying the qualitative composition of gas in the basin establishes a change in the concentration of methane depending on the geological conditions of the bedding, while it is necessary to find the upper boundary of the methane zone, which is defined as the surface, and the methane concentration is 80%.

IOP Conf. Series: Materials Science and Engineering **516** (2019) 012020 doi:10.1088/1757-899X/516/1/012020

The method consists in sampling in a special hermetic vessel from a core raised to the surface with a standard coring projectile or a double coring tube. According to the composition of the gas contained in the sample, the concentration of gas components (methane) at the sampling point is identified.

The simplicity of this method predetermined its wide usage during fieldwork. Currently, the basin accumulated significant material with the nature of changes in the concentration of methane in coal seams.

3. The method of direct determination of methane content of coal seams

The method of direct determination of methane content of coal seams is the main one and is used with the help of a core gas basin with a "bell". The basin uses a type of KG-55 - 2m core gas basin [2].

Interpretation of the results of testing with core gas basins is carried out according to the method [2]. The nature of the increase in methane content with depth is identified according to this method, based on the premise that the ascending increase branch is described by a Langmuir type equation. The processing scheme of the results of testing is reduced to the following operations.

1. Correlation tables of changes in natural gas content (*X*) in a function of sampling depths (*N*) are compiled, for which the number of intervals (m_i) and their optimum value are previously determined depending on the amount of sampling data using the formula:

$$m_i = 1 + \lg \sum n_0 \tag{1}$$

where n_o – the number of sampling points.

The optimal value of the length of the intervals (1) is found as a quotient of dividing the difference between the maximum and minimum values of natural gas content (span) by the number of intervals:

$$R = \frac{X_{\max} - X_{\min}}{m_i} \tag{2}$$

When processing gas sampling in the Karaganda basin, the value of the depth interval is generally taken to be 50–100 m, and by natural methane content - 2m^3 / t gm.

2. The starting point is selected with the coordinates X_i, H_i, so that it could correspond to a stable increase in the natural methane content, i.e. met the condition 3 X_i 10 м^3/t gm.

3. The average values of natural methane content are calculated by depth intervals:

$$X = \frac{\sum X_i}{n_i} \tag{3}$$

where X_i – values of natural gas content in a given depth range, m^3/t gm; n_i – the number of samples

4. The actual value of the methane content stage is determined.

$$a = \frac{H_i - H_1}{X_i - X_1} \tag{4}$$

5. Assuming that the methane content stage *a* varies with depth according to the law of a straight line: $a = A + BH$, by the least squares method from the equation system can be found the values of the coefficients *A* and *B*:

$$\sum a_i = B \cdot H_i + n \cdot A \tag{5}$$

$$\sum a_i \cdot H = B \cdot \sum H_i^2 + A \cdot \sum H_i \tag{6}$$

6. The found values of the coefficients *A* and *B* are used in determining the coefficients *c, b* and H_o of the Langmuir equation:

$$c = \frac{(X_1 \cdot B + 1)^2}{A + B \cdot H_1} \tag{7}$$

$$b = \frac{B \cdot (X_1 \cdot B + 1)}{A + B \cdot H_1} \tag{8}$$

$$H_0 = \frac{H_1 - X_1 \cdot A}{X_1 \cdot B + 1}, \text{ м}^3/\text{т} \tag{9}$$

The numerical values of the coefficients are substituted into the equation:

$$X = \frac{c \cdot (H_i - H_{0)})}{1 + b \cdot (H_i - H_0)}, \text{ м}^3/\text{т} \tag{10}$$

and for given depths, the natural methane content of the seams is determined.

The method for determining the methane content of coal seams according to gas surveys data in preparatory mine workings is indirect. Gas surveys were carried out during coal exploration by the Karaganda Polytechnic Institute and during preparatory works in the mines to clarify the data on methane content obtained during the exploration period and to evaluate the degassing efficiency of the Karaganda branch of VostNII.

The technique developed in work [1,3] is used. The initial parameters for calculating the methane content of coal seam are the results of measurements of the total methane consumption passing through the sections of the peak production pickets $Q^1{}_m$, $Q^2{}_m$, ..., $Q^i{}_m$ m^3/day with the coal wall exposure time at the picket level of t^1, t^2, ..., t^i days.

These data, as well as the thickness of the coal mass of the seam (m), the average speed of the progress of development between the pickets $V^1{}_{np}$, $V^2{}_{np}$,..., $V^i{}_{np}$ allow us to calculate the total outgassing with 1m^2 of coal wall

$$Q_t^i = \frac{1}{2} \cdot \left(\frac{Q_m^1}{V_{np}^1} + \frac{Q_m^2 - Q_m^1}{V_{np}^2} + \ldots + \frac{Q_m^i - Q_m^{i-1}}{V_{np}^i} \right) \frac{\grave{\imath}^3}{\grave{\imath}^2}, \tag{11}$$

If the substantial unevenness of the speed of progress of the development is not fixed, then the measurement is carried out only at the exit from the dead-end development. In this case

$$Q_t = \frac{Q_m}{2 \cdot m \cdot V_m^{cp}} \tag{12}$$

It is known that

$$Q_t = \frac{q_0}{n} \cdot t^n \tag{13}$$

where q_0 – the initial intensity of methane release by the end of the first day of exposure of the coal wall, m^3/m^2 day; n – coefficient characterizing the rate of reduction of methane release over time.

The relatively high accuracy of the determination of methane content compared with the method of core and gas basins promoted the introduction of gas surveys into the practice of work. Currently, the method of gas surveys in the mine workings is used in the exploration of all the mine fields of the basin, where there are necessary conditions for their conduct.

Determining the position of the depth of the upper boundary of the methane gas zone with the required accuracy is one of the main tasks when studying the natural gas content of coal seams. For detailed exploration of a site (deposit), it is necessary to identify its position with maximum accuracy. This problem is solved, basically, by testing coal seams with hermetic cups in the proposed depth range of the transition to the methane gas zone. However, a number of complications in processing the results of testing significantly reduces the reliability of the method and makes it difficult to solve the problem. As a rule, it is not possible to take samples with a methane concentration of 80%. The obtained values of methane concentration in samples vary considerably at the same depth of sampling.

This circumstance often makes linear interpolation impossible when determining the depth of the upper boundary of methane gases.

At the same time, there is often an urgent need to assess the depth of the methane gas zone of the sites where previously testing for methane content for some reason was not carried out or was not carried out to an insufficient extent. In this case, a situation arises in which the total exploration of the site (deposit) is satisfactory, and data on the nature of the occurrence of methane gas zones are missing or do not meet modern requirements. This situation is observed, in particular, with the involvement of lower layers of the Karaganda Formation in the Industrial site of the Karaganda basin.

To solve these on the basis of extensive material on sampling of hermetic cups and core gas basins of coal seams in the zone of demethanization and methane gases, an attempt was made to identify the main geological factors that determined the formation of a zone of gas weathering in the basin, and to assess the quantitative force of their influence. Such an approach will allow, in the first approximation, to calculate the graph-analytical depth of the upper boundary of the methane gas zone for any part of the basin in the known general geological situation.

4. Summary

The role of various geological factors in the formation of the demethanization zone is considered in a number of works [4–9]. The main factors determining the depth of distribution of the gas weathering zone are noted - the history of the geological development of the basin, the angle of incidence of the coal-bearing strata, the presence, thickness and nature of the overburden hydrogeological conditions.

The history of geological development is determined by the nature of the orogenic phases of folding, the oscillating movements of the main stages of erosion and the subsequent demethanization of coal-bearing sediments with positive signs of crustal movement, as a rule, denudation and demetanization are associated, and the coal-bearing strata overlapping by cover sediments slowing demethanization or inhibiting it is associated with negative signs.

When methane migrates to the ground surface, mainly on coal seams, the angle of incidence determines the length of its movement. In connection with this, steeply dipping seams, with other things being equal, demethanize to a greater depth vertically than flat ones.

Cover sediments formed during the majority of negative forms of crustal movement, depending on the thickness, lithological composition, facies consistency and the time of their accumulation affect the distribution of methane in the coal-bearing strata in different ways. Depending on the ratio of these factors, the overlapping of the coal-bearing strata with younger sediments causes a slowdown or almost complete stoppage of methane migration. This often leads to the recovery of methane content in previously demethanized coal seams and in coal-bearing sediments.

Circulating underground water degasses the coal-bearing strata. It is noted that the upper boundary of the methane zone is confined to the bottom of the zone of active water exchange.

References

[1] Kolmakov V.A. 1981 *Methane Emission and Control of it in the Underground Mines* (Nedra: Moscow)
[2] Kizryakov A.D., Khakimzhanov T.E., Kheguy G. 1983 *Methane Content of the Karaganda Basin Mines* (Publishing house «Nauka»: KazSSR)
[3] Sitnikov R.S. 2014 *Improving the Methods of Degassing by the Underground Method when Conducting Mining Operations* (Thesis of Master of Science)
[4] Ayruni A.T. 1981 *Theory and Practice of Mineral Gases Control at Great Depths* (Nedra: Moscow)
[5] Ayruni A.T. *et al.* 1990 *Volume of Gas in Coal Mines in the USSR. Integrated Development of Gas-bearing Coal Deposits* (Nauka: Moscow)
[6] Ayruni A.T. 2002 *Problems of Development of Methane-bearing Coal Seams, Industrial Extraction and Use of Coalmine Methane in the Karaganda Basin* (Publishing House of the Academy of Mining Sciences of Russia: Novosibirsk)

[7] Vasyuchkov Yu.F. 1986 *Physico-chemical Methods of Coal Seam Degassing* (Nedra: Moscow)
[8] Lidin G.D., Sergeev I.V., Zaburdyaev V.S. 1971 *Industrial Labor Safety*
[9] Issagulov A.Z. *et al* 2016 Metalurgija **55(3)** 426-428

IOP Conf. Series: Materials Science and Engineering **516** (2019) 012021 doi:10.1088/1757-899X/516/1/012021

Modeling of synthetic-aperture radar in X and P-band

A N Leukhin[1], A A Voronin[1], V I Bezrodnyi[1], O V Galtseva[2]

[1]Mari State University, Yoshkar-Ola, Russia
[2]National Research Tomsk Polytechnic University, Tomsk, Russia
E-mail: leukhinan@list.ru

Abstract. Synthetic-aperture radar is usually a complex software and hardware system. It allows obtaining images in radio range, comparable in resolution with optical systems. The advantage of radio waves is that the images are of high quality, despite cloudiness and dark time. The development of algorithms for such systems is a rather complex process. Mathematical modeling applied in purpose to reduce costs. In this paper, we give an overview of early created systems. We discuss the methods for calculating the scattered electromagnetic field. We choose methods that are most suitable for simulating a synthetic aperture radar. Combination of different approximation methods allows us to process large scenes. We take into account the various effects that arise when propagating radio waves. Also, we describe algorithms for a synthesis of radar images. In particular, we consider range-migration algorithm and time-frequency processing algorithm. We show that the frequency-time processing algorithm is preferable for synthesis a radio image in the X-band due to its speed. In opposite, the range-migration effect in P-band is too strong to ignore it. The time-frequency algorithm gives not focused image with serious artifacts. It is better to use the range-migration algorithm for P-band.

1. Introduction

Earth remote sensing with radio, infrared and optical waves is an integral part of the modern human activity.

Space and aerial remote sensing are used to solve practical problems of land use, agricultural and forest monitoring, land and sea ecology, control fishing areas, identifying preconditions of dangerous phenomena as: landslides, floods, critical changes in the geometry of natural and artificial objects and engineering structures, as well as technical and military intelligence tasks.

Remote sensing of the Earth's surface started at the beginning of the 20th century and is still ongoing. There are two types of systems depending on the wavelength: optical and radio. The advantages of the optical remote sensing are high resolution and the relative simplicity of the onboard equipment. Disadvantages of this type of systems are dependence on meteorological conditions and time of day. In opposite to that, radar remote sensing is an effective method for obtaining operational and long-term information.

Historically, the implementation and development of space satellites with SAT can be characterized in 4 stages.

The first stage is the research stage, which includes the launch of the American SAR Seasat-A (1978), SIR-A (1981), SIR-B (1982) and Soviet Mech-K, Mech-KU, Kosmos-1870 (1987–1989), Almaz-1 (1991–1992). Also should mention SIR-C/X-SAR (1994) developed by USA and Germany. At this stage, a large amount of experimental material was accumulated. Developers have produced methods and software packages for processing and interpreting radar images. Methods for solving military intelligence tasks have appeared at that time.

The second stage is the refinement of the techniques for the processing of radar images applied to practical remote sensing problems in various fields of human activity. European satellites ERS-1 (1991–2000), ERS-2 (since 1995), Canadian Radarsat-1 (since 1995), Japanese JERS-1 (1992–1998), USA military satellites Lacrosse (launches 1988, 1991, 1997, 2000, 2005) have provided information for improvements of processing methods on this stage.

The third stage is the great leap in radar imaging which includes launches interferometric SAR in 2000. This type of SAR with rigid base allows creating relief height maps. Such SAR systems found its place in full-polarimetric Envisat-1 and PALSAR, that can cover 80% of Earth surface. Commercial use of SAR imaging has developed.

The fourth stage is an information explosion in the field of space radars, which occurred in 2007 when Germany, Italy, China, Japan, and Canada launched into orbit 9 satellites with radar systems.

Modern technologies make it possible to create satellite radars with a spatial resolution about few centimeters. The achieved high resolution of space-based SARs allows solving the problem of radio vision, having advantages not only in all-weather conditions but also in the possibility of detecting and recognizing objects of observation by radar contrasts and selection of moving ground objects.

Still, some important problems of radar imaging remain unsolved today. Tasks such as detecting, tracking, and estimating parameters of airborne targets are unaffordable for modern airborne radar systems.

2. Research Method

2.1 Time-frequency algorithm

The time-frequency algorithm assumes a transition from the calculation of a two-dimensional convolution to the execution of two one-dimensional operations of processing a radio hologram.

To do this, after compression of the signal in range, a transition is made to the frequency domain along the azimuth coordinate [1]. The dependence of the Doppler frequency on travel time is

$$f(t_x) = \frac{1}{2\pi}\frac{d\varphi(t_x)}{dt_x}. \tag{1}$$

Function $\phi(t_x) = \frac{4\pi}{\lambda}(t_x)$ describes the phase modulation of the azimuth signal. Then we get

$$f(t_x) = f_x = \frac{V_n^2 t_x}{\lambda r_0}. \tag{2}$$

The current increment of the slant range relative to the traverse channel as a function of travel time is

$$\tilde{r}(t_x) = r(t_x) - r_0 = \frac{V_n^2 r_x}{2r_0}, \quad t_x = -\frac{f_x \lambda r_0}{2V_n^2}. \tag{3}$$

The time is related to the Doppler frequency f_x linearly. Therefore, substituting (2) into the range function (3), we obtain the following dependence of the current migration on Doppler signal frequency:

$$\tilde{r}(f_x) = \frac{f_x^2 r_0}{8V_n^2}, \quad -\frac{\Delta f_x}{2} \le f_x \le \frac{\Delta f_x}{2}, \tag{4}$$

where Δf_x is spectrum width of the SAR trajectory signal.

Equation (4) expresses the parabolic law of changing the distance from the current frequencies. At the same time, the parabola curvature is proportional to the slant range r_0. With the help of the Fourier transform only along the coordinate x, the transition from the two-dimensional region "slant range –

IOP Conf. Series: Materials Science and Engineering **516** (2019) 012021 doi:10.1088/1757-899X/516/1/012021

path range" to the two-dimensional region "slant range-Doppler frequency f_x" is performed. At the same time, for all points of the surface that have the same traverse range, the dependence of the magnitude of migration on frequency will be the same. Consequently, the subsequent direct rectification allows you to move signals with the same traverse range into one channel and form a two-dimensional signal with no migration.

However, the dependence calculated with the help of (4) will have a different view for different slant ranges, therefore, along with the slant range, the Doppler shift correction will occur with an error. It is believed that it is advisable to perform the correction relative to the center point of the frame with an inclined range r_{0c}, however, at the edges of the frame, the quality of compression will be lower, which leads to distortion and an increase in the level of side lobes. In addition, with the complex law of motion of the carrier, when the support function must be recalculated at each repetition period, the use of this algorithm is impossible.

In this regard, the use of the algorithm described above may be impractical for some applications.

2.2 Range migration algorithm

Range migration algorithm directly compensates the offset of the signal in range channels [2].

Range migration is estimated by

$$\frac{\Delta R}{r_d} = \frac{\sqrt{R^2 + x^2} - R}{r_d}. \tag{5}$$

Range migration at the edge of the antenna pattern is equal

$$\frac{\Delta R_{\max}}{r_d} = \frac{\sqrt{R^2 + (L_s/2)^2} - R}{r_d}. \tag{6}$$

Thus, during the synthesis of the aperture, a signal from a point reflector with a slant range $R = 6000$ m, $r_d = 0.15$ m, $\Delta x = 0.3$ m, $L_s = 350$ m will be observed in 17 channels of the range.

The range-migration effect greatly complicates the task of the synthesis of radar images. This is explained by the fact that the invariance of the sample line of the radio hologram to a shift along the X axis and, consequently, to a time shift is broken. Thus, the solution of the problem of compression in azimuth as a compression of the chirp signal using a matched filter in the frequency domain along a single line of the radio hologram is impossible. Therefore, compression in azimuth must be performed individually for each synthesized point of the image $\dot{I}_{p,q}$ by calculating the scalar product

$$\dot{I}_{p,q} = \sum_{n=-N/2}^{N/2} w_{n+p,m+\Delta m(q \cdot r_d, n \cdot r_a)} \cdot \dot{h}_{n,m}, \tag{7}$$

where

$$\Delta m(R_m x_n) = \frac{\Delta R}{r_d} = \frac{\sqrt{R_m^2 + x_n^2} - R_m}{r_d}, \tag{8}$$

$$\dot{h}_{n,m} = \exp(-j\phi(n \cdot T \cdot R_m)), \tag{9}$$

$$\phi(t, R_m) = \frac{4\pi}{\lambda}\sqrt{R_m^2 + x_n^2} = \frac{4\pi}{\lambda}\sqrt{R_m^2 + (V(t - t_0))^2}, \tag{10}$$

where $w_{n,m}$ is an element of the radio hologram after compression by the slant range, $\dot{h}_{n,m}$ is a filter impulse response for the m-th channel slant range R_m.

3. Results and Discussion

In our case, we decided to use two mathematical models for calculating the scattered field, the results of calculations from each of them we combine according to the principle of superposition.

Vehicles and man-made structures that are of particular interest in SAR are usually composed of metal structures. Thus, we can restrict the exact model of the calculation to only perfect conductors, the scattered field from which is determined mainly by the geometry. A similar situation applies in the military aviation radar. One of the most well-known methods for calculating the scattered field from a polygonal model is the method of the physical theory of diffraction (PTD), developed by P. Ya. Ufimtsev [3]. In the best way, the formulas for calculating the scattered field in accordance with the PTD are shown in [4–6].

On the other hand, the underlying surface usually consists of dielectric materials, calculations for which are more difficult than for an ideal conductor. It remains for us only to apply the statistical description of the reflective properties of materials, taking into account the polarization and angle of irradiation. The entire underlying surface is marked with one of the prepared materials such as asphalt, concrete, dry dirt, wood, etc. Description of materials built on the works [7–10] in which produced the experimental evaluation of scattering characteristics.

In order to compare the previously described approaches to the synthesis of radar images, we produce a mathematical modeling of the synthetic-aperture radar. The radar scene is represented by a forest road with different density of trees, and there are also open areas of terrain. In addition, there are 5 vehicles in various locations on the scene.

Figures 1 and 2 represent the result of work range-migration algorithm. The scene for modeling the P-band is an exact copy of the X-band, with the exception of the material properties of the scene.

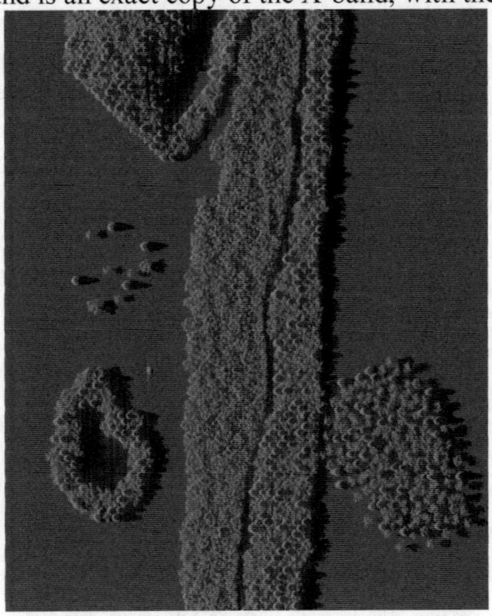

Figure 1. Range-migration algorithm (X-band).

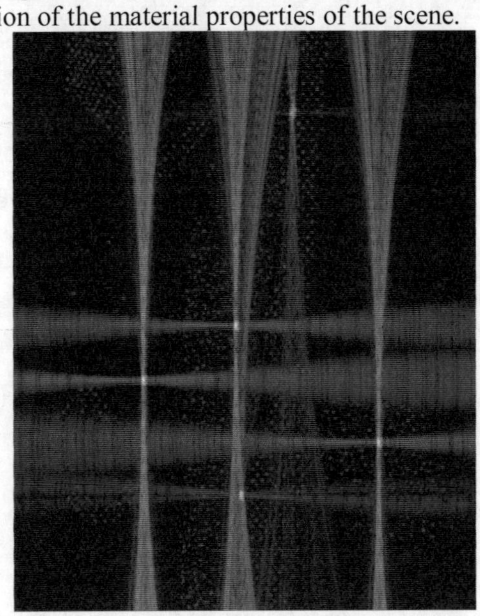

Figure 2. Range-migration algorithm (P-band).

The cross along azimuth and slope range (Figure 3) in both sides of the vehicle appears because of high brightness of vehicles in P-band (Figure 4). Side lobes after matched filtration along with slant range and azimuth are brighter than underlying surface.

IOP Conf. Series: Materials Science and Engineering **516** (2019) 012021 doi:10.1088/1757-899X/516/1/012021

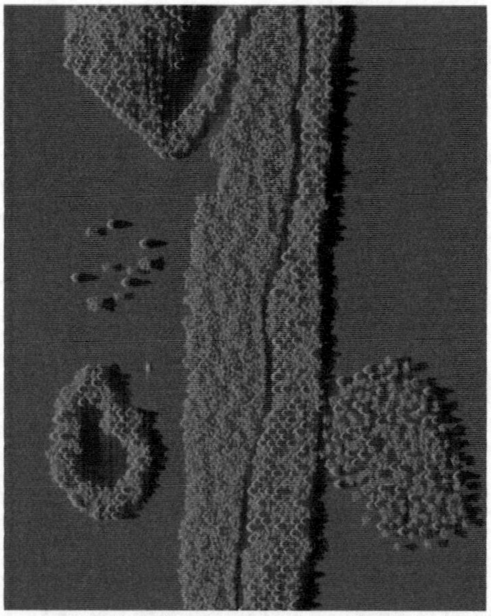

Figure 3. Time-frequency algorithm (X-band).

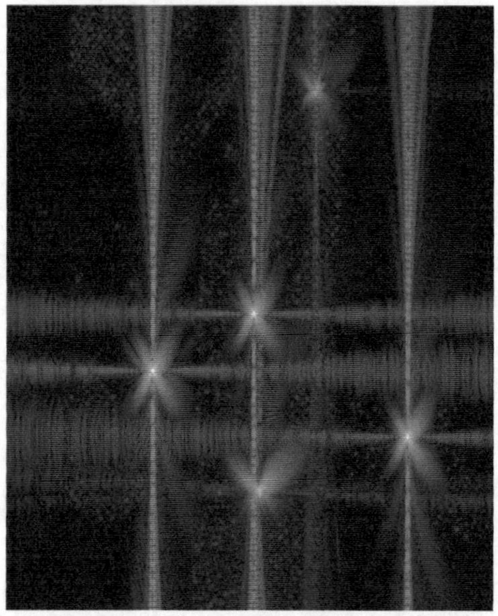

Figure 4. Time-frequency algorithm (P-band).

First of all, it should be noted that with time-frequency processing, vehicles are not focused enough. This is due to the fact that for the formation of radar images in the P-band, a significantly larger synthesized aperture is needed and, therefore, the effect of migration in range will be stronger. It should also be noted that in the P-band, due to the radio transparency of the foliage, it is possible to detect targets under the cover of the forest.

Figures 5 and 6 represent a difference between the two algorithms in different bands. As it can be seen in Figure 5, there is no much difference between range migration and time-frequency algorithm (Figures 1 and 3). But there is a small difference to the left and right sides of the image, which appears as it been discussed in section 2 because the time-frequency algorithm focused only in the middle of the image. Due to a larger size of the aperture in P-band and brightness of the vehicles, this effect yields a stronger difference (Figure 6).

Figure 5. Difference between two methods (X-band).

Figure 6. Difference between two methods (P-band).

It can be seen that the time-frequency processing algorithm creates an image that does not differ

IOP Conf. Series: Materials Science and Engineering **516** (2019) 012021 doi:10.1088/1757-899X/516/1/012021

much from the result obtained using the range migration algorithm. In this case, the method of frequency-time processing is preferred, since it allows you to implement an FFT procedure to increase the speed of synthesis.

4. Conclusion

The calculation of the scattered field usually meets with two requirements in the simulation problem. First, a sufficiently accurate electrodynamic model is needed to observe diffraction, diffuse and specular reflection effects. Secondly, a model is needed that is capable of calculating large areas of the underlying surface in adequate time.

Acknowledgments

The work is executed at financial support of the Ministry of Education and Science of the Russian Federation, project No 2.2226.2017/Project Part and project No 2.9140.2017/Basic Part.

References

[1] Čučej Ž. *and* Gleich D. 2010 *Proc of Int Conf IWSSIP* 40–44 Retrieved from http://www.iwssip.org/archive/2010/proceedings/nav/papers/ip10.pdf
[2] Shiro E. G. 2001 Methods and equipment for digital synthesis of high-quality radar images *PhD thesis* (Scientific Research Institute of Micro-Instruments: Moscow) (in Russian)
[3] Ufimtsev P. Y. 2009 *Theory of Edge Diffraction in Electromagnetics Elektronische Ressource Origination and validation of the physical theory of diffraction* (NC. SciTech Publishing Inc.: Raleigh) doi: 10.1049/SBEW054E
[4] Borzov A. B., Suchkov V. B. and Sokolov A. V. 2004 *Journal Of Radio Electronics* **2** 1–13 (in Russian)
[5] Borzov A. B. *et al.* 2014 2014 *Journal of Communications Technology and Electronics* **59(12)** 1356–1368
[6] Akhiyarov V. V. *et al.* 2015 *Journal of Communications Technology and Electronics* **60(12)** 1297–1304
[7] Cloude S. R., Pottier E. 1996 *IEEE IEEE Trans Geosci Remote Sens* **34(2)** 498–518
[8] Boev S. F. *et al* 2018 *IOP Conf. Ser.: Mater. Sci. Eng.* **457** 012010 doi: 10.1088/1757-899X/457/1/012010
[9] Yakunin M. A., Yurchenko A.V. 2015 *Technical Physics* **60** 141–144
[10] Pottier E., Cloude S. R. 1997 *Proc SPIE* **3120** 132–143 doi: 10.1117/12.278958

IOP Conf. Series: Materials Science and Engineering **516** (2019) 012022 doi:10.1088/1757-899X/516/1/012022

Policy service design of elements and rural distribution mains of 10 Kv on the results of technical position diagnosis

L M Rybakov, N L Makarova, A O Zakhvataeva, A N Basov, D V Rukavishnikova

Mari State University, Yoshkar-Ola, Russia
E-mail: diagnoz@marsu.ru

Abstract. This article considers questions of overhead high-voltage lines working capacity and ways of its improving reliability. Influence of different factors especially climatic on overhead line condition is shown. Zoning plan of the Russian Federation on annual average duration of thunderstorms in hours is presented. Also, it shows data on numbers of thunderstorm activity intensity on the territory of the Republic of Mari El over a period of observation from 2007 to 2016 years. Statistic data on number of overhead line shutdowns in thunderstorm periods from 1985 to 2010 years. For estimation of working capacity of device and elements of distribution mains weather two-factor models are presented. Key reasons of damage of overhead electric lines elements are shown, besides, data on shutdown of overhead lines because of pin-insulators damages and cords on the territory of the Republic of Mari El over a period from 2007 to 2016 years are illustrated. Maintenance work system of overhead lines that accepted in operation are considered. On the results of reasons of overhead lines breakdown analysis it was found that most of overhead lines emergency shutdowns are caused by nature and climate effects that depend on conditions of usage, high quality conduction routine-preventive repairs, consequently on technical staff qualification etc. Service policy considering nature and climate factors influence, material-and-technical support by modern instruments and equipment, rational service and timely preventive measures implementation was designed.

1. Introduction

Electric energy distribution is based on overhead lines (OHL) with voltage of 6–10–35 kV that general length is more 1 mio. km. Overhead transmission lines are more injured electric installation of electrical mains because of their length and location on open countries that's why they are subjected to climatic effect influence. Overhead line failure rate is sequence higher than failure rate of transformers and switches.

In the operating process different climatic and operational factors influence on overhead lines. Prime climatic factor influencing on accident-free operation of overhead lines are atmospheric overvoltage, icing, wind, air temperature.

Security of service of electricity transmission lines depends on a lot of factors designing quality, construction and installation operations, bringing into action and proper operating in different states and atmosphere and climatic conditions.

2. Research Method

Failure insulating elements analysis of power distribution zone of 10 kV shows that most number of shutdown of distribution networks of Mid-Volga region is connected with loose of function of insulating elements. Main part of shutdowns are accrued to summer period that is explained by intensive thunderstorm activity and internal over-voltages.

Thunderstorm activity intensity is characterized by average number of thunderous hours per year. According to chapter 2.5 Electrical Installations Code [1] comprehensive map of thunderstorm activity intensity of RF presents in the Figure 1. As it can be seen Republic of Mari El is IV region of thunderstorm season with duration of 40-60 hours per year.

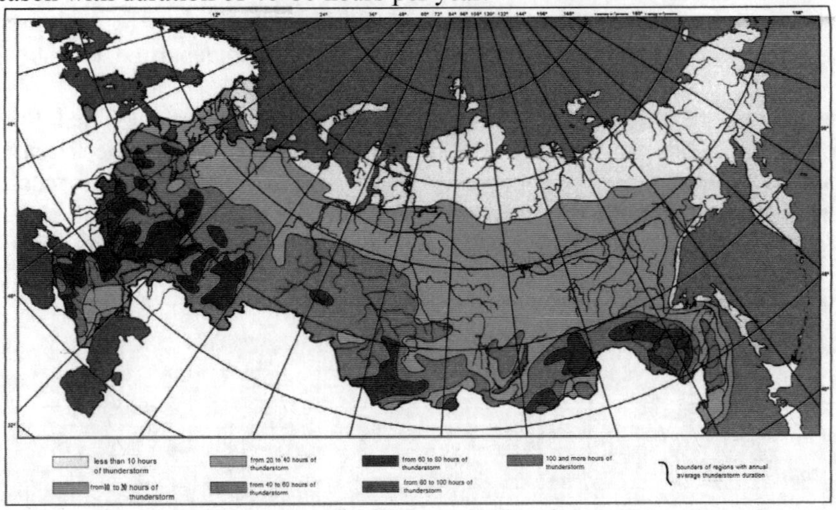

Figure 1. Zoning plan of the Russian Federation on annual average duration of thunderstorms in hours.

Table 1 illustrates received average rates of number of thunderstorm hours by month of the year over last 10 years in the Republic of Mari El (2007-2016 years).

Table 1. Thunderstorm activity intensity in the Republic of Mari El over an observation period from 2007 to2016 years).

	Months						
	April	May	June	July	August	September	October
Number of thunderstorm hours, [h]	0.1	2.5	7.6	13.1	6.7	1.2	0

Figure 2 presents bar chart of distribution of shutdowns of overhead lines in thunderstorm period over a period of 1985-2010 years.

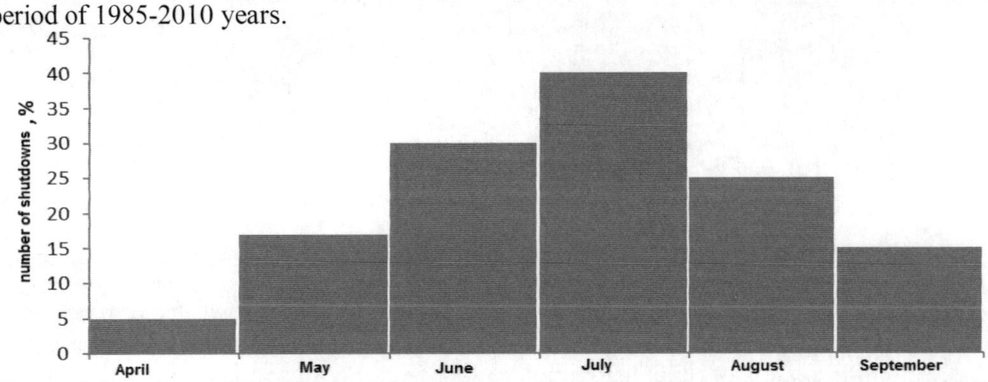

Figure 2. Bar chart of distribution of shutdowns of overhead lines in thunderstorm period by months over a period of 1985–2010 years.

Long-term operating analysis shows that equipment and distribution main elements depend on a seasons. For example, period from December to February is a period of most number of failures caused by cords failures, period from March to November because of insulator failures [2–4]. Among these months there is a period of most emergency shutdowns caused by equipment damage and distribution mains elements that is from April to October.

Prime factor of cord damages is its break from wind actions and gust and gaze influences. Cord break intensity is high over all period of a season, however exceedance of failures caused by its failures falls on winter period because of high wind force, ice loads and snow. That's why it's taken a period of December-February for cords.

Figure 3 shows distribution of number of bar chart of cord damages by months of the year for period of observation 2007–2016 years on the sample of production department Yoshkar-Ola electrical mains of the Republic of Mari El.

Most common cause of insulators failure is mechanical damage from one-side influence in cases of cord break in span because of high temperature cause by covering with electric arc and electrical breakdown caused by lightning strike or surge voltages [5–7]. Figure 4 shows bar chart of distribution of number of shutdowns caused by pin-insulators by months over observation period 2007-2016 on the sample of production department Yoshkar-Ola electrical mains of the Republic of Mari El.

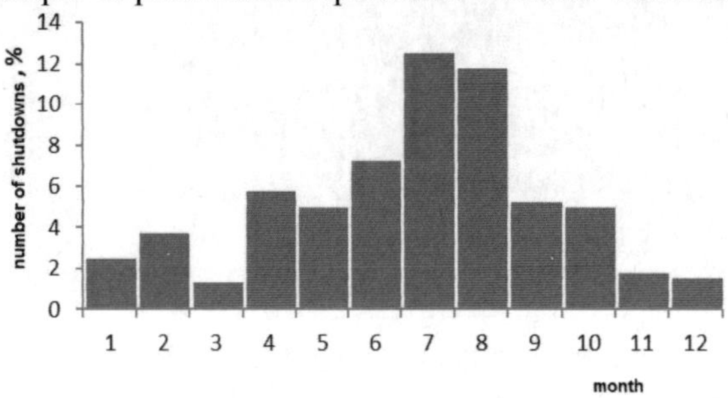

Figure 3. Bar chart of distribution of number of cord damages by months of the year over a period of observation 2007-2016 years on the sample of production department Yoshkar-Ola electrical mains of the Republic of Mari El.

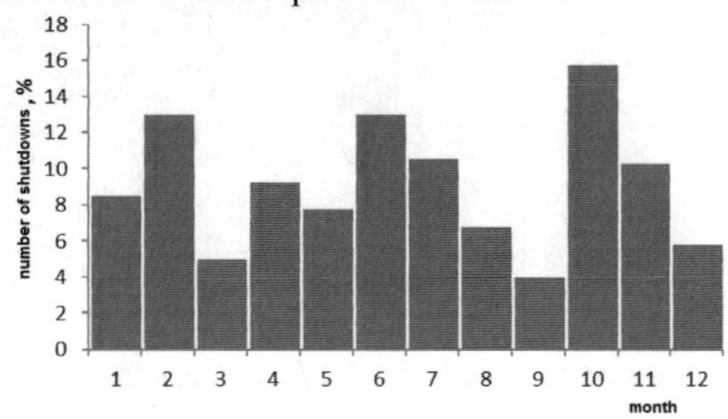

Figure 4. Bar chart of distribution of number of shutdowns caused by pin-insulators by months over observation period 2007-2016 on the sample of production department Yoshkar-Ola electrical mains of the Republic of Mari El.

On the base of experimental researches weather two-factor models were made for estimation of working capacity of pin-insulators presented on the Figures 5 and 6 [8, 9].

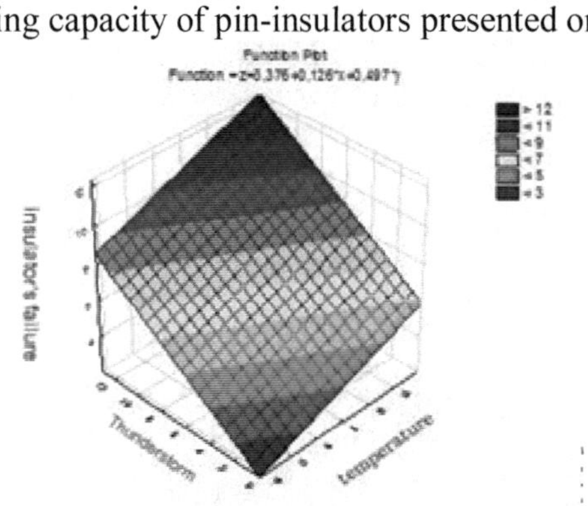

Figure 5. Weather two-factor model (temperature (x) and thunderstorm(y)) for estimation of working capacity of pin-insulators of rural distribution mains.

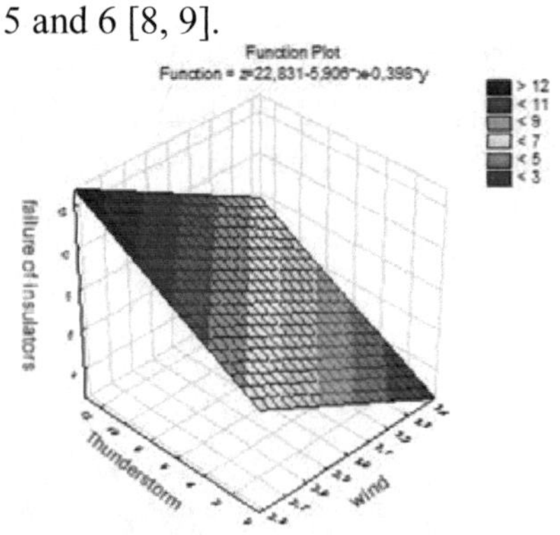

Figure 6. Weather two-factor model (wind (x) and thunderstorm(y)) for estimation of working capacity of pin-insulators of rural distribution mains.

IOP Conf. Series: Materials Science and Engineering **516** (2019) 012022 doi:10.1088/1757-899X/516/1/012022

Figure 7 illustrates prime causes of damages of distribution lines elements of 10 kV. Table 2 illustrates data on distribution of number of distribution mains elements damages of 10 kV of the Republic of Mari El serviced by production department Yoshkar-Ola electrical mains of the Republic of Mari El over last 10 years (given average rates by months of the year for researching 10 years) because of which failures happened.

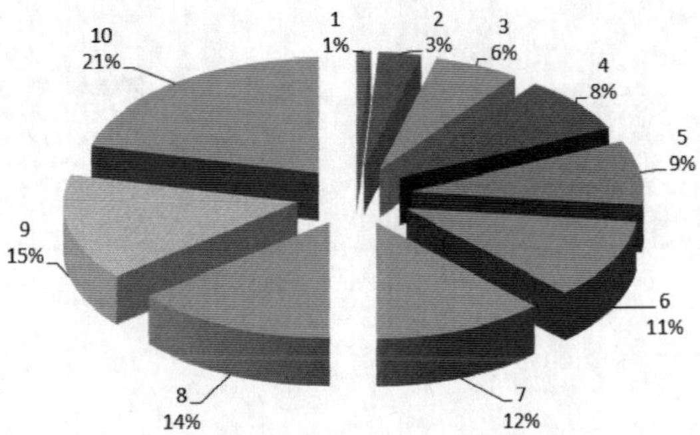

Figure 7. Prime causes of damages OHL elements, where 1 is carrying capacitance, wear, corrosion; 2 is gust-and-glaze loading; 3 is insulation pollution; 4 is vandalism; 5 is design fault, building and assembly faults; 6 is defects of production; 7 is reasons are not found out; 8 is process defects; 9 is off-design condition of main (internal overvoltage); 10 is atmospheric overvoltage.

3. Results and Discussion

Overhead transmission lines because of their big length have determined number of identical elements that have own reliability index. As it was established weather factors influence on proper overhead transmission lines functioning and failure rate of overhead transmission lines elements depends on features of construction, operation conditions and outside environment. It's revealed that key reasons of overhead transmission lines basic elements failures are atmospheric, natural and climatic and other effects. Generally, proper overhead transmission lines functioning depends on quality preventive maintenance, early diagnostics and forecasting of researched object [10–12]. Currently, designing, operation and diagnosis of overhead transmission lines are conducted in according with such normative documents as GOST839-80, GOST 6490-93, GOST 18322-2016, GOST 25866-83, GOST 26656-85.

In operating practice use following systems of holding maintenance works overhead transmission lines preventive maintenance, that are conducted after the expiry of operation definite period. Such method is not always optimal as it tends to unjustified shutdowns of working items. Emergency recovery repairs and repairs on the results of technical diagnosis when it is necessary to know real state of researched object. Solving problems of technical diagnosis it is possible to support normal functioning and in particular cases improve overhead transmission lines reliability that more over liable to nature and climatic effects. Necessity of moving to maintenance system appears. Herewith, scope and terms are determinate by real state of electric equipment that necessitates creation of effective diagnostic system of overhead transmission lines [13–15].

Using of means of technical diagnosis of overhead transmission lines that function in a real time makes it possible to increase essentially reliability of functioning of overhead lines owing to early defects detection, take-down for repairs , failure inhibition.

Reliability of overhead lines is determined by using conditions, their timely and right accounting bring about change of reliability index.

On the results of calculations statistical model that is dependence of mathematical expectation failures of distribution mains from time by months during a year was received. Received model is presented as a graph on the Figure 8.

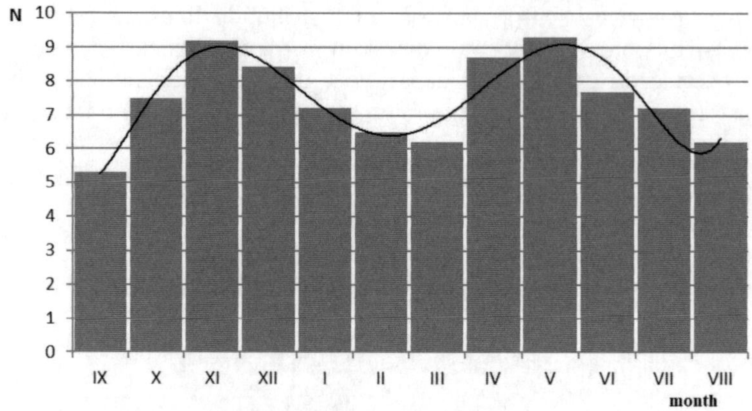

Figure 8. Distribution of intensity of failures of insulator elements of distribution mains.

4. Conclusion

On the results of reasons of overhead lines breakdown analysis it was found that most of overhead lines emergency shutdowns are caused by nature and climate effects that depend on conditions of usage, high quality conduction routine-preventive repairs, consequently on technical staff qualification etc. Service policy considering nature and climate factors influence, material-and-technical support by modern instruments and equipment, rational service and timely preventive measures implementation on the base of recommendations [16].

References

[1] Collection of Normative Documents 2016 *Electrician's Bible* (Normatics: Novosibirsk) (in Russian)
[2] Cimador A. *et al* 1994 *Proc of 35th CIGRE Session* 3-04
[3] Thione L. 2000 *Proc of 38th CIGRE Session* 1-02
[4] De Tourrel C., Ishivari M. 2000 *Proc of 38th CIGRE Session* 1-04
[5] Boev S. F. *et al* 2018 *IOP Conf. Ser.: Mater. Sci. Eng.* **457** 012010 doi: 10.1088/1757-899X/457/1/012010
[6] Rybakov L. M., Volkov S. V. 2006 *Mechanization and electrification of agriculture* **2** 4–6 (in Russian)
[7] Rybakov L. M., Ivanova Z. G. 2015 *Bulletin of the Chuvash University* **1** 04–110 (in Russian)
[8] Gmurman V. E. 2004 *Guide for solving problem on the theory of probability and mathematical statistic* (Visshaya Shkola: Moscow) (in Russian)
[9] Gunst R. F., Mason R. L. 1980 *Regression analysis and its application: a data-oriented approach* (Marcel Dekker: New York)
[10] Shilin A. N., Shilin A. A. 2011 *Bull of the VSTU* **8** 84–88 (in Russian)
[11] Lilien J. L., Destine J. 2012 *Patent US* 8184015 URL: http://patft.uspto.gov/netacgi/nph-Parser?Sect1=PTO1&Sect2=HITOFF&d=PALL&p=1&u=%2Fnetahtml%2FPTO%2Fsrchnum.htm&r=1&f=G&l=50&s1=8184015.PN.&OS=PN/8184015&RS=PN/8184015
[12] Olsen R. G., Edwards K. S. 2002 *IEEE T POWER DELIVER* **17(4)** 1142–1152
[13] Yager R. R. 1982 *Fuzzy Set and Possibility Theory: Recent Developments* (Pergamon Press: New York)
[14] Mamyrbek A. B. *et al* 2016 *International Journal of Applied Engineering Research* **11** 11237-11240
[15] Aijaz A. 2013 *J Am Chem Soc* **135(44)** 16356–16359
[16] Alekseev B. A. *et al* 2004 *The scope and norms of electrical equipment testing* (Publishing House of the NTS ENAS: Moscow)

Formation of informative characteristics of network traffic states using its dynamic filtration

A Krasnov[1], D Nikol'skii[1], D Repin[1], O Galtseva[2]

[1]Federal State Autonomous Educational Establishment of Additional Professional Education "Center of Realization of State Educational Policy and Informational Technologies", Moscow, Russia
[2]National Research Tomsk Polytechnic University, Tomsk, Russia
E-mail: krasnovmgutu@yandex.ru

Abstract. The approach to the task of detecting DDoS attacks based on the formation of secondary informative characteristics of its states on the observed parameters of data packets headers of network traffic is considered. Secondary informative characteristics characterize the traffic time structure or its behavioral character. The problem is solved by the method of dynamic filtering of traffic using various operators: the causal transformation operator and the evolution operator.

1. Introduction

Traditionally, the traffic anomaly detection is carried out according to the results of identification of deviation of its current state from patterns of normal states. At the same time, various methods of intellectual analysis are used such as method of principle component analysis [1], wavelet analysis [2], histogram analysis [3], support vector machines [4, 5] and detection of shifts in spatial-temporal traffic patterns [6]. Secondary informative features which are formed as logical fingerprints for various network protocols [7], entropic [8], correlational [9] and structural functions of primary informative features of traffic [10] are used for more in-depth analysis. Various difference measures and metrics are used to detect DDoS attacks against the background of changing legitimate traffic, including taking into account the "flash crowds" effect [11, 12].

The use of signatures to detect DDoS attacks requires the accumulation of a significant amount of data for all types of attacks; it is used in a similar way to virus detection methods [13]. Neural networks are widely used as part of complex systems for detecting and protecting against DDoS attacks [14–16]. Another approach is based on identifying group anomalous behavior, which is typical for botnets and differs from the usual actions of legitimate users. IP addresses of the traffic originator [17] and packet identification numbers [18] can be used to detect DDoS attacks. A detection system was proposed which effectively identifies the anomalous state of the system using patterns of group activity of network nodes [19]. Known clustering methods were used to detect malicious packets in traffic and identify botnets [20–23].

It should be noted that there are no publications in the network security field in which issues of formation and productive use of the space of secondary informative features of traffic from the standpoint of statistical dynamics are. Network traffic in the works mentioned above is considered as a set of static data (for example, the number of packets per second) without taking into account their dynamic structure, which is determined by the statistics of time consecution and the corresponding stable connections of the primary features. However, in fact any network channels and traffic passing through them are the dynamic systems [24]. It is necessary to take into account both the speed of changes in traffic and the instant values of network loads for their adequate description.

The aim of the this work is to summarize the results of theoretical and experimental studies of the authors on the analysis of network traffic [10, 24, 25] and to substantiate the methods of formation of secondary informative features of network traffic on the basis of its consideration as a dynamic system.

2. Research Method

Dynamic filtering is performed by scanning the vector $\boldsymbol{P}_n=(P_{1n}, P_{2n}, \ldots, P_{Mn})^T$ of the stream M of parameters of network traffic data packet headers (n = 1, 2, …, N) by dynamic window $w_{(n, \Delta N)}$ with an aperture ΔN. The formation of vector of secondary informative features of network traffic $\boldsymbol{F}_n= (F_{1n}, \ldots, F_{kn})^T \cdot K$ occurs at every n-m window position. In the general case, each k-th secondary informative attribute is formed by the action of the corresponding functional or operator $\boldsymbol{F}_k$ on the set of vectors of parameters of the packet headers of network traffic data packets included in the window $w_{(n, \Delta N)}$:

$$\boldsymbol{F}_{kn}= F_k(\boldsymbol{P}_n, \boldsymbol{P}_{n-1}, \ldots, \boldsymbol{P}_{n-\Delta N+1}),\ k = 1, 2, \ldots, K. \quad (1)$$

The formation of vectors of secondary informative features is necessary for account the statistical relationship of vectors traffic parameters ΔN (primary informative features) to take into account the temporal structure of its behavior. Normalized frequency distributions $\omega(F_1, \ldots, F_K|r)$ formed from a set of observations N of vectors $\boldsymbol{F}_n$ in different states of St_r of traffic (r = 0, 1, …, R-1) are used for a stable description of such connection. Normalized frequency distribution was considered as the conditional probability of values of the secondary informative features of traffic under specified states. Some trajectory in the phase space of its secondary informative features corresponds to traffic dynamic filtering in such a formulation of the problem. The trajectory passes through various subspaces corresponding to different traffic states St_r (r = 0, 1, …, R-1). In this case, the subspaces can overlap, which is typical for real network traffic containing a mixture of attacking packets. Therefore, the detection of fact of entry of the traffic trajectory into any subspace and the classification of the subspace type are probabilistic procedures.

3. Results and Discussion

The following operators were used for dynamic filtering.

3.1 The causal transformation operator

This operator converts auxiliary features X_{mn}, calculated from the primary informative features of traffic $P_{mn}(m=1, \ldots, M)$, into its intermediate features Y_{mn}. Hilbert's FIR filter with finite impulse characteristic is used for transformation h_n [24]:

$$Y_{mn}=h_1 (X_{(m, n+1)} - X_{(m, n-1)})+h_3 (X_{(m, n+3)} - X_{(m, n-3)})+h_5 (X_{(m, n+5)} - X_{(m, n-5)})+h_{\Delta N} (X_{(m, n+\Delta N)} - X_{(m, n-\Delta N)}). \quad (2)$$

Filter coefficients $h_n \in (1, 1/3, 1/5, \ldots, 1/\Delta N)$ provide orthogonality of sequences X_{mn} and Y_{mn}:

$$\textstyle\sum_n X_{mn} Y_{mn}=0 \text{ for } \forall m =1, 2, \ldots, M. \quad (3)$$

Intermediate features of traffic describe the generalized rates of its change, because they depend on higher-order derivatives of auxiliary features ΔN, which correspond to generalized coordinates. Wherein odd central differences should be used for linear independence of intermediate features in (2) so that each Y_{mn} does not depend on X_{mn} $\forall n$=1, 2, …, N.

The set of intermediate and auxiliary features of traffic allows to consider the functioning of network data transmission channel as the behavior of multidimensional dynamic system of the Δ-th order with generalized coordinates X_{mn} and generalized speeds Y_{mn}. It is enough to use ΔN=7 ... 13 for practice.

In accordance with the approach of Ludwig Boltzmann, Henri Poincare and Willard Gibbs [26], the states of multidimensional dynamic system with generalized coordinates X_m and generalized velocities Y_m are described by a family of phase portraits which do not contain time explicitly, but describe the change of the states of the system, i.e., its dynamics in the form of phase trajectories in the corresponding phase spaces $\{X, Y\}_m$. Figure 1 shows examples of phase portraits corresponding to the normal state of traffic and its abnormal state (the case of TCP Syn Flood attack) [24]. The ratio of parameters (number of octets to the number of packets) of the load part of the headers of traffic data packets received from the edge router of the network via the Net Flow Protocol is taken as a generalized coordinate *X*.

It is quite difficult to perform the numerical calculations, which are necessary to detect attacks with predetermined probability at describing traffic conditions in the form of phase portraits. Therefore, generalized coordinates and pulses were converted into one-dimensional secondary informative features by hashing to simplify the calculations [24]:

$$F_{mn}=X_{mn}+Y_m,\ m = 1, 2, \ldots, M;\ n = 1, 2, \ldots, N. \quad (4)$$

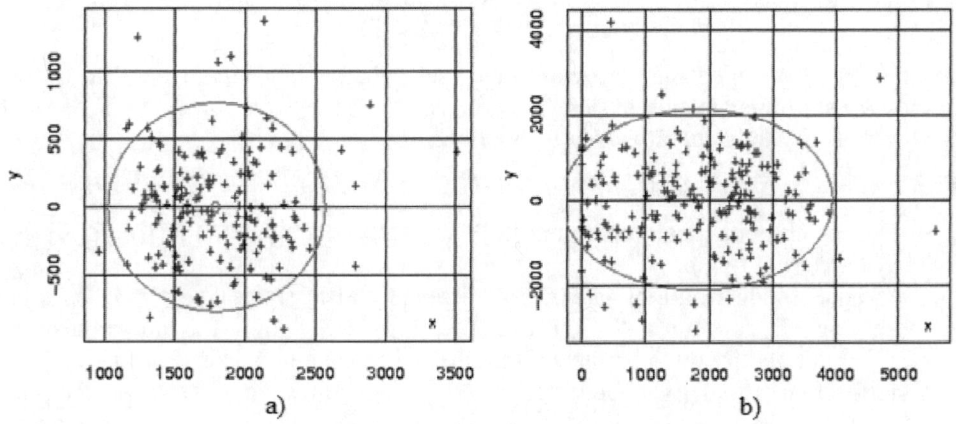

Figure 1. Examples of phase portraits of normal state of traffic (a) and its abnormal state (b) (TCP Syn Flood attack) [24].

Normalized frequency distributions $\omega(F_1, \ldots, F_K|r)$ of hash functions were formed. Figure 1 shows example of normalized one-dimensional frequency distributions ω (FK|r), formed in relation of the parameters (number of octets to the number of packets) of the load part of the traffic for N=4000 counts of observations at different states $r \in$ {Normal; HTTP Flood; Slow Loris}.

The frequency distributions of all traffic states differ significantly on different intervals of values *F* of its informative feature (Figure 2).

3.2 The evolution operator

It is assumed that the continuous vector $\boldsymbol{P}(t)= [P_1(t), \ldots, P_M(t)]^T$ M of traffic characteristics $Pm(t)$ (m=1, …, M) at forming this operator is described by the differential equation [10, 25]:

$$d\boldsymbol{P}(t)/dt=(1/\Delta t)\ \boldsymbol{H}(t)\cdot\boldsymbol{P}(t), \quad (5)$$

where the dynamic operator $\boldsymbol{H}(t)$ is unknown, and only the values of the components of the vector $\boldsymbol{P}$ of the parameters of the packet headers of traffic data observed at various discrete moments of time $t_n \in (t_0, t_N]$ are known. The time interval Δt is introduced into (5) because of dimension reasons. It is also assumed that there is an operator $\boldsymbol{S}(t, \tau)$ for $\boldsymbol{H}(t)$, which gives the solution of the Equation (5):

$$\boldsymbol{P}(t)=\boldsymbol{S}(t, \tau)\ \boldsymbol{P}(\tau),\ \boldsymbol{S}(t, t)=\boldsymbol{E}, \quad (6)$$

where $\boldsymbol{E}$ is unit matrix.

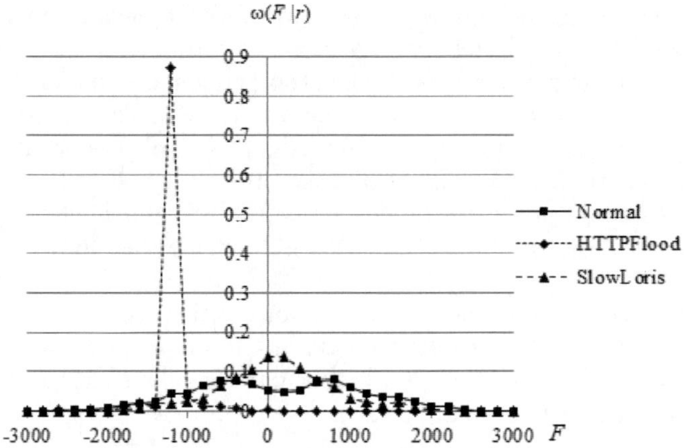

Figure 2. Example of normalized one-dimensional frequency distributions ω(F|r) for three traffic states (N = 4000, ΔN=7).

The operator ***S****(t, τ)* is called the matriciant or Cauchy matrix in mathematics and the evolution operator or the time shift operator in physics. The connection of the evolution operator of the dynamic system ***S****(t, τ)* with its dynamic operator ***H****(t)* is described as a series [25]:

$$\boldsymbol{S}(t,\tau)=\boldsymbol{E}+1/\Delta t\cdot\int_{\tau}^{t}\boldsymbol{H}(t_1)dt_1+(1/2!\cdot(1/\Delta t))^2\cdot\int_{\tau}^{t}dt_1\cdot\int_{\tau}^{t}dt_2\cdot J\,[\boldsymbol{H}(t_1)\cdot\boldsymbol{H}(t_2)]+\cdots, \quad (7)$$

where J is Dyson chronological operator [27] (J $[\boldsymbol{H}(t_1)\cdot\boldsymbol{H}(t_2)]=\boldsymbol{H}(t_1)\cdot\boldsymbol{H}(t_2)$, if $t_1>t_2$ and J $[\boldsymbol{H}(t_1)\cdot\boldsymbol{H}(t_2)]=\boldsymbol{H}(t_2)\cdot\boldsymbol{H}(t_1)$, if $t_1<t_2$).

Normalized frequency distributions ω(F|r) for different traffic states St_r(r=0, 1, 2, …, R) were formed by observing all its counts N. For example, Figure 3 shows the normalized frequency distributions ω(F|r), formed by two parameters (number of octets and packets) of the load part of the traffic for N=4000 observations of adjacent counts t and τ at different states r ∈ {Normal; HTTP Flood; Slow Loris}.

Frequency distributions of all traffic states differ significantly on different intervals of values F of its informative feature (Figure 3). However, the window for the formation of secondary informative features by the evolution operator method was used; its size is significantly smaller than the size of the window at the causal operator method.

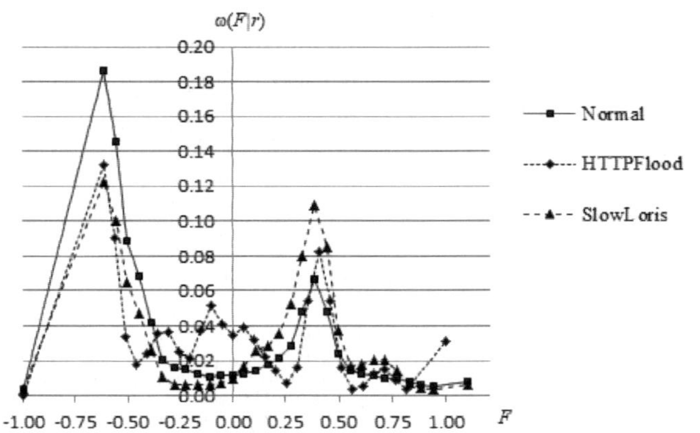

Figure 3. Example of normalized one-dimensional frequency distributions ω(F|r) for three traffic states (N = 4000, ΔN=2).

Normalized frequency distributions ω(F|r), formed by any known method of dynamic filtering, are peculiar images (patterns) of traffic states Sr(r = 0, 1, …, R-1). Additional experiments using the tools developed by the authors [28, 29] have shown that the use of frequency distributions allows to adjust automatically the decision-making thresholds at given probabilities of detection and errors based on the use of the method of Wald's sequential analysis [30,31].

4. Conclusion

The method of formation of secondary features of network traffic states, which is necessary to detect DdoS-attacks; it is based on its dynamic filtering is developed. The causal transformation operator and the evolution operator are used for dynamic traffic filtering. Experimental studies shown the principal possibility of using the developed method in the intelligent systems for detection and identification of the main types of DDoS attacks. The advantage of the developed method is unified approach to the description of traffic states and the ability to detect anomalies by automatic setting of thresholds of decision making at given probabilities of detection and errors. These provisions are fundamental for solving of the tasks of network traffic analysis of super-large volume (more than 100 Gbit/s). Thus, the proposed methods can complement and significantly expand the methods used in real systems of protection against DDoS-attacks.

Acknowledgments

The investigation was carried out with financial support of the state represented by the Ministry of Education and Science of Russia on the lot with the code 2017-14-579-0002 on the topic "Design of effective network attack detection algorithms based on detecting deviations in the traffic of super-large volumes coming to the data transfer network border routers, and creation of specimen of software complex on their basis for detecting and preventing threats to the security of information aimed at rejection in the service", agreement No. 14.578.21.0261 on the provision of a grant from 09/26/2017, a unique identifier of work (project) RFMEFI57817X0261.

References

[1] Liu Y. *et al 2010 Proc. of IEEE Int Conf ICDCS* 807–816 doi: 10.1109/ICDCS.2010.45
[2] Srihari V., Anitha R. 2014 *Proc of Int Symposium SSCC* 291–303 doi: 10.1007/978-3-662-44966-0_28
[3] Kind A. *and* Stoecklin M. P. 2009 *IEEE Trans on Netw Serv Manag* **6(2)** 110–121
[4] Catania C. A. *et al* 2012 *Expert Syst Appl* **39(2)** 1822–1829
[5] Ye J. *et al* 2018 *Secur Commun Netw* **2018** 9804061 doi:10.1155/2018/9804061
[6] Yuan J. and Mills K. 2005 *IEEE T Depend Secure* **2(4)** 324–335
[7] Crotti M. *et al* 2007 *ACM Sigcomm Comp Com* **37(1)** 5–16
[8] Reddy V. S. *et al* 2012 *IOSR J Comput Eng* **7(1)** 13–18
[9] Yu Sh. *et al* 2012 *IEEE T Parall Distr* **23(6)** 1073–1080 doi: 10.1109/TPDS.2011.262
[10] Galayev V. S. *et al* 2017 *IJAER* **12** 10781–10790
[11] Ke L. *et al* 2009 *Proc of IEEE Int Conf NNS* 9–17
[12] Chawla Sh. *et al* 2016 *IJCSIS* **14(10)** 382–389
[13] Malina L. *et al* 2015 *Proc of Int Conf SPI* 113–128
[14] Chen J. H. *et al* 2012 *Proc of IEEE Int Conf GrC* 38–43
[15] Gupta B. B. *et al* 2012 *Int J Comput Sci Netw* **14(2)** 61–70
[16] Saied A. *et al* 2016 *Neurocomputing* **72** 385–393 doi: 10.1016/j.neucom.2015.04.101
[17] Baishya R. C. *et al* 2017 *IJINS* **19(6)** 929–939 doi: 10.6633/IJNS.201711.19(6).09
[18] Thang T. M. *and* Nguyen V. K. 2017 *Software Networking* **1** 213–228 doi: 10.13052/jsn2445-9739.2016.012
[19] Chen Ch. – M. *and* Lin H. – Ch. 2015 *JISA* **21** 42–51 doi: 10.1016/j.jisa.2014.05.002
[20] Dietrich Ch. J. *et al* 2013 *Computer Networks* **57(2)** 475–486 doi: 10.1016/j.comnet.2012.06.019
[21] Terzi D. S. *et al* 2017 *Proc of IEEE Int Conf UBMK* 592–597 doi: 10.1109/UBMK.2017.8093473
[22] Lu W. *et al* 2011 *Computer Communications* **34(3)** 502–514 doi: 10.1016/j.comcom.2010.04.007
[23] Wang K. *et al* 2011 *Computer Networks* **55(15)** 3275–3286 doi: 10.1016/j.comnet.2011.05.026
[24] Krasnov A. E. *et al* 2018 *IJAER* **13(8)** 5647–5654
[25] Krasnov A. E. *et al* 2018 *The Bulletin of Udmurt University. Mathematics. Mechanics. Computer Science* **28(3)** 407– 418 doi: 10.20537/vm180310 (in Russian)
[26] Nolte D. D. 2010 *Physics Today* **63(4)** 31–33 doi: 10.1063/1.3397041

[27] Dyson F. J. 1949 *Physical Review* **75(11)** 1736–1755 doi: 10.1103/PhysRev.75.1736
[28] Repin D. S. *et al* 2018 Program Comlex "DDoS Attacks Simulator *The Certificate on Official Registration of the Computer Program* 2018615691 (in Russian)
[29] Krasnov A. E. *et al* 2018 *Proc of All-Russian Pospelov' Conference Hybrid Synergistic Intelligent Systems* 349–354
[30] Serikov T. G. *et al* 11th International Forum on Strategic Technology (IFOST 2016) 7884168 510-514
[31] Wald A. 1947 *Sequential analysis* (J. Wiley & Sons, Inc.: New York)

IOP Conf. Series: Materials Science and Engineering **516** (2019) 012024 doi:10.1088/1757-899X/516/1/012024

Numerical modeling of ultrasonic waveguide with piezoceramic element

A Kuzovova, T Muksunov

Tomsk State University, Tomsk, Russia
E-mail: sdy@mail.tsu.ru

Abstract. A model of an ultrasonic waveguide with a cos junction and optimization of the bolt attachment point and the length of the reflector was proposed. A numerical simulation of the proposed waveguide design was carried out in the Comsol Multiphysics program. According to the results of numerical simulation, the resonant frequency of the waveguide with the cos junction was found, the optimal length of the reflector and the point of attachment of the bolt ensuring its minimum deformation. Also the design of this waveguide was simulated in a developed program that provides parallel computations using OpenCL technology.

1. Introduction

Ultrasonic waveguides are used in technologies of ultrasonic welding [1, 2], cutting various materials [3–5], crushing [2], and also used in surgery [6, 7]. However, when developing ultrasonic instruments for medicine, it is necessary to take into account the specifics of the field of application. In particular, to evaluate the biological effect of ultrasound on living tissue and the possibility of obtaining the desired biological effect and the requirements of the doctor for ultrasonic instruments: the duration of continuous operation, reliability and efficiency, weight, heating, ease of operation. In medicine, ultrasonic waveguides are most often used with an operating frequency of 20 to 44 kHz. Already there are technologies for calculating and manufacturing ultrasonic instruments with various features of application [8–10]. There are two main ways to excite ultrasonic vibrations: based on the magnetostriction effect [11] and based on the piezoelectric effect [12]. Ultrasonic waveguides based on piezoelectric elements, as a rule, show a greater efficiency [12].

Waveguides are made in the form of bodies of rotation of variable diameter with stepped, conical or exponential transitions [8, 13, 14]. The most beneficial in terms of the possibility of obtaining significant amplitudes of displacements at low load are stepped waveguides, in which the amplitude gain factor is equal to the ratio of the areas of the input and output sections. Particularly promising are waveguides with smooth, exponential or radial transitions. The presence of a transitional exponential sector reduces the stress concentration and provides more favorable conditions for the propagation of ultrasonic vibrations, improves the strength properties. In addition, the presence of an exponential segment allows you to transform the load without a significant change in the resonant mode of the oscillatory system. Piezo elements are stacked to into Langevin packets [15], which allows multiply the amplitude of oscillations by number of elements. Langevin packets are connected to the ultrasonic resonator using a bolt tie [16] or cylinder [15].

One of the problems of ultrasonic instruments based on waveguides with a Langevin package is overheating of the package or its destruction due to deformation. The principle of constructing an ultrasonic tool based on a waveguide and a Langevin package with a bolt tie with optimization of the bolt attachment point and reflector length is proposed. At the same time it is proposed to consider the use of cos-like transitions for the form of a waveguide.

2. Principled model of ultrasonic waveguide

A model of an ultrasonic waveguide (Figure 1), consisting of a reflector, a piezoelectric element and a working part, was considered. We have solved the Helmholtz equation for considered structure. As a result of solving a system of differential equations, a mathematical model of an ultrasonic tool was obtained, which allows to estimate the resonant frequencies and select the optimal length of the reflector to maximize the amplitude of oscillations of the working part. By variation of parameters we have found optimal size of it's parts to minimize bolt deformation while maximizing vibrations of the waveguide end. The bolt attachment point in the working part was chosen so that its offset relative to the outer edge of the reflector was zero at the operating frequency. Thus, the minimum deformation of the tightening bolt during the operation of the tool is ensured, and the oscillation energy is mainly directed to the working part.

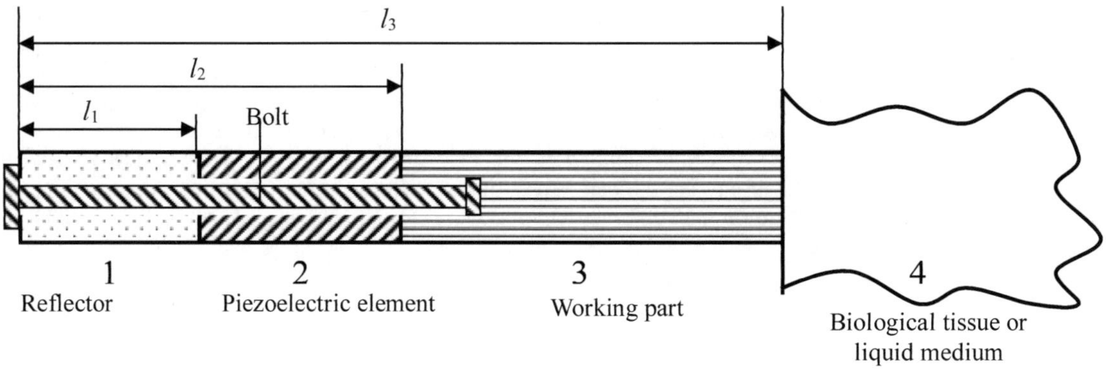

Figure 1. Considered construction of a composite waveguide.

3. Numerical modeling of the ultrasound waveguide by finite elements method in the Comsol Multiphysics program

A numerical modeling of the ultrasonic instrument was carried out in the COMSOL program, which showed close resonant frequencies with an analytical model. Figure 2 presents the image of the longitudinal shift of the aluminum waveguide at a frequency of 23750 Hz. (Figure 1).

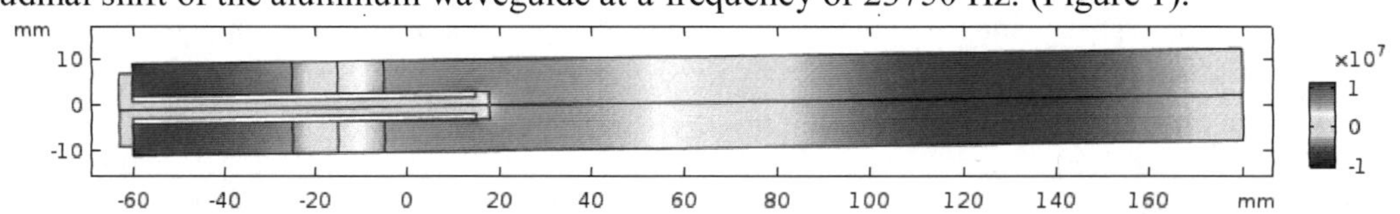

Figure 2. Deviation in the longitudinal direction at a frequency of 23750 Hz according to the results of numerical simulation of a waveguide in COMSOL medium for a cylindrical waveguide.

After changing the shape of the waveguide to a smooth transition in the form of a cos-like function 60 mm long, with a reflector length of 35mm bolt 78mm and piezoelectric elements 20mm long, the total length of the waveguide was 180mm, there was a slight change in the resonant frequencies. The resonant frequency under consideration has shifted to 25250 Hz. This is due to the fact that any changes in the dimensions of the waveguide geometry inevitably lead to a shift of the resonant frequencies. Figure 3 shows the longitudinal deviation of the aluminum waveguide at a given frequency.

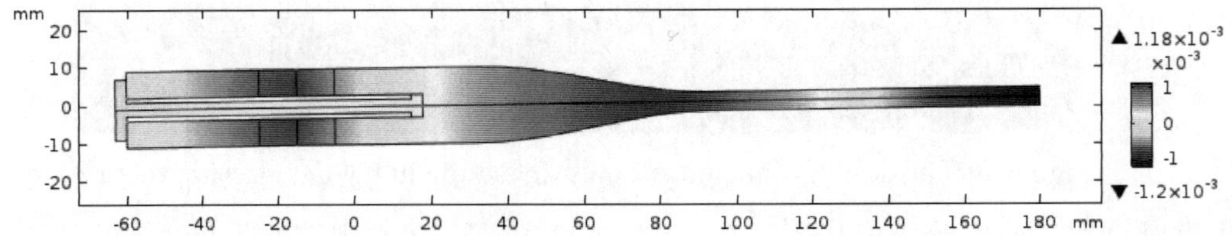

Figure 3. Deviation in the longitudinal direction at a frequency of 25250 Hz according to the results of numerical simulation of a waveguide in a Comsol software for a waveguide with a cos(x) transition.

According to the results of numerical simulation, we can conclude that the selected dimensions of the design of the ultrasonic waveguide are optimal. From Figure 3 it can be seen that with this design of the waveguide at a frequency of 25250 Hz, the maximum amplitude of oscillations of the end of the working part is achieved. This ensures minimal deformation of the bolt during the operation of the tool, and the oscillation energy is mainly directed to the working part.

4. Numerical simulation of the ultrasound waveguide by finite difference time domain approach

A numerical simulation of the design of the ultrasonic waveguide presented in Figure 3 by finite difference time domain approach was carried out. We have developed program that provides parallel computations using the OpenCL technology. In the simulation, the considered ultrasonic waveguide is represented as an array of particles in a cubic body-centered crystal grid. In addition, each particle has its own mass and speed. Particle dynamics is described by Newton's equations of motion. The forces of interaction between particles are characterized by a force function, which is the dependence of the force of attraction between particles by the distance. Near the point of equilibrium, the force of attraction is zero, which corresponds to a point of stable equilibrium of particles. The angle of inclination of the force function near the equilibrium point is proportional to the Young's modulus. Particles in the lattice are placed periodically, but there are two different distances between the neighboring particles: d and $d\sqrt{3}/2$. For each particle, we calculate the velocity and coordinate by numerical integration of the acceleration by time. The acceleration of a particle is calculated according to Newton's second law as the ratio of the force acting on the particle to its mass. The mass of the particle and the force of interaction determine the type of material. Such force dependence is characterized by the formula.

$$F_1(x)=\begin{cases} k\dfrac{d}{(3-2\sqrt{2})}\left(1-\dfrac{d}{x}\right)\left(\sqrt{2}-\dfrac{x}{d}\right)^2 & if\ \left(x<d\sqrt{2}\right) \\ 0 \quad if\ \left(x\geq d\sqrt{2}\right) \end{cases}$$

$$F_2(x)=\begin{cases} k\dfrac{d\sqrt{3}/2}{(3-2\sqrt{2})}\left(1-\dfrac{d\sqrt{3}/2}{x}\right)\left(\sqrt{2}-\dfrac{x}{d\sqrt{3}/2}\right)^2 & if\ \left(x<d\sqrt{3/2}\right) \\ 0 \quad if\ \left(x\geq d\sqrt{3/2}\right) \end{cases} \tag{1}$$

Where $F(x)$ – interaction force between particles, x – the distance between particles, k – coefficient of elasticity. Mass and force function determine the density and speed of sound in a particular material, which allows you to associate modeling with processes in real materials. Figure 4 shows the design of the ultrasonic waveguide with ultra wide band vibrations of particles speed represented by color. This construction is considered as monolithic. In numerical modeling, the speed of sound and the density of the material were set. The speed of sound in aluminum is set to 5050 m/s, the density is 2712 kg/m^3. Ceramic PZT-8 was used as piezoelements, sound speed 3200 m/s, density 7600 kg/m^3.

IOP Conf. Series: Materials Science and Engineering **516** (2019) 012024 doi:10.1088/1757-899X/516/1/012024

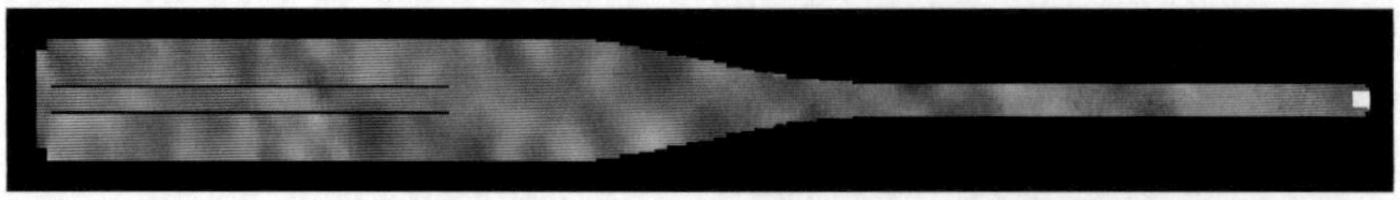

Figure 4. The design of the ultrasonic waveguide in the developed program.

A short pulse was applied as the exciting signal. Different colors represent the speeds with which the particles oscillate. When a short pulse propagates in the system, the system begins to respond at all its frequencies. At the end of the waveguide a monitoring particle was placed, which allows to record deviations of the coordinate and velocity of the particle. As a result of the numerical modelling, a resonance frequency spectrum of the structure being modeled was obtained. During the simulation, the coordinate X (t) of the monitoring particle is controlled. As a result of numerical simulation, we obtained the function X (t) and calculated its spectrum. The spectrum of edge spatial oscillations is presented in Figure 5.

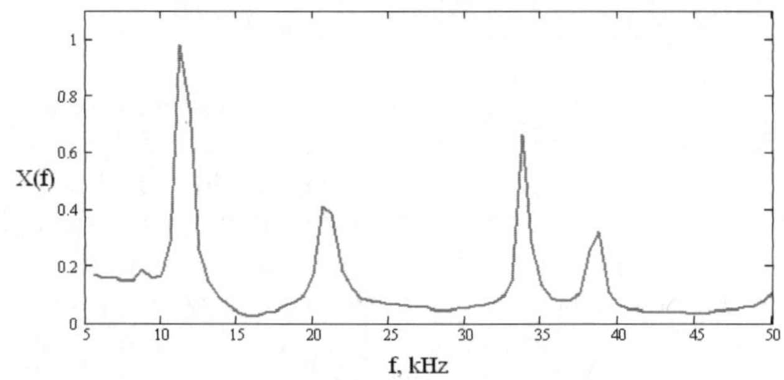

Figure 5. The resonant frequency spectrum of a monolithic ultrasonic waveguide with a cos (x) transition in the developed program.

In the numerical modeling of the monolithic structure of the ultrasonic waveguide, Figure 3 by the Comsol Multiphysics program, the spectrum of the resonant frequencies of the working part of the ultrasonic waveguide was close to the spectrum of resonant frequencies obtained in the developed program Figure 6.

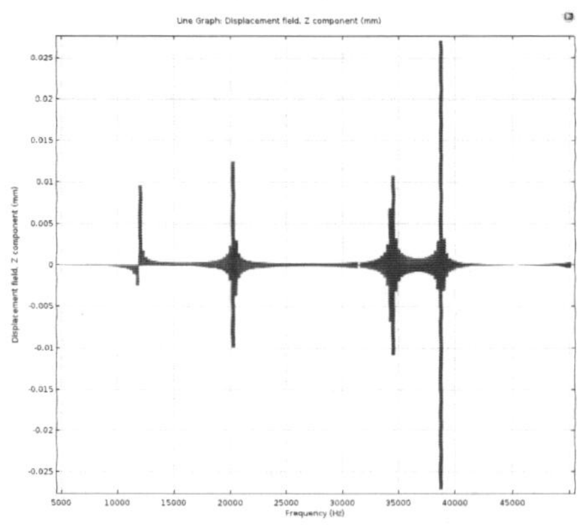

Figure 6. Resonance frequency spectrum of a monolithic ultrasound waveguide with a cos (x) transition in the Comsol Multiphysics program.

That allows to judge the reliability of the results of numerical modeling in the developed program.

5. Conclusion

As a result of the work, the proposed design of a waveguide with a cos junction was considered. A numerical simulation was carried out, in the program Comsol Multiphysics, based on the results of which, the resonance frequency, the optimal reflector length and the bolt attachment points were checked to ensure minimal deformation. For comparison, we modeled this waveguide design using finite difference time domain approach. As a result a resonance frequencies spectrum was obtained close to the spectrum obtained in the Comsol Multiphysics program.

Acknowledgments
The work was supported by the Ministry of Education and Science of Russia within the framework of the project "Creation of a high-tech complex of ultrasound surgery" (Unique identifier of the project RFMEF157517X0163).

References

[1] Gershgal D.A, Friedman V.M. 1976 *Ultrasonic Equipment* (Energy: Moscow)
[2] Donskoy A.V., Keller O.K., Kratysh G.S. 1982 *Ultrasonic Electrotechnological Installations* (Energy: Moscow)
[3] Severdenko V.P., Klubovich V.V. 1967 *Application of ultrasound in industry (*Science and Technology: Minsk)
[4] Lebedeva D.A., Novikov A.A. 2013 *Resource-efficient systems in management and control: a look into the future in 4 volumes* **1** 121 – 124.
[5] Lebedeva D.A. Novikov A.A. 2012 *Modern technologies and management in the energy and industry: resp. ed. V.N. Goryunov* 195 –198
[6] Graham P.G., Hashmi M.S.J., Finbar D., McGuinness G. B. 2005 *II International conference on computational bioengineering* 14–16
[7] Gossot D., Buess G., Cuschieri A., Leporte E., Lirici M. 1999 *Surgical Endoscopy* 48–51
[8] Khmelev V.N., Levin S.V., Khmelev S.S., Tsyganok S.N. 2013 *South Siberian Scientific Herald* **2 (4)** 42
[9] Ambrozevich E.G. 1982 Development of technology and equipment for ultrasonic welding and antimicrobial treatment of peritoneum in peritonitis: Diss. Cand. tech. Sciences
[10] Kvashnin S.E. 1997 *Topical issues of creating biotechnical systems* (2) 160–168
[11] Kikuchi E. 1972 *Ultrasonic Transducers* (WORLD: Moscow)
[12] Kvashnin S.E. 1997 *Conversion* (10) 30–31
[13] Stepanenko D. A., Minchenya V. T. 2010 *Ultrasonics* (50) 424 – 430 doi:10.1016/j.ultras.2009.09.033
[14] Novik A.A. 2007 *News spggatu "Laty".* 25–31
[15] Kvashnin S.E. 1999 *Ultrasonic electroacoustic transducers and waveguides-tools for medicine.* (MSTU: Moscow)
[16] Khmelev V.N., Tsyganok S.N., Demyanenko M.V., Levin S.V. 2003 *South Siberian Scientific Bulletin* (3) 67–71

IOP Conf. Series: Materials Science and Engineering **516** (2019) 012025 doi:10.1088/1757-899X/516/1/012025

Selection and substantiation of cultivator adjustment parameters for differential soil treatment on potato based on the rheology state of soil horizons

A B Kalinin[1], I Z Teplinsky[1], A A Ustroev[2], P P Kudryavtsev[2]

[1]Saint Petersburg State Agrarian University, Saint Petersburg, Russia
[2]Institute of Agro Engineer and Ecological Problems in Agriculture, Saint Petersburg, Russia
E-mail: andrkalinin@yandex.ru

Abstract. The article presents results of theoretical studies on the determination rational parameters of placement tines on the frame interrow cultivator for differential tillage of potato field. Differential tillage means depth adjustments of tines according soil condition and their placement on the cultivator frame. The main goals of differential tillage by inter row cultivators are the elimination of soil compaction between ridges, formed by potato planters, as well as loosening of the zone of location of the potato tubers. Deep loosening after planting provides good conditions for the complete assimilation of water by soil after rain or irrigation, protects the top soil layer from erosion and also eliminates the over moistening in the autumn. Loosening the soil in the zone of the tuber placement ensures the obtaining of a fine-grained soil structure, which helps to create best conditions for the growth of plants and development their root system, as well as to ensure favorable conditions for soil separation during harvesting. The substantiation of rational parameters for tine placement on the frame of row cultivator was carried out on the base of the regularities of soil rheology, the heterogeneity of soil composition in different layers of arable and subsoil horizons, as well as the variability of their physic-mechanical properties. These regularities influence the development of deformation and cracked zones during the interaction ripper tines with the soil. As a result of the theoretical studies was developed a method for calculating the rational placement of ripper tines on the frame of the row cultivator for fulfil differential tillage on the potato based on the rheological state and heterogeneous structure of the soil horizons. Using of this method allow ensure high quality structure of tilled soil.

1. Introduction

Removing of compacted zones during hilling of potato is an important factor in increasing crop yields and product quality. Investigations of traditional European technology of potato growing have established [1] a significant increase in soil density occurs after potato planting. Increasing of soil density in this case is a result of man-made impact after action of the wheel systems of heavy tractors and planters (Figure 1).

In order to eliminate the compaction after planting it is proposed to use the differentiated tillage that allow provide deep loosening inter row space during hilling of potato [2, 3]. For the implementation of this tillage method on the row cultivator was installed rigid tines (ripper tines) with chisel points (Figure 2), which performs deep loosening between rows, and two side spring tines with chisel points that moves along borders of protective zone c [4]. All installed chisel points are adjustable in height and lateral ones

IOP Conf. Series: Materials Science and Engineering **516** (2019) 012025 doi:10.1088/1757-899X/516/1/012025

can also move relative to the center. For final ridge forming behind all these tines are installed ridging mouldboard or ridging hood [5]. The proposed scheme of tines placement on the frame of row cultivator in addition to destroying of compacted zone make it possible to create the best conditions for the development of plants and spreading their roots into soil as well as providing fine crumbling of soil layers around potato tubers. In order to fulfill these actions by row cultivator during differential tillage of the potato were justified rational location of all offered type of tines on the cultivator frame based on the carried out theoretical studies.

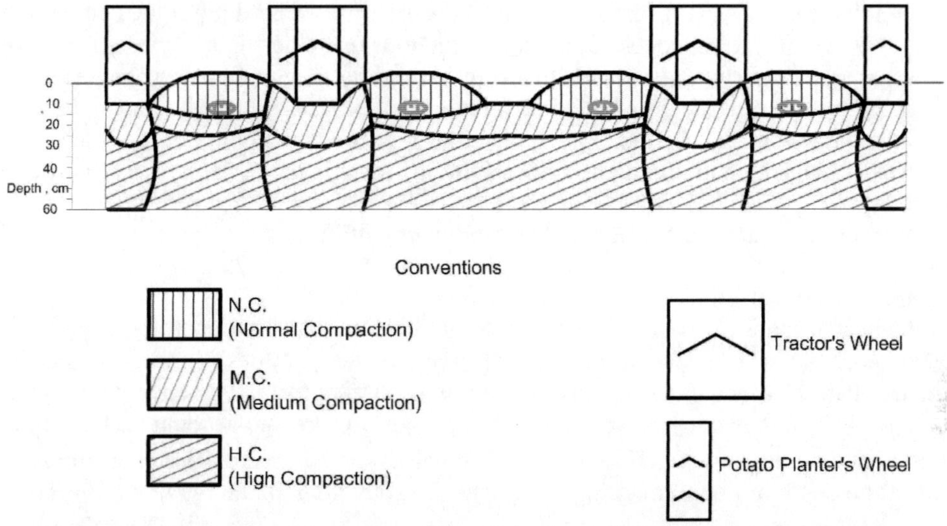

Figure 1. Distribution of compacted zones in the soil after the potato planting.

Objective of our researching is determination of technological parameters for the rational placement of ripper and spring tines on the frame of inter row cultivator carry out on the basis of calculation zones of soil deformation, taking into account the rheological state of soil horizons having a heterogeneous structure and various physical and mechanical properties.

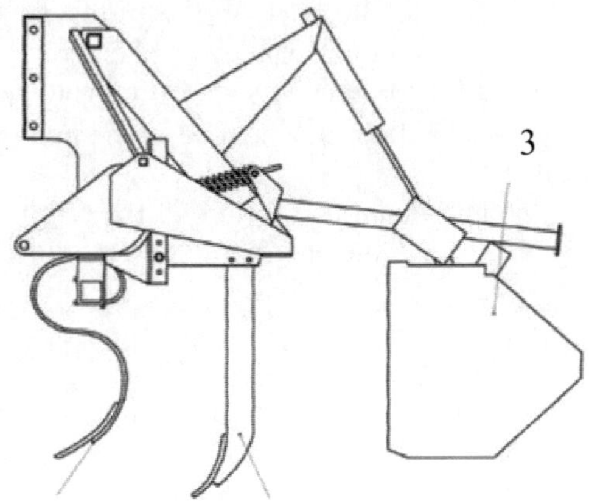

Figure 2. Scheme of placement of working elements on the row cultivator section, where 1 is side spring tine; 2 is ripper tine; 3 is ridge former plate with spring suspension.

2. Experimental

According to the generally accepted theory [6, 7], the cultivator's tines are placed taking into account the possible zone of deformation of the soil that occurs during their interaction with soil. When tines move in the soil, their points cause deformation of the treated horizons, which leads to the formation of a network of cracks after cultivator passes [8]. In turn, the fracture network creates the conditions for the crumbling of the soil in the upper horizons and the formation in them of small soil particles. Taking into account the main provisions of the soil rheology it is assumed that for a horizon with homogeneous soil structure the boundary zone of deformation in the longitudinal direction is limited by a line of the friction angle φ "steel-soil" from the normal to the point of the tines. In the transverse direction the boundary zone of soil deformation is limited by line of internal friction with angle ψ. It is assumed that a homogeneous soil structure is characterized by certain physical-mechanical properties such as density, hardness, humidity, internal and external coefficients of friction. According to the data obtained by [9] the depth of the potato's roots penetration zone, which ensures the highest yield, is at least 70 cm. Therefore working area of ripper tine will consist of several heterogeneous horizons, each of them has own structure with characteristic parameters mentioned above. This working area can be conditionally divided into three horizons A_1, A_2 and B [10]. Layer A_1 represents the upper part of the arable horizon, which, as a rule, coincides with the planting depth of potatoes; layer A_2 is the lower part of the arable horizon, and layer B placed below plowing depth. As rule, horizon B needs to be loosened for better penetration of the root system of the potato and good water infiltration.

3. Results and consideration

Scheme of boundary zones of soil deformation after interaction with by ripper tines of a row cultivator in heterogeneous soil horizons during differential tillage of the potatoes is shown in the Figure 3. In view of the fact that the row cultivator works on the ridge surface formed by the potato planter, then, in addition to the width of the main row spacing, it is necessary to take into account such parameters as the ridge height h_0 and the furrow bottom width b_0 . We will assume that the position of the seed tubers in the ridge is at the depth of potato planting *a*, measured in the XOY plane relative to the flat surface of the soil.

We define the zone of possible spreading of soil deformation in the longitudinal direction. When the ripper tine with the point width b_c (Figure 3) aimed for deep loosening is interaction with the soil a crack is formed [11, 12], which propagates in the longitudinal direction at an angle φ_B to the normal to the share at the point *m*. The angle φ_B is the angle of friction "steel-soil" valid for subsoil horizon B. When the crack reaches the boundary of horizons B and A_2 its further direction will depend on the angle of internal friction "soil-soil" ψ_{A_2} valid for horizon A_2. Therefore, at the point *n*, the line direction of spreading crack in the longitudinal direction is changed. When crack reach the point *p* at the boundary of the horizons A_1 and A_2, the line direction of its spreading changes again according the angle ψ_{A_1}, which is valid for the horizon A_1. The deformation zone in the longitudinal direction can propagate to the point *q* at the top of the ridge.

Thus, taking into account the individual parameters of the physics mechanical properties valid for each soil horizons, treated by ripper tine, the length of a deformation zone in the longitudinal direction Lc is determined by the expression:

$$\mathrm{L_c} = \mathrm{L_c^0} + tg(\varphi_B + \propto_c) \cdot \mathrm{B} + \mathrm{A_2} \cdot tg\psi_{A_2} + (\mathrm{A_1} + h_0^t) \cdot tg\psi_{A_1}, \quad (1)$$

where $\mathrm{L_c^0}$ is the protrusion of bottom end of point; $\propto_c$ is the inclining angle of point; h_0^t is height of the ridge relative to the initial level of the soil.

To ensure the operation of the ripper tines without clogging, it is necessary that the parameter L of their placement on the cultivator's frame in the longitudinal direction is greater than Lc. Therefore, when placing the side tines on the frame of row cultivator for differential tillage of potatoes, it is necessary to fulfill the requirement L ≥ Lc.

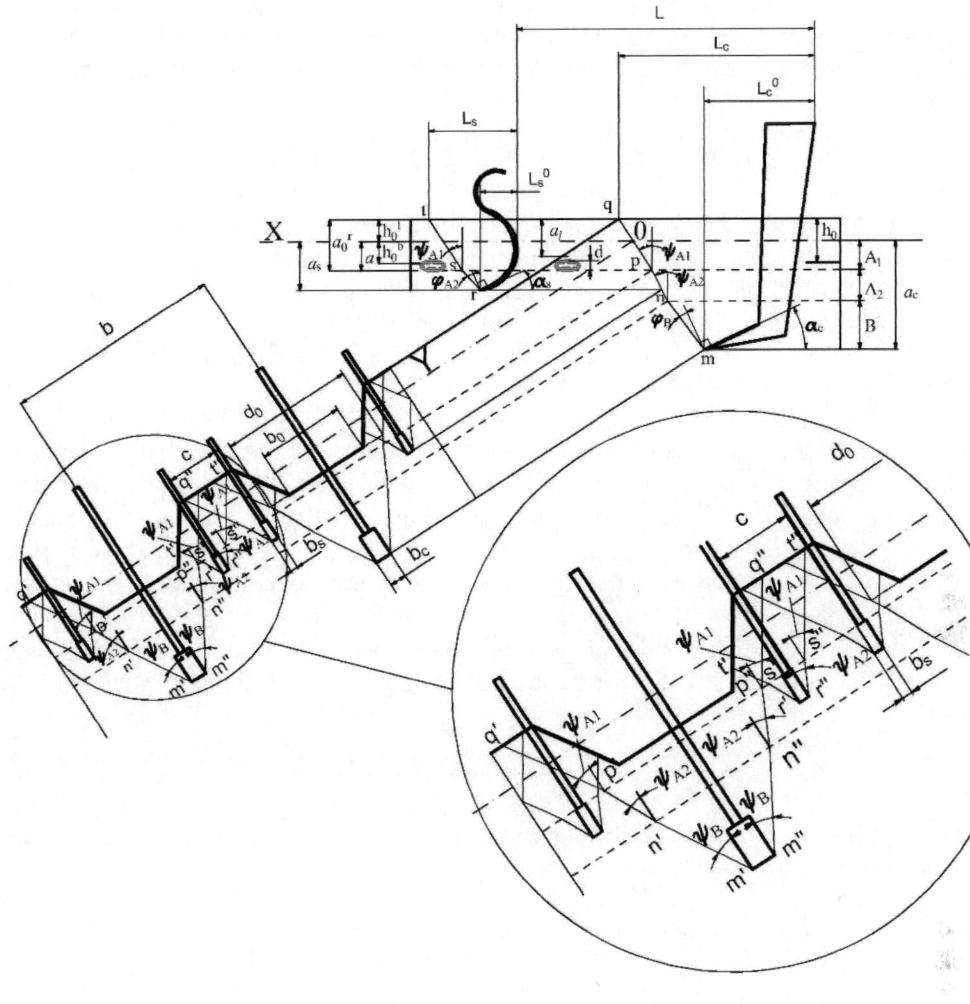

Figure 3. Scheme of spreading of soil deformation caused by ripper and side spring tines of a row cultivator in heterogeneous soil horizons during differentiated tillage of potatoes.

Since the row cultivator forms a ridge from the soil taken from the middle rows, it is important to fulfill two conditions to ensure the high quality soil crumbling:

✓ the crumbling zone d_0 after the passage of the ripper tine must exceed the furrow bottom width b_0;

✓ the zones of the deformation soil from the side tines must intersect directly above the mother tuber at a depth of representing the distance from the top of the ridge to tuber upper part.

To fulfill the conditions for forming ridges with fine loosed soil, will consider the zone of propagation of soil cracking in the transverse direction. This zone is limited by the projection of the line *mnpq* in the direction of spreading of the crack and is determined by the internal friction angles ψ_B, ψ_{A_2} and ψ_{A_1} in the horizons B, A2 and A1, respectively. Thus, in the transverse direction, the width d_0 of propagation the soil deformation after passes ripper tine for deep loosening of potatoes, is bounded at the level of the furrow bottom by the lines *m'n'p'q'* and *m"n"p"q"* is determined as:

$$d_0 = b_c + 2\left[\frac{B \cdot tg\psi_B}{cos(\varphi_B + \alpha_c)} + \frac{A_2 \cdot tg\psi_{A_2}}{cos\,\psi_2} + \frac{(A_1 - h_0^b) \cdot tg\psi_{A_1}}{cos\,\psi_1}\right], \quad (2)$$

where $h_0^b = h_0 - h_0^t$ is distance from bottom of furrow to the flat soil surface.

To provide a sufficient zone of loosening of rows by ripper tines we determine the required working depth a_c. Since $\mathrm{B} = a_c - (\mathrm{A}_1 + \mathrm{A}_2)$, the condition for depth setting will be determined by the expression:

$$a_c \geq (\mathrm{A}_1 + \mathrm{A}_2) + \left[\frac{d_0 - b_c}{2} - \frac{(\mathrm{A}_1 - h_0^b) \cdot tg\psi_{A_1}}{\cos\psi_{A_1}} - \frac{A_2 \cdot tg\psi_{A_2}}{\cos\psi_{A_2}}\right] \cdot \frac{\cos(\varphi_B + \alpha_c)}{tg\psi_B}. \quad (3)$$

When calculating the adjustment values for the side spring tines placement relative to the ridge center, we make the assumption that their arrangement is completely determined by the width of the protective zone *c*, where the deformation zones from the interaction of adjacent tines must intersect above the mother tuber at a distance a_l from the ridge top. Assume that $a_l = a_0^r - 1{,}5d$. Here $a_0^r = a + h_0^t$ is the distance from the top of the ridge to the depth of planting and *d* is an average diameter of the mother tuber.

Since the setting depth of the side spring tines is below the planting depth, they will treat the soil layer with layers A_1 and A_2. When the side spring tines are moved in the soil at the point *r* appears a crack, which propagates in layer A_2 at the angle of external friction "steel-soil"ψ_{A_2} from the normal to the chisel point. When the crack caused by side spring tine reaches the point *s* at the boundary of the layers A_2 and A_1, direction of its spreading line changes depending on the value of the angle of internal friction "soil-soil" $\psi_{A_1}$and further it reaches the surface of the ridge at the point *t*. In the transverse direction the boundary of the soil deformation will be limited by lines $r's't'$ and $r''s''t''$.

To determine the depth of the side spring tines we need take into account the width of the protective zone *c*, which is characterized the complete crumbling of the soil above the mother tuber by the expression:

$$\mathrm{c} = 2 \cdot \left[\frac{(a_s - \mathrm{A}_1) \cdot tg\psi_{A_2}}{\cos(\varphi_{\mathrm{A}_2} + \alpha_s)} + \frac{1{,}5 \cdot d \cdot tg\psi_{A_1}}{\cos\psi_{A_1}}\right]. \quad (4)$$

By solving the inverse task we determine the value of the side tines depth when adjust it on the frame of row cultivator by expression:

$$a_s \approx \mathrm{A}_1 + \left(\frac{c}{2} - \frac{1{,}5 \cdot d \cdot tg\psi_{A_1}}{\cos\psi_{A_1}}\right) \cdot \frac{\cos(\varphi_{\mathrm{A}_2} + \alpha_s)}{tg\psi_{A_2}}. \quad (5)$$

4. Conclusion

A method based on the rheological state and the heterogeneous structure of soil horizons A_1, A_2 and B has been developed the for calculating the rational placement on the frame of row cultivator the ripper and side spring tines for differential tillage of potatoes.

It has been obtained the theoretical substantiation of the technological parameters of tines depth adjustments that takes into account physical and mechanical of soil properties and allow to provide high quality soil treatment.

It was obtained regularities of interaction tillage tool with soil during differential tillage can be used to calculate its technological parameters while improving existing and designing new machines for potato growing.

References

[1] Kalinin A. B. *et al* 2016 *Potatoes and vegetables* **2** 35–36 (in Russian)
[2] Kalinin A. B. *et al* 2016 *Izvestiya of the St. Petersburg State Agrarian University* **43** 327–330 (in Russian)
[3] Stark J. and Love S. 2003 Potato production systems: A Comprehensive Guide for Potato Production (University of Idaho Extension US)
[4] Kalinin A. B. *et al* 2016 *Patent RU* 169780 **19** (in Russian)
[5] Dozorova T. A. *et al* 2016 *International Research Journal* **6(48-1)** doi: 10.18454/IRJ.2016.48.204
[6] Srivastava A. K. *et al* 2006 *Engineering Principles of Agricultural Machines* URL https://fcaib.edu.ng/books/Engineering/%5BAjit_K._Srivastava__Carroll_E._Goering%5D_Engineer(BookFi.org).pdf

[7] Tarverdyan A. P., Tonapetyan P. A. 2016 *Annals of Agrarian Sciences* **14(4)** doi: 10.1016/j.aasci.2016.09.014
[8] Bietresato M. *et al* 2018 *Int J of Safety and Security Eng* **8(1)** 77–89 doi: 10.2495/SAFE-V8-N1-77-89
[9] Gliessman S. R. *et al* 1981 Agro-Ecosystems **7(3)** 173–185 doi: 10.1016/0304-3746(81)90001-9
[10] Wale S (ed) 2018 *Achieving sustainable cultivation of potatoes* **2** (Burleigh Dodds Science Pub: UK)
[11] Boev S. F. *et al* 2018 *IOP Conf. Ser.: Mater. Sci. Eng.* **457** 012010 doi: 10.1088/1757-899X/457/1/012010
[12] Gudehus G. 2011 *Physical Soil Mechanics* (Springer: Berlin) doi: 10.1007/978-3-540-36354-5

IOP Conf. Series: Materials Science and Engineering **516** (2019) 012026 doi:10.1088/1757-899X/516/1/012026

Digital 3d x-ray microtomographic scanners for electronic equipment testing

S A Klestov, D N Kuznetsov, S B Suntsov

National Research Tomsk State University, Tomsk, Russia
E-mail: klestov_simon@mail.ru

Abstract. The article studies the operating procedures of an X-ray microtomographic scanner and the module of reconstruction and analysis 3D-image of a test sample in particular. An algorithm for 3D-image reconstruction based on image shadow projections and mathematical methods of the processing are described. Chapter 1 describes the basic principles of X-ray tomography and general procedures of the device developed. Chapters 2 and 3 are devoted to the problem of resources saving by the system during the X-ray tomography procedure, which is achieved by preprocessing of the initial shadow projections. Preprocessing includes background noise removing from the images, which reduces the amount of shadow projections in general and increases the efficiency of the group shadow projections compression. In conclusion, the main applications of X-ray tomography are presented.

1. Introduction

Modern science allows analyzing the internal microstructure of objects by means of various methods. The method of X-ray microtomography is considered to be one of the best ways of nondestructive imaging. X-ray microtomography allows visualization of the internal structure of nontransparent objects in three dimensions (3D) with high spatial resolution. There is a need to examine the internal structure of nontransparent objects, especially biological ones, in the visible range of the electromagnetic radiation with micron resolution. The development of X-ray microscopy methods has allowed looking inside nontransparent objects with resolution that exceeds the capabilities of optical microscopy. Today, computer microtomography is a primary method of 3D visualization of the internal microstructure of organic and inorganic objects using X-rays. The method is similar to medical imaging, but it has significantly higher spatial resolution. Scanning visualizes the entire internal 3D structure of the object and completely preserves the sample for other studies [1, 2].

Methods of digital X-ray imaging allow carrying out studies of both organic and inorganic objects and materials, identifying statistical features of composition and structure of the samples [3, 4].

2. Basic principles of x-ray microtomography

The work process is as follows: the X-Ray unit irradiates the object that is placed on the controlled operating area. Penetrating the object, X-rays are received by the detector element unit, which provides element-by-element recognition the full frame image of the internal structure of the object.

The analog signal from the detector is digitized by an analog-digital converter and is led to the electronic computing machine. A digital signal from the second analog-digital converter, modifying the signal of the control unit of power spectrum of X-ray radiation, enters the second input of the electronic computing machine. The control unit of power spectrum of X-ray radiation measures the spectral components of the signal received from the detector unit [5, 6].

Let us consider a kinematic scheme of the intelligent 3D X-ray microtomographic scanner (XRMTS). Figure 1 shows the kinematic diagram of the device. During the survey, the sample rotates by 180 or 360 degrees with a fixed pitch. A shadow (transmission) image of the sample is fixed for each micro-rotation. The system saves all these projections as 16-bit tiff files. After scanning, the data array represents a set of normal transmission X-ray images. The number of files in the array depends on the step size and the value of the selected total rotation angle. For example, for a 180 ° rotation angle with 0.9 ° step, the data array will contain 200 images and a small number used to transform images to compensate for the correctness of the X-ray beam.

Completion of the sample shooting is followed by reconstruction of its image. The obtained 16-bit shadow tiff images are used to reconstruct virtual sections of the object. Further, using the reconstruction algorithm, a preliminary array of sections is generated. These data are not yet images; it is a matrix containing the absorption values in the section under reconstruction.

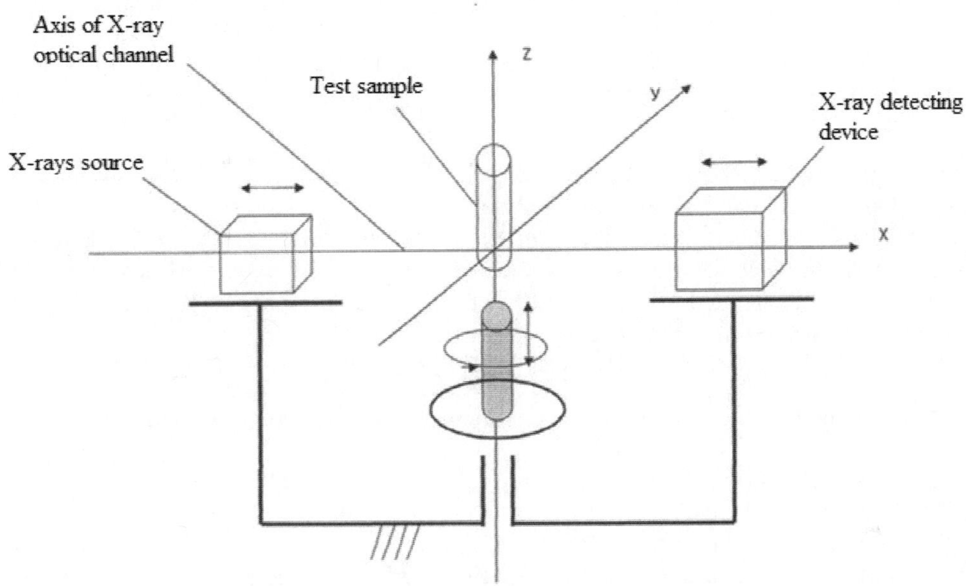

Figure 1. Kinematic diagram of the device.

The size of the matrix is similar to the number of pixels inside the section or the line of the CCD matrix (*n* is a number of pixels in the line of the shadow image or the CCD matrix). Now we can save the reconstructed section as a floating-point matrix containing attenuation values after reconstruction [2].

Completion of formation of the preliminary array of sections is followed by creation of a 3D image of the sample.

3. 3d-reconstruction analysis

Using the 3D image reconstruction unit (3D IRU), a three-dimensional image of the internal structure of the object is reconstructed by its shadow projections (Figure 2). Next, a defectoscopy unit detects discontinuities (defects, cracks, inclusion of garbage, etc.). Then, a pseudo color image-shaping unit (PCISU) presents a pseudo-color image on the monitor screen using information from the 3D IRU. Via the Internet connection unit, the results of the device operation are transferred to the Internet system from the defect detection unit and PCISU.

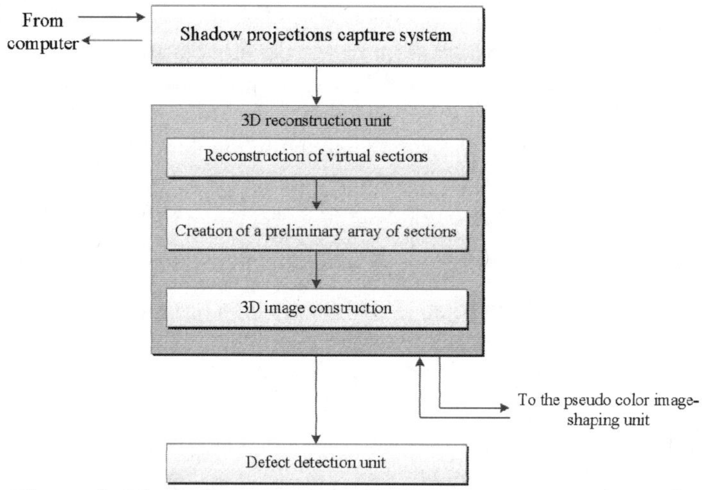

Figure 2. Block diagram of the 3D image reconstruction unit.

Algorithmic support of the defect detection unit is based on the gradient, correlation and fractal methods as well as wavelet and Fourier analysis described below.

An array of density values at each point of the sample is initial data for the gradient analysis method. Based on these values, the density function of the material $\rho(x, y, z)$ is determined followed by construction of a gradient field of the sample. Defects will be detected by the field nonuniformity, that is, by the presence of density gradients:

$$\nabla\rho(x, y, z) = \left(\frac{\partial\rho}{\partial x}, \frac{\partial\rho}{\partial y}, \frac{\partial\rho}{\partial z}\right).$$

It is possible to determine the size and nature of the defects by the volume arrangement of these gradients.

Advantages of this method include:

-possibility to determine the defect type;

-possibility to determine the defect location;

-possibility to determine geometric and physical characteristics of the defect.

The gradient of the image $f(x,y)$ is defined at the point (x,y) as a two-dimensional vector:

$$G[f(x,y)] = \begin{bmatrix} G_x \\ G_y \end{bmatrix} = \begin{bmatrix} \frac{df}{dx} \\ \frac{df}{dy} \end{bmatrix}.$$

From the vector analysis, it is known that the vector G indicates the direction of the maximum change in the function f at the point (x, y). However, when determining defects, the modulus of this vector is of interest; it is usually called as a gradient and is denoted as $G[f(x,y)]$, where

$$G[f(x,y)] = [G_x^2 + G_y^2]^{1/2} = [\left(\frac{\partial f}{\partial x}\right)^2 + \left(\frac{\partial f}{\partial y}\right)^2]^{1/2}.$$

In practice, as a rule, the gradient is approximated by absolute values

$$G[f(x,y)] \cong |G_x| + |G_y|.$$

This approximation greatly simplifies implementation of the method.

The gradient calculation is based on finding the first derivatives $\frac{\partial f}{\partial x}$ and $\frac{\partial f}{\partial y}$. There are several ways to do it for digital imaging. One of the approaches is to use the difference between neighboring points

$$G_x = \frac{\partial f}{\partial x} = f(x,y) - f(x-1,y),$$

IOP Conf. Series: Materials Science and Engineering **516** (2019) 012026 doi:10.1088/1757-899X/516/1/012026

$$G_y = \frac{\partial f}{\partial y} = f(x,y) - f(x,y-1).$$

Another method, which is more complicated, involves points in the neighborhood of 3x3 size centered at the point (x, y):

$$G_x = \frac{\partial f}{\partial x} = [f(x+1,y-1)+2f(x+1,y)+f(x+1,y+1)]-[f(x-1,y-1)+ \quad (1)$$
$$+2f(x-1,y)+f(x-1,y+1)] = (g+2h+i)-(a+2b+c),$$

$$G_y = \frac{\partial f}{\partial y} = [f(x-1,y+1)+2f(x,y+1)+f(x+1,y+1)]-[f(x-1,y-1)+ \quad (2)$$
$$+2f(x,y-1)+f(x+1,y-1)] = (c+2e+i)-(a+2d+g),$$

Where the letters from a to i denote neighboring points of the center (x,y). Calculation of the gradient in the 3×3 area is more advantageous than using equations (1) and (2) because of a larger averaging, which makes the gradient less noise-sensitive. In principle, use of wider neighborhoods in finding the gradient is also possible.

Currently, two main directions of the wavelet transform have become widespread. The first direction is continuous wavelet analysis, the main applications of which include localization and classification of the signal singular points, as well as calculation of its various characteristics and frequency-time analysis. The second direction is discrete wavelet analysis, the main application of which is compression of video information, as well as image processing. The result of the wavelet transform with high information content is characterized by a large amount of computation, and, as a rule, by the redundant representation of the results (in comparison with the Fourier transform). This is primarily because of the fact that the wavelet transform allows calculating the relative contribution of frequencies at each point of time (by finding convolution with different-scale versions of the wavelet).

Consequently, it is possible to observe evolution of a spectrum analogous to Fourier, but not for a selected period of time, as in the case of the Fourier transform, but over the entire time interval. Thus, for informative decomposition of the original signal it is sufficient to know its wavelet transform on some rare grid in the time-frequency plane. Analysis of literature data shows that even high-quality and high-contrast images obtained by means of various information and measuring equipment carry a large number of various distortions, noise fields, etc. This interferes with the operation of image processing algorithms and requires additional efforts for suppression and elimination of distortions.

The wavelet spectrum of the images is also exposed to various distortions related to noise properties of signals. It should be noted that the concept of “noise” is connected with the type of image processing and the same effect can be interpreted differently in various image-processing tasks such as tracing, compression, texture processing, etc. In this paper, the concept of “noise” is associated with the factors that prevent proper detection of a useful signal (forming deformation relief on the surface). Let us list the main “disturbing” factors:

- “white noise” (associated with transformation of the light field into an electrical signal);
- distortion of the constant component in the signal (low-frequency noise);
- high-frequency filling (high-frequency noise);
- texture noise, not related to the method of obtaining a two-dimensional signal.

The main (basic) ways of solving the noise-filtering problem include the following operations. The “white noise” can be eliminated by a linear signal filtering (for example, a linear average filter) or a nonlinear “median” filtering. Distortions of the constant component can be eliminated either by spectral methods or by reduction to zero of the constant component in the bounding sliding window.

Let the resolving power of the tomographic scanner be the quantitative measure of accuracy of tomographic images reconstruction. We will define it as the minimum distance between two opaque objects, which allows us to distinguish these objects on the reconstructed tomogram.

Let us consider the main factors, which determine the resolving power of reconstructed tomographic images. Obviously, the limiting value of the resolving power (the minimum distance resolved by the given tomographic scanner) is determined by the width of the diffraction half-shadow area. Figure 3 shows the dependence between the modulus of the Fresnel integral (which determines the ratio of the level of the diffraction field beyond the obstacle to the level of the field in free space) and excess of the observation point above the obstacle edge by u. The value u is measured in the dimensions of the first Fresnel zone. The maximum value of $|F(u)|$ equal to 1.85 is reached when $u \approx 1.2$ (this value is marked by a dashed line in Figure 1). The total diffraction curve obtained when illuminating two objects, will contain two distinguishable maxima if the distance between objects is more than Δ_1. The value Δ_1 is determined by the level of $\frac{1}{\sqrt{2}}$ from the maximum value and is equal to $\Delta_1 = 1.2\sqrt{\lambda d}$, where λ is the radiation wavelength, d is the distance between the radiation source and the receiver. When $d = 0.2$ m, $\lambda = 2.3 \cdot 10^{-10}$ m, the value Δ_1 is about 13 μm.

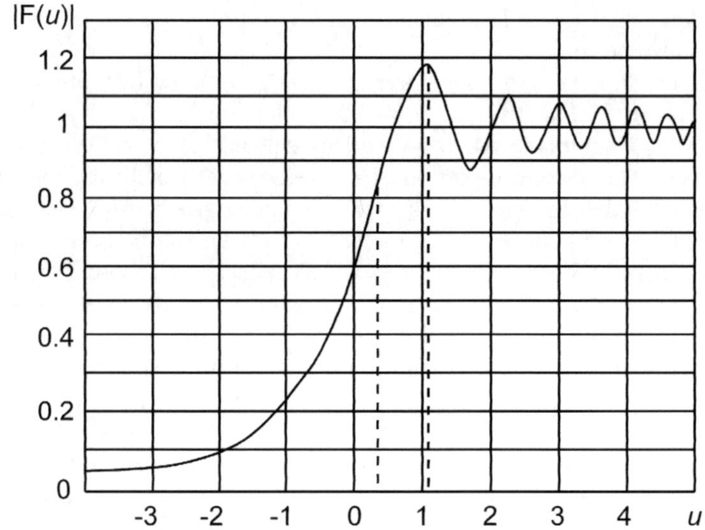

Figure 3. The values of the Fresnel integral modulus determine the ratio of the diffraction field amplitude to the level of the field in free space. Dashed lines indicate the width of the area, which determines the minimum separation of distinguishable objects.

Spatial distribution of the diffraction field intensity is recorded at the receiving aperture. A set of registered distributions is the Radon transform of the spatial distribution of the optical density of the object under investigation. This density is restored by the registered projections by means of inversion of the Radon transform. To estimate the accuracy of the reconstruction of the optical density distribution (accuracy of the reconstructed tomogram), we use simplified representations about the inversion procedure of the Radon transform [3, 4].

Traditionally, when calculating the fast Fourier transform (FFT) according to the Cooley-Tuckey algorithm, a technique that significantly reduces the level of quantization noise in the spectral region is used. This method consists in dividing the computation results in each step by 2. Using this technique, estimation of the quantization noise at the output of the calculator can be obtained in the form $\sigma^2 = \frac{2}{3} \cdot 2^{-2b}$. When $b = 16$, $\sigma = 1.2459 \cdot 10^{-5}$.

The value $\sigma_{\Delta} = \dfrac{\sqrt{\sigma_{\xi}^{2} + \sigma^{2}}}{\dot{S}(X_{0}')} \approx \dfrac{1.25 \cdot 10^{-5}}{0.4341} = 28.8 \cdot 10^{-6}$ m characterizes deterioration of the system resolution because of the thermal noise of the receiving system and the quantization noise of the analog-to-digital converter (ADC).

The overall resolution of the entire system δ can be estimated by summing the values Δ_1, Δ_2 and σ_{Δ}:

$$\delta = 1.2\sqrt{\lambda d} + \frac{2 \cdot \Delta y}{\cos(\Delta \varphi)} + \sigma_{\Delta}$$

In the general form, the operating scheme of the XRMTS algorithm can be written as follows:

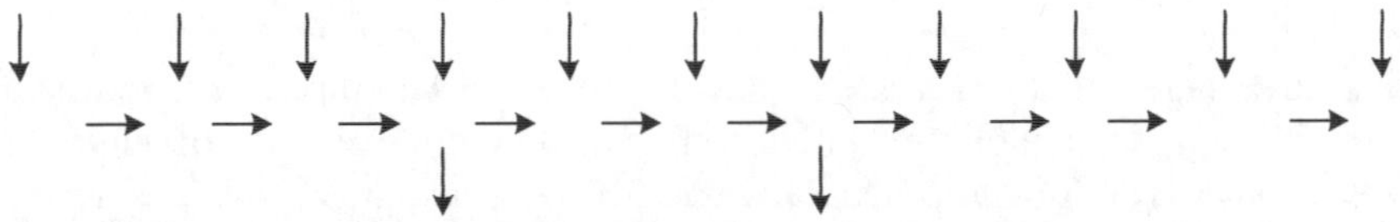

Let us explain the meaning of the operators introduced.

1. The operator **Start** means start of the XRMTS operation: the nodes and units of the XRMTS are prepared for operation (resetting, clearing the memory, etc.), and instruction is given to the synchronizer that determines the sequence of operations.

2. The operator **I** means installation of a sample (the object of investigation) in the working area (the surface of the table). This operation is preceded by preparation of a sample (assurance of the sample size).

3. The operator **E** includes an electromechanic system and a monitoring system for the operation parameters by displaying on the monitor the metrological support unit.

4. Operator **M** includes a metrological provision unit, including a color television control system, as well as the XRMTS security system. Here, fulfillment of the condition q is required: the object of investigation does not exceed the required dimensions and physical characteristics (for example, hardness, because external vibrations do not affect the measurement), otherwise, control functions pass to one of the operators Start, I, E.

5. Operator **H** includes a high-voltage power supply and a system for monitoring stability of its operation by means of displaying it on the screen of the metrological provision unit.

6. Operator **D** includes a detector that perceives the x-ray signal passing through the object of investigation. The detector converts the X-ray signal into an analog electrical, and then into a digital one. Here, the power supply and cooling units of the detector are turned on.

7. The operator **T** generates test signals for testing all modules (electronic, mechatronic and software), evaluates their state and gives an enabling signal in case of compliance of the XRMTS with the technical requirements.

8. The operator **IS** forms an array of two-dimensional images of the object under investigation when it is moved and rotated in the working area; (i.e. the operator provides image sensing when scanning the object of investigation). Here, fulfillment of the condition p is required: the power of X-ray radiation is sufficient for conducting measurements, all XRMTS modules function correctly, otherwise, the control functions pass to one of the operators Start, I, E, M, H, D.

9. Operator **A** analyzes a three-dimensional image and performs the following operations:

a) restores the 3D image of the internal structure of the material;

b) processes the image (filters, poses);

c) analyzes discontinuities (defects) of the material;

d) represents a color three-dimensional image (colors discontinuities).

10. Operator **Disp** displays and transmits the received information.

11. Operator **End** indicates the end of the XRMTS operation.

4. Conclusions

The developed intelligent X-Ray microtomographic scanner for testing of materials of different origin has the following distinctive advantages:

1) high precision positioning system that is able to position objects under study with the accuracy of ± 1 micron;

2) complete automation of the X-ray micro tomography process and reconstruction of a 3D-model of the object;

3) built-in algorithms and classification analysis of the internal structure and defects of objects;

4) built-in algorithms for pre-processing of initial lossless compression data in order to save computing resources of the system;

5) high operating speed of both hardware and software components because of the use of restructurable control algorithms providing significant accuracy of the X-ray micro tomographic scanner [7].

Competitive advantages of the developed device include portability, compactness, ability to test different materials (organic, nonorganic, and constructive) and elements of electronic equipment, compatibility with other types of equipment, and competitive price [8, 9].

Acknowledgement

The paper is prepared within the framework of the competitiveness improvement programme of National Research Tomsk State University and grant of the Russian Foundation for Basic Research No 16-29-04388.

The authors wish to thank Tatiana B. Rumyantseva from National Research Tomsk State University for English language editing.

References

[1] Elliot A. J. *Application of X-ray Microtomography and Image Processing to the Investigation of a compacted Granular System* URL: http://people.pwf.cam.ac.uk/jae1001/CUS/research/pfizer/Fu_etal_PPSC_2006.pdf.

[2] Panin S., Syryamkin V., Glukhikh A. 2003 *Avtometriya* (4) 79–92

[3] Marusina M. Ya., Kaznacheeva A.O. 2016 *Modern methods of tomography* (Mechanics and Optics Press: St. Petersburg)

[4] Maire E. *et al On the Application of X-ray Microtomography in the Field of Materials Science*, URL: http://onlinelibrary.wiley.com/doi/10.1002/1527-2648%28200108%293:8%3C539::AID-ADEM539%3E3.0.CO;2-6/abstract

[5] Kornilov S., Rempe N., Shidlovskiy S. 2016 *Technical Physics* **61**(6) 841–848 doi: 10.1134/S1063784216060116

[6] Bureev A.Sh. 2015 *Digital x-ray tomography* (Red Square Scientific, Ltd.: London)

[7] Bogomolov E.N. *et al* 2012 *Modern methods of materials and nanotechnologies research* (Laboratory course) (Tomsk State University Press: Tomsk)

[8] Bubenchikov M.A. *et al* 2010 *Modern methods of materials and nanotechnologies research* (Tomsk State University Press: Tomsk)

[9] Okhotnikov V. *et al* 2016 *IOP Conference Series: Materials Science and Engineering* **116** doi: 10.1088/1757-899X/116/1/012001

Mechanized application of ameliorants for preservation of soil moisture on cultivated lands

I I Maksimov[1], S A Vasilyev[2], A A Vasilyev[3]

[1]Mari State University, Yoshkar-Ola, Russia
[2]Chuvash State Agricultural Academy, Cheboksary, Russia
[3]Novgorod State Engineering-Economic University, Knyaginino, Russia
E-mail: alexei.21@mail.ru

Abstract. During the drought season agronomic crops get into a critical situation when a lack of moisture and nutritional chemicals may lead to a wilt of plants and crop loss. For preventing of the following situation it is proposed to apply liquid ameliorants in the process of subsurface tillage which can accumulate moisture in a soil. The introduction of ameliorants offered to carry out work on liquid ameliorants directly into the root soil layer, to create favorable conditions for the development of the plant. Also, when making soil ameliorants preventing the leaching of their streams generated from rainfall and meltwater. Experimental studies have shown that zero tillage with simultaneous application of liquid ameliorants (in the process of experiment was used urea-ammonium nitrate) can increase soil moisture at cultivation depth from 22 to 47%. Moisture of soil determined after two months after cultivation and applying of liquid ameliorants according to the National Standard 28268-89 "Soils. Methods for determination of moisture, maximum hydroscopic moisture and moisture of steady plant fading". In this case, received values of filtration filters and soil porosity show qualitative soil cultivation for crop growing.

1. Introduction

Greater focus should be placed on preparation of field in the process of cultivation. The most important is preserving of moisture and nutritional chemicals in the soil as during the drought season this affects process of sprouting and vegetation that directly influences on the yield. For this it is necessary to carry out soil cultivation oriented on preservation of moisture, mineral and nutritional chemicals for creation of favorable condition of developing withstanding root system. Also, topsoil can be affected drying that can lead to destruction of a plant. In this case applied ameliorants placed in the substrate allow the roots to feed moisture and nutritional chemicals.

It's of great current interest application of liquid ameliorants have a high sorption capacity to accumulate moisture [1].

Ameliorants occur as:

- powder or gel mass (hydrogel, polyacrylimide) can absorb a large volume of liquid and grow in size [2, 3];
- solution and emulsion (cryogel, carbamide- ammonium nitrate, liquid complex fertilizers) high-nitrogen and phosphate [4].

IOP Conf. Series: Materials Science and Engineering **516** (2019) 012027 doi:10.1088/1757-899X/516/1/012027

It's the best of all to apply ameliorants to a depth of cultivation by deep tillage cultivator-25…30 cm. As in this case it is prevented illuviation of ameliorants from surface by runlets formed by rainfall and melt waters.

Quality of applying of liquid ameliorants is determined by soil water conductivity as thereat may be liquid stagnation [5]. This may lead to overwatering of roots or running off along a slope and consequently creates erosion process. Filtration factors are a part of this process and threshold is moisture absorption when soil pores gradual full of liquid.

2. Experimental

Experimental researches were aimed at the determination of moisture after applying of liquid ameliorants and its impact on filter factors and soil porosity.

In Figure 1*a* is shown work tool for applying of liquid ameliorants for subsurface tillage set at the frame of KPG-250.

Work tool for applying of liquid ameliorants [6, 7] is a hat hoe, under the space of it set a device allowing to pulverize liquid ameliorants during the process of soil cultivation(Figures 1*b* and 1*c*).

a *b* *c*

Figure 1. Work tool for applying of liquid ameliorants, where a is work tool state on the frame KPG-250; b is open furrow after using of work tool; c is spreading of ameliorants on the furrow sole.

Experimental researches were conducted in Integrated Agricultural Production Centre "Trud" in Baturevskiy region of Chuvash Republic. Tilth-top soil is presented by heavy loamy chernozem.

Experimental was conducted on the place with smooth relief and minimal field slope at the speed of aggregate 6…8 km/h.

Applied liquid ameliorants are carbamide-ammonium nitrates.

Samples of soil for analysis were chosen as at the parcel cultivated by work tool for applying of liquid ameliorants so at the parcel cultivated by hat hoe without applying of liquid ameliorants.

Moisture of soil determined after two months after cultivation and applying of liquid ameliorants according to the National Standard 28268-89 "Soils. Methods for determination of moisture, maximum hydroscopic moisture and moisture of steady plant fading".

Measurement of moisture was carried out with a help of soil samples taken every each 10 cm to a depth of 0 to 30 cm. Each sample was placed into numbered aluminum small glasses weighed and then delivered to laboratory where it was kept for 7 hours with the temperature of 105 ºC in drying closet. After drying and cooling soil was weighed. Further on the base of received measurements was determined soil moisture of study sample in percentage correlation.

Determination of filtration factors and porosity was conducted according to National Standard 25584-90 “Soils. Laboratory methods for determination of filtration factors”.

3. Results and Considerations

After mechanized application of liquid ameliorants based on the results of experimental research of soil moisture W_w in different layers a graph of moisture variance was constructed (Figure 2).

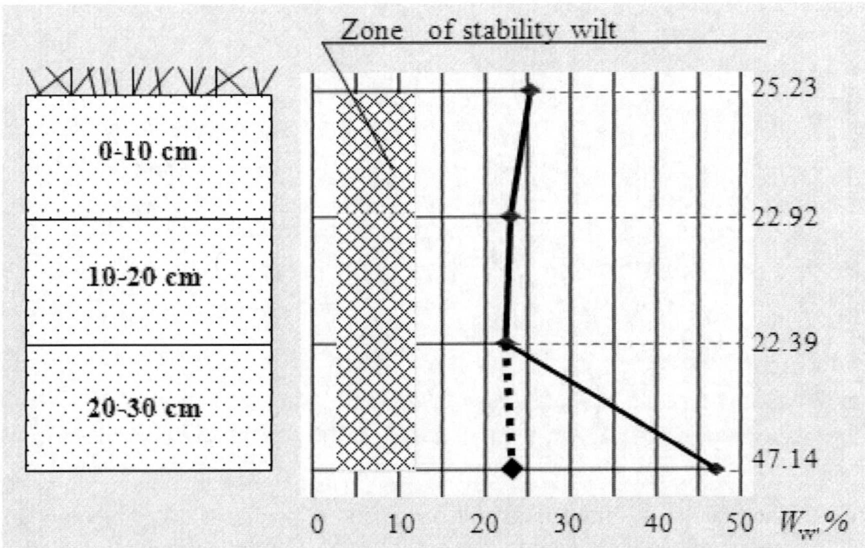

Figure 2. Soil moisture after cultivation by work tool for applying of liquid ameliorants.

Moisture values of soil samples on the surface decrease from 25.23% to 22.92% after cultivation in layers of 10÷20 cm (Figure 2). It so happens because of appearance of new pores filling out with air. High values of moisture are noticed in soil layer 20-30cm more than 47% as liquid ameliorants applied just in this zone. Cultivation by work tool without applying of liquid ameliorants values at the same depth no more than 23% (the Figure 2 shows it with the dashed line).

Also Figure 2 demonstrates with mark-making a zone of steady wilt of plants. So, sandy loam appears if moisture is 3.5…12%, for sandy clay is 1.5…4% and for loam is 12…20% [8].

According to research carried out on definition of filtration factors and soil porosity let to judge about qualitative soil preparation with applying of liquid ameliorants [9, 10].

Filtration factors allow characterizing hydraulic conductivity of soil under hydrostatic pressure force. Herewith a liquid permeated in a ground spreads and provides necessary access of moisture to roots.

The Figure 3 shows data of filtration factors in different soil layers after pass of tool for applying of liquid ameliorants.

Filtration factors taken from results of soil samples have more high values in the depth of 20-30 cm- from 90.6 to 106 cm/d where direct tillage and damage of soil layer by work tool happens (Figure 3). This process is characterized by crumbling of lower layer of soil. Also in formed space liquid ameliorants are pulverized and so affect with stream on the layer.

For comparison half values of filtration factors in different sources are presented for different types of soil. For sand it is 300…800 cm/d, sandy loam is 20…100 cm/d, loam is 1…50 cm/d [11].

Porosity is characterized by existence of pores which fill out with liquid or air. Their presence creates stable deposit of moisture concurrent with good air exchange [12–15].

The Figure 4 presents data about porosity of different soil layers before and after using of work tool for applying of liquid ameliorants. Soil layers were taken in 5 sample points.

The best porosity on the graph (Figure 4) pointed after soil cultivation in depth of 20÷30 cm. Right in this depth applied liquid ameliorants in the process of soil cultivation without soil overturning. Formed porous spaces fill out with air and nutrient elements of ameliorants. In higher slices of soil horizon values down fall because of subsurface tillage. Herewith tillage goes with cutting of layer and action of shank of the tool.

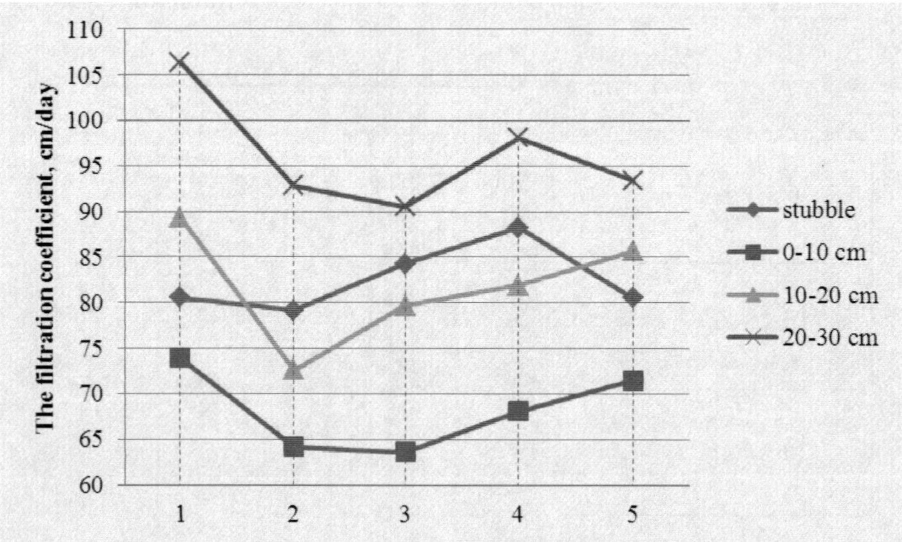

Figure 3. Filtration factors of soil samples taken from different places of experimental parcels before and after pass of work tool for applying of liquid ameliorants.

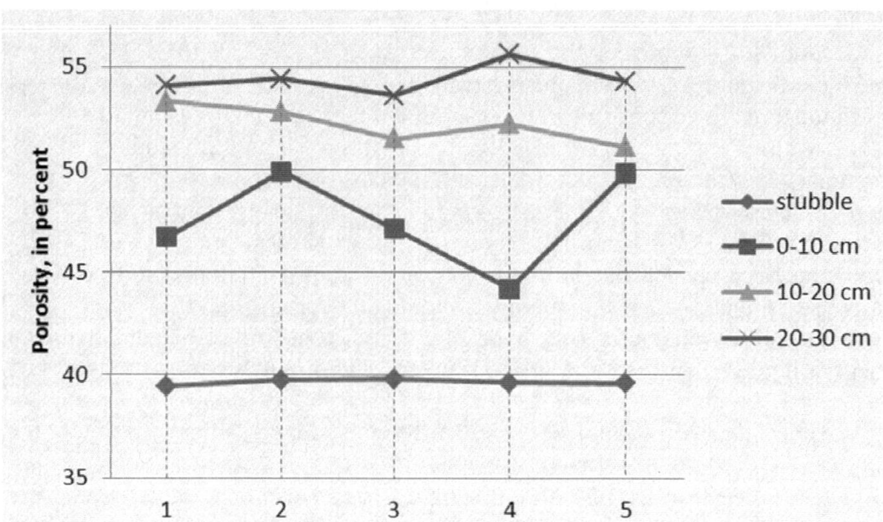

Figure 4. Porosity of different soil layers before and after using of work tool for applying of liquid ameliorants.

4. Conclusion

Applying liquid ameliorants created conditions for preventing plants from critical situation which can appear during the drought season characterized by a lack of past precipitations and temperature rise from a norm.

In this case, received values of filtration filters and soil porosity show qualitative soil cultivation for crop growing.

Thus, applying of liquid ameliorants lets save and increase soil moisture that helps plants to survive during the drought seasons.

References

[1] Sojka R. E. *et al* 2007 *Adv Agron* **92** 75–141 doi: 10.1016/S0065-2113(04)92002-0
[2] Williams B. G. *et al* 1966 *Aust J Soil Res* **4** 131–143
[3] Wang S. *et al* 2018 *Appl Surf Sci* **428** 1070–1078 doi: 10.1016/j.apsusc.2017.09.225
[4] Turchin F. V. 1972 *Nitrogenous nutrition of plants and applying of nitrogen fertilizers* (Russia: Kolos) (in Russian)
[5] Six J. *et al* 2002 *Plant and Soil* 241 155–176
[6] Maksimov I. I. *et al* 2010 Patent RF 24248829 *Byull Izobret* **26** (in Russian)
[7] Vasilyev S. A. *et al* 2014 *Bulletin of SSAU* **1** 55–58 (in Russian)
[8] Eze V. C. *et al* 2017 *Chem Eng J* **322** 205–214
[9] Semenov E. I. *et al* 2016 *Res J Pharm Biol Chem Sci* **7(1)** 1860–1868
[10] Mohayeji M. *et al* 2016 *J Taiwan Inst Chem Eng* **60** 76–82
[11] Chen Z. *et al* 2015 *Chem Eng Sci* **130** 254–263
[12] McDonough J. R. *et al* 2015 *Chem Eng J* **265** 110-121 doi: 10.1016/j.cej.2014.10.113
[13] Zhang J. *et al* 2011 *Bioresour Technol* **102** 7407-7414
[14] Mamyrbek A. B. *et al* 2016 International Journal of Applied Engineering Research **11** 11041–11045
[15] Hodgkinson J., Tatam R. P. 2013 *Meas Sci Technol* **24** 012004 doi: 10.1088/0957-0233/24/1/012004

The use of artificial intelligence`s elements to block the manifestations of individuals` behavioral activity going beyond the quasi-stable states

E L Loginov[1], V V Grigoriev[2], A A Shkuta[3], V Y Bortalevich[4], D D Sorokin[3]

[1] International Research Institute for Advanced Systems, Moscow, Russia
[2] Moscow State Institute of International Relations, Moscow, Russia
[3] Financial University under the Government of the Russian Federation, Moscow, Russia
[4] Central Economics and Mathematics Institute of the Russian Academy of Sciences, Moscow, Russia
E-mail: dima.dd.sor@mail.ru

Abstract. The monitoring of the metastable states of individuals as a bifurcation point of development of intellectual dynamics of behavioural activity allows us to identify a set of characteristics of coherent-resonant clusters of manifestations of biophysical factors and information-cognitive nature. Conscious changes in the functioning modes of brain and human nervous system can be achieved by electromagnetic influence or purely informational influence as well as by the combined effect of both these factors. The arrangement of complex processes of informational influence is required to penetrate to the level of conscious (semantic) and unconscious (mental, emotional, meditative) interpretation of events when it is necessary to ensure sufficiently stable cognitive-behavioural stereotypes of individuals and their groups. Convergent informational and electromagnetic effects facilitate the effect on the brain, including the use of neuro-linguistic programming for this element with correction or even a complete change of the reflexive matrix (matrix of key reflexive reactions) with a corresponding change and fixation of the personality-style in the personality events to block the release of behavioural activity of individuals beyond quasi-stable states.

1. Introduction

It is necessary to forecast the appearance of behavioural manifestations beyond quasi-resistant states as a source of increased risks of emergency situations. The use of factors of biophysical and informational-cognitive character creates opportunities for blocking the output of manifestations of behavioural activity of individuals beyond the limits of quasi-stable states, pushing the personality to active destructive actions. The assessment of the probability of critical intellectual dynamics in behavioural activity can be realized based on an analysis of the dynamics of the functional interrelationships of the psycho-semantic subjectivity of a person and various political, economic, social and other factors based on a multiparametric analysis of diverse data.

2. Highlighting of a set of characteristics of coherent resonance clusters of manifestations of biophysical and informational-cognitive factors

In fact, the behavioural activity of individuals and their groups is enlarged by a combination of a group of factors:

- biophysical factors that form the modes of the brain and the human nervous system;
- psycho-semantic characteristics of a person, determining the activity and forms of realization of the interests (basic attitudes) of a person in relation to external irritants;
- imprinted reflexive matrices - matrices of key reflexive reactions of a conscious and unconscious nature, determining the interpretation of events (incoming information);
- informational irritants - information coming from the external in relation to the individual, the environment in conjunction with the communicative characteristics of the channels of receipt of information [1].

Authors suppose monitoring of metastable states of personalities as a bifurcation point of development of the intellectual dynamics of behavioural activity and the resulting higher level processes (politics, economics, etc.) that are crucial for the state and society in identifying the multiparameter structure behavioural activity, allows to select a set of characteristics of coherence-resonant clusters of manifestations of biophysical and information-cognitive factors as part of a certain The ongoing activity of the supersystem "the bioelectrical activity of the brain - the psycho-semantic subjectivity of the personality - imprinted reflexive matrices - informational irritants".

The study of the properties of these associations (clusters) of manifestations of biophysical and informational-cognitive factors, which form the conditions for successful-perceived and unconscious influences of an individual, makes it possible to predict the appearance of behavioral manifestations beyond quasi-stable states as a source of increased parties of external forces or groups of intrasystem actors.

3. Evaluation of the state of intellectual dynamics of behavioral activity and the probability of its going beyond the limits of critical conditions

Especially important is the prediction and proactive exclusion (initiated failure) of the possibility of resonance of these factors based on technologies of distant influence, psycho-correction and psycho-sensing, taking into account the psychosemantic personal qualities of an extended nature (official and real political orientation, quality of its professional training, cultural level, interests, volitional qualities, internal motivation, etc.).

It is necessary to divide the transitional activity of the supersystem "brain bioelectrical activity — psychosemantic personality subjectivity — imprinted reflexive matrices — informational stimuli" so that each aggregated segment (personality group) of an ordered interconnection of functioning and interaction of distributed information objects, information networks and information consumers (work, leisure, personal life, participation in public or political activities, collective networks s communications, intellectual leisure, create, honours, etc..) was presented as a kind of one macro-object.

This macro-object includes many evolving subsystems, each of which corresponds to a local state of short-term relative equilibrium of the super-system, which can be characterized by a consolidating "convolution" of manifestations of biophysical and information-cognitive factors into cluster-forming structures.

To forecast a possible emergency situation due to peak behavioral activity of large groups of the population, it is necessary to form a package of monitoring models, including It is necessary to assess the state of intellectual dynamics of behavioral activity and the probability of its going beyond the limits of critical conditions as a source of increased risks of emergency situations. Estimation of the probability of the critical intellectual dynamics of behavioral activity can be realized on the basis of an analysis of the dynamics of the functional interrelationships of the psycho-semantic subjectivity of an individual and various political, economic, social and other factors based on a multi-parametric analysis of heterogeneous data obtained as part of constant monitoring. dynamically changing situation.

The complexity or impossibility of using a physical (or other substantive) or statistical approach in assessing the criterion indicators of aggregates of a possible database of the processes in question is most pronounced in relation to hard-to-form objects (psycho-semantic subjectivity of an individual), as well as in complex multifactor relationships of the assessed characteristics, in particular, indicators of the expected adequacy of the model of a complex system constructed using standard mathematical models and algorithms chesky units. Neural network modelling using information obtained during the analysis of [information and telecommunications] events that may be related to a particular person allows predicting the intellectual dynamics of behavioural activity, incl. allows you to get a summary of the summary information about the object (person), which is not explicitly available in information sources by analysing large arrays of complex structured data. This data can be obtained when monitoring the Internet and the dynamics of various telecommunication services (the dynamics of using channels and television programs, the dynamics of mobile message traffic in conjunction with current political processes, query keywords when searching the Internet, atypical activity of social networks communication etc.), as well as the dynamics of electromagnetic fields of natural and man-made origin in relation to the territorial mobility of a particular person.

4. Programming cognitive-behavioral stereotypes as a basis for event-based chains of personal actions

Identification of the psycho-semantic qualities of an individual based on an analysis of their interests and preferences regarding the viewing of information programs, activity in social networks, the choice of computer games, etc. as a set of electronic content data, it allows to form a cognitive-reflexive model adapted to a specific personality with embedded elements of neuro-linguistic programming for identifying and interpreting what is happening, serving as a source of actions of this personality within the framework of world-view clusters that are virtually and organizationally organized for this case. and professional patterns of interpretation of the surrounding reality. That is, to ensure the identification of qualities and personality programming.

Management of the considered information-cognitive processes necessitates the development of new approaches to the development of information systems based on the convergence of the process of perception of information stimuli. In these approaches, the manipulative concentration of the attention of the individual and the interpretation of information should be “combined” into a single composition for operating with semantic structures, translating the semantic transcription from the conscious to the unconscious level using neuromodulation — the vector stimulating activity of various subsystems of the brain and nervous systems through convergent information and communication and electromagnetic effects.

There is a problem of “input” and “output” of data from a specific (identified in the framework of monitoring electronic content) identity, which can be implemented using a software neural interface used in any intelligent electronic communicative devices.

Such an interface creates a basis for programming cognitive-behavioral stereotypes as a basis for event chains of personal actions for proper functioning of social control systems in difficult conditions with a large component of uncertainty in order to eliminate the possibility of interception of control of social elements from external forces or groups of intrasystem actors [2].

The arrangement of complex processes of informational impact is required for penetration to the level of conscious (semantic) and unconscious (mental, emotional, meditative) interpretation of events when it is necessary to ensure sufficiently stable cognitive-behavioral stereotypes in individuals and their groups.

It is required that the incoming information supplant one of the future event-based nets (actions) of a specific person in the network of causal relationships between interpreted event chains. This approach is implemented in the framework of the configuration of interactive communication processes used for the exchange of information between people with the establishment of feedback and taking corrective measures based on neuro-information systems and services for various purposes.

5. Using neuromodulation to obtain the desired cognitive effect

Selecting the parameters of neuromodulation individually for an individual, one can cause inhibition or activation of brain structures, thereby achieving the same cognitive effect. Especially high efficiency of neuromodulation manifests itself with electromagnetic effects on the brain and the human nervous system [3].

As an example: in order to relieve a patient from hand-shake in Parkinson's disease, high-frequency impulses (100 Hz and more) are used for stimulation, and in order to ease the severity of phantom limb pain - low-frequency pulses (25—30 Hz) .

The use of elements of artificial intelligence allows for dynamic adjustment of neuromodulation based on monitoring the electrical activity of the brain of a particular person [4, 5]. When electrophysiological signs appear, manifestations of aggression as a reaction to the person receiving an identifiable information package containing information about the surrounding reality, a stimulator of electromagnetic effects (for example, built-in an electronic communication gadget: computer, smartphone, smart TV, etc.) starts automatically and immediately sends stimulating electromagnetic impulses that suppress a sharply negative reaction, pushing the person to active destructive effects.

When adjusting the electromagnetic effect to the activity of the key points of the brain, similar effects can be achieved by remote influence, for example, from a cell phone, with a less pronounced effect on segments of the nervous system - - from an intellectual medical bracelet, intellectual paired wristwatches, etc. At the same time, a point dynamic electromagnetic or laser effect on individual biologically active points of the nervous system (for example, being on the wrist) can also quite effectively suppress the psychological irritation of the personality, switching its active attention from the interpretative (conscious, semantic) reaction to the information received - on the physiological (unconscious, reflexive) reaction to changes in the activity of a specific nerve point (nerve ending) [6].

6. Results

Conscious change in the modes of functioning of the brain and the human nervous system can be achieved by electromagnetic action or by purely informative influence as well as by the combined influence of both these factors. An informational impact is the receipt of an information package that causes the brain to temporarily “crash” by analogy with a computer.

At the same time, the brain enters a trance state in which it is exposed to an increased uncontrolled perception of information received from outside (analog: “Gypsy hypnosis”), which forms a specific vector of personal activities of the person, realized in the near or distant period [7].

That is, convergent informational and electromagnetic effects facilitate the impact on the brain, including the use of neuro-linguistic programming for this element, with correction or even a complete change of the reflexive matrix (matrix of key reflexive reactions) with a corresponding change and fixation events.

Acknowledgment

The theses were prepared with the financial support of the Russian Foundation for Basic Research (Project No. 19-07-01066 “Creating an artificial intelligence system as a component of a digital platform for monitoring the behavioral activity of large groups of people based on the application of methods for analyzing large weakly structured data, building thematic models with cognitive and multi-parameter semantic interpretation, exploratory search and collaborative filtering with convergent control”.

References

[1] Loginov E.L., Raikov A.N., Eriashvili N.D. 2015 *Public Service and Personnel* 1 89–95
[2] Ageev A.I., Loginov E.L. 2017 *Economic strategies* **19** 124–139
[3] Marchenko V.G., Zaichenko M.I. 2014 *Journal of Higher Nervous Activity of I.P. Pavlova* **64** 255
[4] Golubev V. 2016. *Information as a mapping of objects of the world in the cerebral cortex. Information technologies* **22** 233–239
[5] Novikova S.I. 2015 *Modern foreign psychology* **4** 91–108
[6] Loginov E.L., Raikov A.N., Shkuta A.A. 2018 *Neurocomputers: development, application.* 9 34–45
[7] Bogdanov A.V., Galashina A.G., Tukaev R.D. 2016 *Psychiatry, psychotherapy and clinical psychology* 1 (12) 133–144

IOP Conf. Series: Materials Science and Engineering **516** (2019) 012029 doi:10.1088/1757-899X/516/1/012029

Detection of alive people with the Radiodozor-M device

S E Shipilov, R N Satarov, I S Tseplyaev , V L Khmelev, I S Fedyanin
Tomsk State University, Tomsk, Russian Federation
E-mail: s.shipilov@gmail.com

Abstract. The article discusses the approach to the restoration of radio images of objects behind dielectric barriers. The approach is based on the method of mathematical focusing of reflected ultra-wideband signals. The sounding system is a radar with a synthesized aperture. Here we present a method for improving the quality of radio images based on additional processing of received signals. The article also presents the results of numerical simulation and experiment.

1. Introduction

Dozens of catastrophes happen every year (traffic accidents, industrial disasters (accidents at nuclear power plants, oil pipelines)) and natural disasters (earthquakes, landslides, avalanches, fires, etc.). All had victims, the number of which is thousands and hundreds of thousands of people. In addition, mining mine collapses and avalanches happens almost every week, as well as more and more often happens terrorist attacks. Regardless of the scale of the incident, there is always the problem of saving lives. Searching and rescuing people in disasters, accidents, or emergencies is a very difficult task. Obviously, the task for creating locating devices for alive humans under collapse is relevant.

Most preferred are radio wave tomography techniques. The solution of object detection requires a development of special software and hardware with active location tomography. We believe that the use of clocked antenna arrays is promising, as was shown in the articles [1, 2, 4].

The developed device "Radiodozor-M" detects moving and motionless humans in real time behind reinforced concrete, brick walls, multi-layered structures and other building structures in smoke conditions, under the rubble. The device gives tactical advantages in rescue operations: accurate determination of location and distance, both to moving and stationary objects. The two-dimensional image shown on the screen allows you to quickly perceive in the situation and make decisions about further actions. Measuring the breathe rate of immobile people allow you to take emergency actions for saving people with low signs of life (weak breathing).

2. Signal processing

Algorithm for recovering a radio tomographic image based on focusing UWB signals reflected from objects. The main idea of data processing is the synthesis of aperture and focusing. We sums all received signals from different points of a scanning space for each given point of the image in phase.

This device implements the differential-difference method for detecting moving people, based on filtering signals from stationary objects. The essence of the method is calculating a difference of signal in the adjacent periods.

$$dS(t, T_i) = S(t, T_i) - S(t, T_{i-1}),$$

where $dS(t, T_i)$ – filtered signal, $S(t, T_i)$-registered signal, $S(t, T_{i-1})$ - a signal, which one was received in the previous period of time. The signals, reflected from motionless, background objects of the scene, will be the same for different periods and self-compensate after subtracting. The difference of signals *dS* received from a moving object will non-zero. Next, the resulting set of perspectives will use for building a radio image of moving objects.

In cases where a person is stationary, you can determine his presence behind an obstacle by observing the movement of his chest and stomach. Look to Figure 1. The signal reflected from the chest in a breathe out position S_2, is different from the signal, which one reflected from a breathe in position S_1 by a time delay $T = 2\Delta x / c$, where Δx - chest movement during change breathe in/out position.

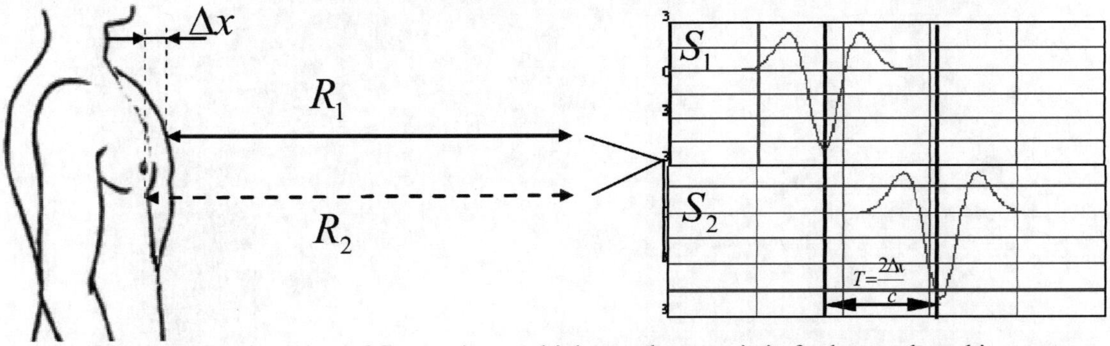

Figure 1. Registration of fluctuations, which are characteristic for human breathing.

The device provides the transformation of data, which are obtained in UWB probing, to detect characteristic fluctuations for human breathing.

The complex amplitudes of the spectra of neighboring signals will differ from each other only by the multiplier $\exp(i\omega T)$:

$$SS_2(\omega) = SS_1(\omega)\exp(i\omega T)$$

The selected phase from the resulting expression will contain information about fluctuations of the chest.

3. Device description

Figure 2 shows developed the radio-tomograph UWB "RadioDozor-M". The device includes an UWB clocked antenna array [3], a minicomputer-based hardware module for interfacing an antenna array with a RF emission module, and registration of signals. The RF module has a wide dynamic range and provides a range for detection of moving objects up to 20 meters. The developed software for the radio-tomograph computes a radio image of moving and motionless people with a frequency about 5-10 frames per second. For motionless humans, in addition to their location, is possible to determine the frequency of their breathing. The device provides for the computing of the radio image of the scene with frequency 10 frames per second.

IOP Conf. Series: Materials Science and Engineering **516** (2019) 012029 doi:10.1088/1757-899X/516/1/012029

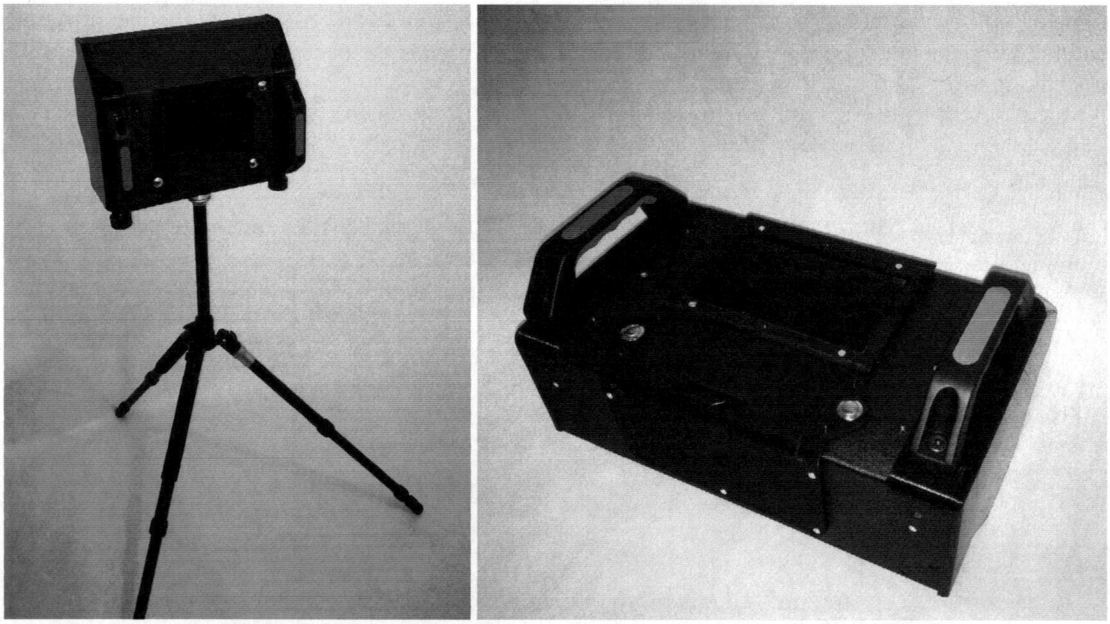

Figure 2. The appearance of the radio-tomograph "Radiodozor-M".

4. Results

Figure 3 shows results of sounding for two motionless humans. 30 seconds data collection is enough to accumulate an information for calculation locations and breathe rates of each person.

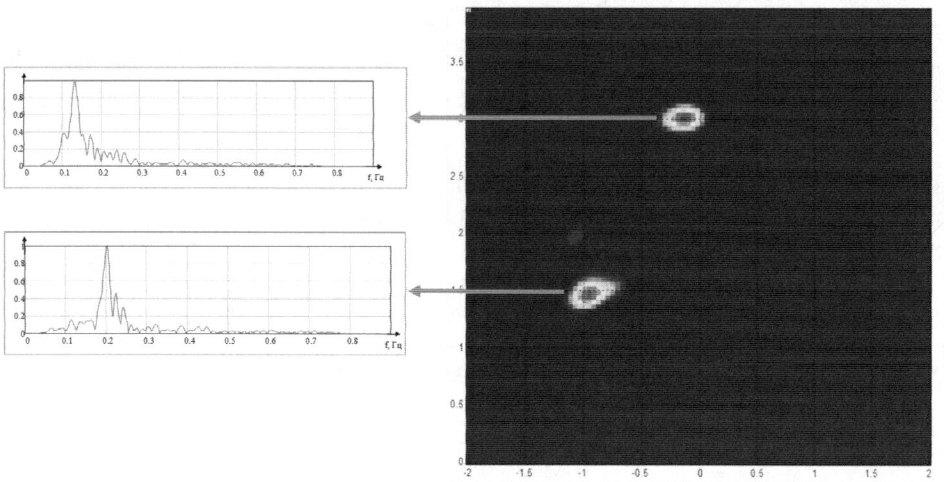

Figure 3. Radio-image of two motionless humans and their breathe rates.

5. Summary

Developed methods of registration and tomographic processing of radio wave measurements of objects, which hidden behind dielectric barriers by using large-aperture synthesis focusing. The performance of the proposed approach verified by experimental data processing, that were obtained with using ultra-wideband radiation. The experiment involved two volunteers. In this article, we presented a developed method of filtering signals reflected from fixed objects to select moving objects behind dielectric barriers.

Acknowledgments
This work was supported by the grant of the Russian Science Foundation №18-79-00294

References

[1] Shipilov S.E. *et al* 2018 *Russian Journal of Nondestructive Testing* **54**(6) 443-447 doi: 10.1134/S1061830918060086
[2] Shipilov S.E. *et al* 2017 *Optical and Quantum Electronics* **49**(10) 1–12 doi: 10.1007/s11082-017-1172-7
[3] Satarov R.N. *et al* 2013 *Russian Physics Journal* **55** (8) 884–889 doi: 10.1007/s11182-013-9896-8
[4] Gallagher K.A. *et al* 2017 *Electronics (Switzerland)* **6**(2) 884–889 doi: 10.3390/electronics6020030

IOP Conf. Series: Materials Science and Engineering **516** (2019) 012030 doi:10.1088/1757-899X/516/1/012030

Information and analytical provision of the satellite monitoring system

A Kadochnikov[1], O Yakubailik[1,2]

[1]Institute of Computational Modeling SB RAS, Krasnoyarsk, Russia
[2]Siberian Federal University, Krasnoyarsk, Russia
E-mail: oleg@icm.krasn.ru

Abstract. This scientific work discusses the problems of formation of information and analytical support of satellite monitoring tasks. The information and computational technologies for regional level monitoring systems, peculiarities of their organization and software architecture are considered. A brief description of the completed work, aimed at development of software for the satellite image catalogue, formation of a regional satellite data center is given.

1. Introduction

Intensive current development of natural resources in Russia inevitably forms the problem of developing new modern means of appropriate environmental control, monitoring of changes in the state of the environment. The existing nowadays ideas about what kind of information support system should be used to solve these problems, need to be clarified and reinterpreted due to the significant progress in the field of Internet technologies during the recent years. We should also note the dramatic changes in the situation with high-resolution satellite imagery: the information becomes available almost daily, which opens up unprecedented earlier opportunities [1–4].

The traditional monitoring system today includes an information and measurement network, a data transmission network, a monitoring center and a network of user terminals, including mobile terminals. As a rule, extensive databases of spatial data, remote sensing data (RS), and analytical support tools are used, all this means allow real-time mapping of current and predicted changes in state, resources and objects of the controlled territory, based on mathematical modeling methods and by solving direct and inverse transfer problems. A modern automated information system should also provide its users with effective means of information interaction [5, 6].

Web technologies can provide a new quality of monitoring systems, bring them into a new level, and improve dramatically the task solving effectiveness. On the other hand, modern geoinformation systems (MIS) provide a huge range of tools, including measurement tools, advanced tools for topological and spatial analysis, as well as tools for working with external databases. In addition, GIS provide work with any reporting materials, remote sensing data (aerial photographs, space images, seismic sounding data, etc.) [7, 8].

On the basis of the analysis of the current information and computational technologies and the corresponding functional requirements of the modern monitoring systems, it can be concluded that the most appropriate technical model of such systems is a multicomponent software and technology complex, i.e. a distributed geoinformation system based on the modern web interfaces and services, it provides high-level interactive user and program interfaces to geospatial data (web GIS). The client software, forming the basis of automated workstations for users (AWs), is created on the basis of a standard web browser that can work on personal computers, tablets, and smartphones. It is also necessary to create specialized software that provides, among other things, automatic loading of monitoring data from remote sensors and observation posts [9]. The system server components are usually built on the basis of relational DBMS, business logic servers, tools for providing spatial data stores. Given the current realities, we should consider the distributed storage systems, specialized data centers of the regional level in the "cloud" architecture. The specific configuration of the systems depends on the tasks that should be solved [10, 11].

This scientific work studies the experience of research and development carried out on the topic under consideration at the Federal Research Center "Krasnoyarsk Scientific Center of the Siberian Branch of the Russian Academy of Sciences" (FRC KSC SB RAS) is considered.

2. The system software architecture

As a program basis of this GIS, a geoinformation web server was used, the modular architecture of which allows to develop the system already in the process of its exploitation, by adding new means of access to data, and also improving the existing ones. The entire development is performed on the basis of free and open source software – both in the GIS part and in all other development components.

The basic functional components of the considered system are:

- database of characteristics of monitoring objects;
- catalog of information resources and spatial metadata;
- a set of auxiliary web services.

In particular, the catalog of information resources contains objects of various types:

- structural elements (organization, server, folder);
- the security system elements (user, role);
- information resources (cartographic layer, map, attribute data, analytical service with web access, publication, etc.);
- elements of classification of information resources;
- information and navigation elements (HTML documents).

This GIS provides the user with means for filling and editing the system data and metadata, searching and classifying of cartographic resources, web services for direct access to the data based on standard WMS / WFS protocols, and analytical processing possibilities [12, 13].

The server part of the system is implemented using the MVC design pattern (model-view-controller) [14]. Using this architecture involves splitting of application data, the user interface, and control logic into three separate components, i.e. a model, a view, and a controller, so that each component can be modified independently. In conditions of the system constant modernization, continuous specification of technical requirements and task statement, these possibilities become very actual [15, 16].

In the process of the considered system development, several new software components and libraries, that can be used (replicated) in other projects, were created. These are the elements of the user interface, services for working with geospatial DBMS, the applied cartographic web services, etc. [17].

The UMN Mapserver and MapGuide Open Source software tools provide the basic web-GIS functionality. Geospatial DBMS PostgreSQL/PostGIS, Drupal CMS-based web content management system and DokuWiki wiki system, a whole family of libraries of building the user interface and other system components for JavaScript programming languages, PHP – JQuery, Fusion, TinyMCE, ExtJS, MapScript, and many others – are also used [18].

3. Peculiarities of satellite data catalogue implementation

Priorities for research and development carried out at the FIC of the KSC of the SB RAS at this stage were related to creation of software tools for working with satellite data catalogs. The developed software tools provide the solution of the priority tasks for the operational processing of remote sensing data from the new UniScan satellite reception facility at FIC KSC of the SB RAS, which was put into operation in spring, 2017, as well as via a dedicated data transmission channel from the Siberian Regional Center for Remote Sensing of Roscosmos based on JSC "Information Satellite Systems" named after academician M.F. Reshetnev in Zheleznogorsk (JSC ISS).

The basic software of the receiving complex UniScan forms a set of standard products of Level 1, i.e. channel images (brightness of spectral channels). Further tasks related to the extraction of the useful information, thematic data processing, are solved separately, by means of special software.

The priority task of the first stage was creation of a system of web-visualization of satellite information available in the formed regional archive of satellite data. It is based on sets of specially created collections of multiscale snapshots, with possibility to select in the web interface the combinations of displayed channels on a small scale, and, at the same time, with the availability of detailed data at the maximum possible spatial resolution. The limitation in the choice of channel combinations at a detailed level is associated with the disk space economy. Thus, a compromise is realized between the class system "we work with rough Quik Looks" and the level system "anything with maximum detail". The created web interface provides viewing of the archive of satellite images. It provides the ability to select a spacecraft sensor, to select a predefined set of channels and products from each snapshot for simple data analysis. The advantages of the created software module allow to combine any combination of channels available in the image without additional configuration of the server software.

To create the color images, LUT (Look Up Table) is used, this is a kind of "table of corrections" for making changes to each of three channels. Previously, to improve the image quality, a spectral transformation was used, it is based on work with a spectral diagram showing the relationship between the number of the image pixels and the spectral brightness values.

With spectral transformations, such parameter as the contrast changes. To increase the contrast, a linear histogram stretching has been used, meaning that all brightness values are assigned with new values in order to cover the entire possible range from 0 to 255. LUT allows to change the brightness value of the image points when converting a picture to 8 bits of 16 bits by linear interpolation between them. This allows to make the picture more vivid. An example of a web interface is shown in Figure 1.

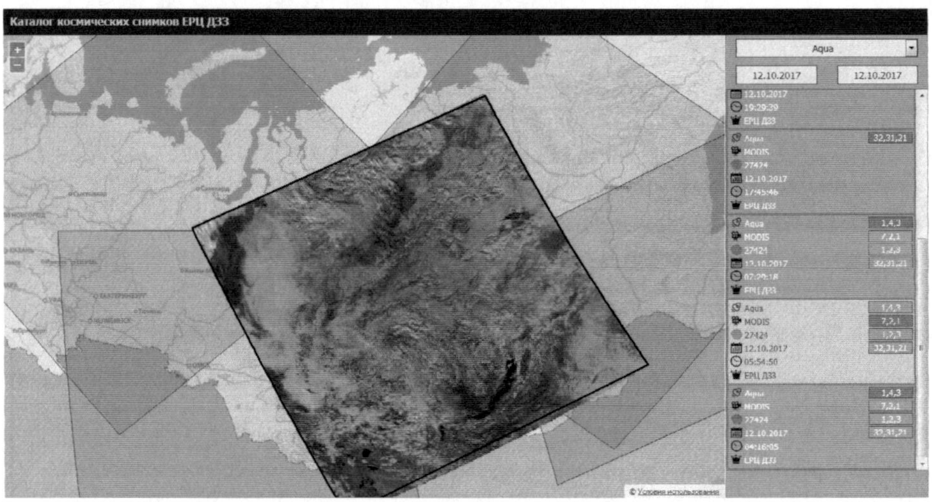

Figure 1. Web-based directory of satellite imagery.

Tactical development tasks are associated with implementation of methods and algorithms for processing satellite data. Technological features of working with raster data are predetermined by the need to quickly display large files. Experiments with various formats and their parameters have shown that the optimal choice in terms of minimizing the display time in a web browser is the TIFF format with georeferencing (Geo-TIFF), with application of the TILES structure (the image consists of a set of independently stored one-type fragments), with previously computed pyramidal (survey) layers OVERVIEW. All the above processing steps are performed by the GDAL library utilities [19].

To generate a number of information products based on satellite imagery, it is assumed that indexed rasters are widely used, where each image pixel contains an index, and the file header, i.e. a table with color values for all palette indices (Figure 2).

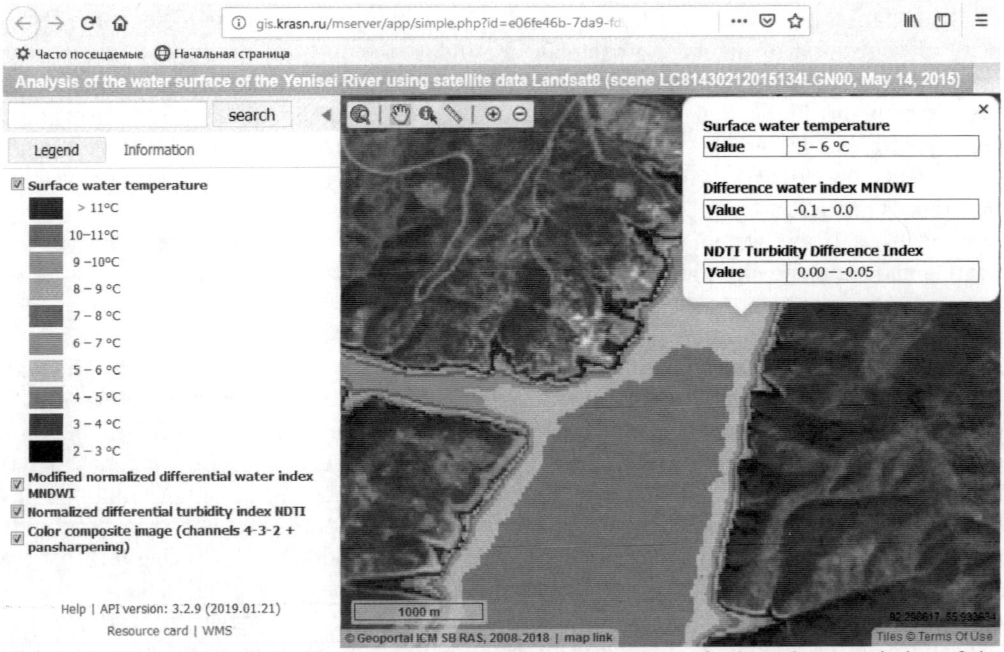

Figure 2. Thematic maps of water temperature and a number of other characteristics of the Krasnoyarsk reservoir, based on Landsat-8 satellite data.

The developed software tools provide the solution of the priority tasks for operational processing of the incoming remote sensing data, their cataloging, and interactive visualization with use of a web application. The following data were considered: MODIS TERRA / AQUA, Suomi NPP, Meteor-M No. 2 CMCC.

To make a quick and easily search in the satellite data catalog, to minimize the load on the server software and hardware, a set of server applications for preliminary processing of the satellite data, including the following processing steps is prepared:

1. Converting the original raster data to GeoTiff format with transformation of the original projection into the Lambert Azimuthal Equal Area (EPSG code: 3576), which is mandatory for functioning of the OGC services (Open Geospatial Consortium). EPSG Geodetic Parameter Dataset is a structured set of data, coordinate systems and coordinate transformations accessible through the Internet registry (www.epsg-registry.org).
2. Creation of a color image for different display scales, consisting of three spectral channels. Such images will be used to view the satellite image in detail, maintaining the original resolution. For MODIS, day and night shots are combined as a basis for a color RGB image: in the daytime shooting time, channels 7-2-1 are used in the range from 0 to 100% albedo with a gamma of 1.8; at night channels 21-31-32 in the negative display are used from 235K to 305K with gamma 1. For the spacecraft Meteor-M No. 2, the base color image contains a red, green and blue channel, calculated on them and the near infrared channel of the CMMS device.
3. Creation of a multi-channel image to show Quicklooks in a web application for different display scales, but with a lower resolution (up to 1 km per point). Unlike the basic color image, the "Quicklinks" in a web application can be displayed in groups, depending on the user's choice,

IOP Conf. Series: Materials Science and Engineering **516** (2019) 012030 doi:10.1088/1757-899X/516/1/012030

whereas the basic color image will be active only for one selected picture. For MODIS spectroradiometer, the multichannel image contains 1-4, 7, 21, 31, 32 channels. Combinations of these channels allow the construction of images with natural colors for the human eye, pseudocolor images for classification of snow and ice, images for determining fires and smoke, etc.

4. Products for the spacecraft Meteor-M # 2 are in the form of a raster multichannel image in GeoTIFF format with NDVI indices (Normalized Difference Vegetation Index) and NDWI (Normalized Difference Water Index). The NDVI index is designed to detect and assess the intensity of vegetation of plants. To calculate the index, the spectral brightness values in the red and near infrared spectral ranges are used. The NDWI index shows the moisture content in the soil and leaves of plants. To calculate the index, the spectral brightness values in the green and near infrared spectral ranges are used. The NDWI index shows the moisture content in the soil and leaves of plants. To calculate the index, the spectral brightness values in the green and near infrared spectral ranges are used.
5. Conversion from 16-bit format to 8-bit, more suitable for displaying satellite imagery in a web application and requiring significantly less resources in the storage system.

4. Prospects for Implementation

Creation of an effective software and technology tool for objectives of the regional satellite monitoring has significant prospects. The strategic task of research and development in this context at FIC KSC SB RAS is solving the actual applied problems, related to the current and future needs of the Krasnoyarsk Territory, based on the use of the remote sensing data, the modern spatial data infrastructure, for effective socio-economic and innovation development, improving the level of competitiveness and safe living. The regional center for satellite monitoring, in conjunction with other regional and federal information systems, should contribute to solving the following tasks of the state authorities and local self-government of the Krasnoyarsk Territory:

- creation of information and analytical tools for making well-founded management decisions based on an analysis of operational and objective information from space monitoring and other sources;
- increase of the efficiency of the exchange of geospatial data between the registered monitoring center and the government of the Krasnoyarsk Territory;
- provision of operational work with the order, collection, storage, and display of space monitoring materials on the territory and individual objects of the Krasnoyarsk Territory;
- implementation of possibility to monitor the land resources, carry out transport management, comple with the terms of licensing agreements for extraction of mineral resources, and also environmental management systems, detection of unauthorized disposal of domestic waste, selection of investment sites;
- realization of possibility to collect and display data on incidents and events in the region on a specialized geoportal;
- providing a quick search and ordering the archival materials of satellite imagery, stored in the archive of space and attributive information;
- display by means of the geoportal of the ready geospatial data sets about various territories of the regional infrastructure of the Krasnoyarsk Territory;

The expanded activities cover a wide range of areas, i.e. water and forestry, prevention and liquidation of emergencies, agriculture, ecology and nature management, land and real estate, territorial planning [20].

During the 2017, specialists from the FIC KSC SB RAS and JSC ISS conducted consultations with the sectoral ministries of the Krasnoyarsk Territory and subordinate institutions. As a result, a preliminary list of priority tasks have been identified, these objectives are supposed to be made "pilot projects" in the Krasnoyarsk Territory. Among them are:

- Monitoring of flood and ice conditions in the areas of transport highways, water transport routes of the Yenisei Basin;

- Monitoring system of flood and ice conditions in the area of the 2nd stage of the Novo-Angarsk concentrator;
- Monitoring systems of specially protected natural territories of the Krasnoyarsk Territory; areas affected by the oil and gas industry and landfills of solid domestic waste;
- Operational twenty-four-hour monitoring of forest fires in selected (agreed) areas in the Krasnoyarsk Territory.
- Mapping the agricultural lands for the purpose of precision farming.

One of the possible mechanisms for infrastructural support of these tasks is formation of a "regional priority project", secured by targeted financing in accordance with the Krasnoyarsk Territory law on priority projects (programs).

5. Conclusion

The system of satellite monitoring opens new opportunities in solving important tasks of the regional level, which it is impossible to settle by other means today. The effectiveness of the proposed approaches is based on modern computer technologies, software, new satellite data, and methods for large data processing. The use of the remote sensing data and technologies can radically improve the efficiency and quality of decision making.

References

[1] Bastin L. *et al* 2013 *Ecological Informatics* **14** 9–16
[2] Christian Heipke 2010 *ISPRS Journal of Photogrammetry and Remote Sensing* **65** 550–557
[3] Huayi Wu *et al* 2011 *Computers & Geosciences* **37** 485–494
[4] Patroumpas K. *et al* 2015 *Web Semantics: Science, Services and Agents on the World Wide Web* **35** 53–62
[5] Díaz L. *et al* 2012 *Applied Geography* **35** 448–459
[6] Foerster Th. *et al* 2010 *Computers, Environment and Urban Systems* **34** 79–88
[7] Maguire D. J. *et al* 2005 *Computers, Environment and Urban Systems* **29** 3–14
[8] Blower J.D. *et al* 2013 *Environmental Modelling & Software* **47** 218–224
[9] Pickard B. R. *et al* 2 015 *Ecosystem Services* **14** 45–55
[10] Rubén Béjar *et al* 2012 *Computers & Geosciences* **46** 66–72
[11] Ling Xue *et al M 2013 Mathematical and Computer Modelling* **58** 480–488
[12] Yakubailik O.E., Popov V.G. 2009 *Computational Technologies* **6** 116–126
[13] Kadochnikov A.A., O.E. Yakubailik, 2015 *NSU Journal of Information Technologies* **1** 37–45
[14] Rautenbach V. *et al* 2013 *Computers, Environment and Urban Systems* **37** 107–120
[15] Ayushi Vyas *et al* 2016 *MATEC Web of Conferences* **57** 05002
[16] Xifang Jin *et al* 2015 *MATEC Web of Conferences* **22** 04027
[17] Kadochnikov A.A. 2014 *InterCarto* **20** 188–196
[18] Matveev A.G., Yakubailik O.E. 2012 *Vestnik of SibGAU 2* 48–54
[19] Kadochnikov A. *et al* 2015 *Scientific GeoConference SGEM* **1** 487–494
[20] Yakubailik O.E. *et al* 2009 *Vestnik of SibGAU* **25** 61–66

Hydraulic drive of vibration stand for testing the robotic systems units by random vibration method

A I Nizhegorodov[1], A N Gavrilin[2], B B Moyzes[2], K A Kuvshinov[2]

[1] Irkutsk National Research Technical University, Irkutsk, Russia
[2] Tomsk Polytechnic University, Tomsk, Russia
E-mail: nastromo_irkutsk@mail.ru

Abstract. This paper reviews the development of hydraulic drive of vibration stand for testing the robotic systems units (technical products) for vibration resistance by random vibration method. The physical laboratory model was developed for studying the dynamic properties of a test stand which allows to realize smooth sweep excitation frequency f1 within the range of 8...56 Hz and detect resonant modes. The new design allows adequate simulation of vibration during transportation and using gapless narrow-band spectrum of random vibrations from power units. The stand allows to implement a large number of test modes that demonstrate the prospects of its application for testing technical products by random vibration method.

1. Introduction

It is known in mechanical engineering that a significant number of products failures during operating are caused by vibration during transportation and from power units. Dynamic loads caused by operation are significant and can lead not only to functional disorder (loss vibration resistance) of hydro-mechanical, pneumatic, electrohydraulic and similar devices, but can also result mechanical failure of structural components of such equipment (loss of vibration strength). Before launching the production, devices must be tested in more extreme conditions than the actual operational ones [1–8].

One method of vibration tests is carried out as follows: the smooth flat pattern of the excitation frequency in the specified range is used to determine the present frequency of the vibrations of the test elements [9, 10]. After such scanning, the product shall be subject to a vibrational load in the local resonance mode. By the result of the tests it is decided whether changes to design or modifications should be fulfilled [1]. Another efficient method is the method of amplitude-modulated oscillations [1, 3] with assistance of which the harmonic structure can be presented by a wide line spectrum [12, 13] thus adequately simulating the transport vibration and the vibration of operating power units of technological machines. Amplitude-frequency modulation test is the most efficient when amplitude fluctuation is supplemented by changed carrier excitation frequency [14–17]. The maximum vibration testing effect [9, 10] can be achieved when implementing the continuous spectrum with a given width range. Any technical object is characterized by the spectrum of natural frequencies, which can have a wide frequency range depending on its size and structure. At transport vibration, the elements of various technical devices of transport and operational machines are impacted simultaneously by the dynamic within the given range. Therefore, the testing method of random vibration is the most difficult from the technical point of view [1, 2, 18–20].

IOP Conf. Series: Materials Science and Engineering **516** (2019) 012031 doi:10.1088/1757-899X/516/1/012031

2. Test stand for excitation of random vibration

Random vibration test method can be carried out in a testing laboratory by using a hydrostatic vibratory drive of linear excitation. Vibration test stand (Figure 1) differs from the construction described above [21], by the presence of:

- test platform *4*, hydrocylinder *3*, which rod is connected with a spring *5* providing a reverse motion of the platform and its setting to its original position when the initial pressure P_0 changes in the hydraulic system of oscillating circuit: the piston – a shell – the test platform;
- reverse two-position direction valve *9*;
- servo-motor No.2 controlled by a random signal generator through the direction valve *9* which provides parameters change a_1 and a_2 of generator signal w_{x1} (1).

Hydraulic power drives the eccentric *6* which acts on the generator piston. In this case the alternating fluid flow w_{x1} is displaced from the shell 1 into the hydraulic cylinder *3* of platform *4*, raising the platform vibration.

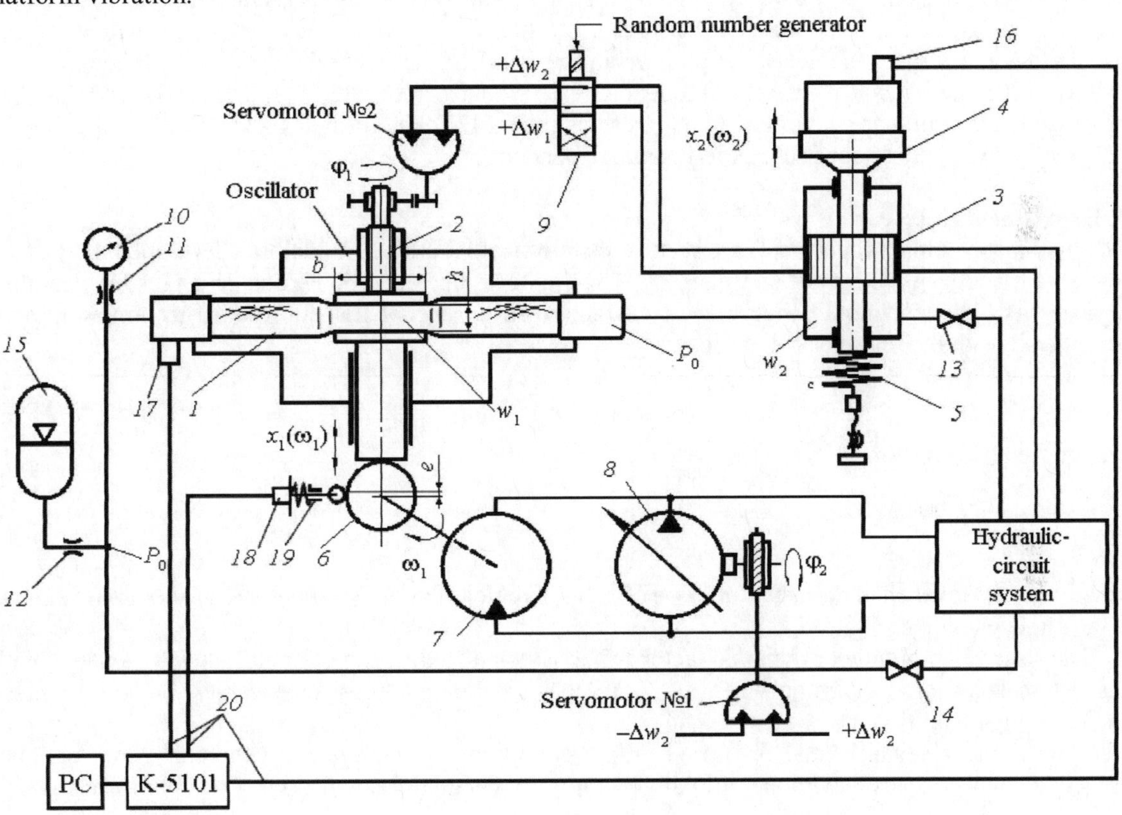

Figure 1. Schematic diagram of the test stand for random vibration excitation:
1 – preloaded elastic shell; *2* –preloaded regulator; *3* – hydro-cylinder; *4* – test platform; *5* – spring stiffness; *6* – eccentric; *7* – hydraulic motor; *8* – pump; *9* – reversing two-position valve; *10* – pressure gauge; *11*, *12* – throttles to prevent pressure pulsations in pressure gauge; *13*, *14* – valves; *15* – hydro-pneumatic accumulator; *16*, *18* – accelerometers; *17* – oil-pressure sensor; *19* – device for accelerometer *18*; *20* – data line;
K-5101 – vibration measurement module; PC – a laptop;
b и *h* – shell preload options; *c* – spring stiffness *5*; w_{xi} – fluid volume in the cavities;
Δw_{xi} – alternating fluid flow; $x_1(\omega_1)$, $x_2(\omega_2)$ – the laws of generator pistons motion of oscillations and hydrocylinder; φ_1, φ_2 – rotation angles of the preload regulator *2* and the regulator of displacement volume of the pump *8*, accordingly; ω_1, ω_2 – angular speed of shaft rotation of hydro-motor *7* and test platform *4* oscillations.

IOP Conf. Series: Materials Science and Engineering **516** (2019) 012031 doi:10.1088/1757-899X/516/1/012031

The law of actuating fluid displacement is described by the quadratic function [21]:

$$w_{x1} = a_1 \cdot x_1(\omega_1) + a_2 \cdot x_1^2(\omega_1) \tag{1}$$

where $x_1(\omega_1)=e{\cdot}\sin\omega_1 t$ – input signal (generator piston movement); e – the generator shaft eccentricity; a_1 and a_2 – geometric constants of the elastic shell [21].

The dynamic structure of stand in question shown by the diagram is [21].

Accelerometers, the data line, the vibration measurement module and the laptop are the part of the mobile complex for rapid diagnosis of the technological system elements, which principle of operation is described in detail. The pressure measurement module has been added.

The accelerometers are installed on the device for accelerometer *19* and on the test platform *4*, oil-pressure sensor is installed on the preloaded elastic shell *1*. Data on pressure and movement from sensors *16-18* come in from the data line to measurement modules (analog-to-digital converter) and after digitization arrive on the laptop equipped by special software «Vibroregistrator-M2», created in a graphical programming environment National Instruments LabVIEW [22]. Thus, the values of the oscillations x_1 (mm) of the generator plunger and x_2 (mm) of the hydraulic cylinder *3*, the pressure P (MPa) in the shell *1*, and hence the frequencies rate f_1 and f_2 (Hz) are obtained. Registered data are processed in the software environment "Vibroregistrator-M2". The formulas given in this paper and in [21] made it possible to construct experimental dependences.

3. Experimental Research

To study the dynamic properties of a test stand the physical laboratory model was developed. It allowed to realize smooth sweep excitation frequency f1 within the range of 8...56 Hz and detect resonant modes.

The first resonant mode (Figure 2) is characterized by abrupt transition to larger amplitudes in accordance with the relative shell preload:

$$\Delta = \frac{h}{d_0} \tag{2}$$

and by frequency ratio:

$$\frac{f_1}{f_2} = 1, \tag{3}$$

where d_0 – the inner shell diameter, m; h – elastic shell preload, m; f_1, f_2 – input $x_1(\omega_1)$ and output signals $x_2(\omega_2)$ frequency, Hz (Figure 1).

The relative amplitude or dynamic factor which is equal to the ratio of amplitude in resonance with respect to the static amplitude when $\omega_1=0$, reaches in the first resonance values $k_d=10\ldots11$. This indicates high quality of the system.

The second resonant mode is carried out in conditions of capture the generator f_1 drive signal frequency by frequency f_2 with the ratio of generator drive signal frequencies and test platform equals:

$$\frac{f_1}{f_2} = 2, \tag{4}$$

Equation 4 corresponds to the resonance with the parametric excitation of oscillations, so we can use the Mathieu equation [23–30].

Relating to this system it can be shown that the value range of its parameters, where second (parametric) resonant mode is implemented in the second (parametric) resonant mode can be described by the expressions [23, 24]:

$$c_{\text{sh}} = k_1 + \left[\frac{m(a_1 k_2 e f^2)}{2\alpha} \cdot \left(1 \pm \sqrt{1 - \frac{2\alpha^2 k_2 f^3 A_{x2}}{m(a_1 k_2 e f^2)}} \right) - c - k_2 f^2 \right] \tag{5}$$

where α – viscous friction coefficient, N·s/m; d – outer diameters of the shell, mm; c_{sh} – volumetric shell rigidity; k_1, k_2 – empirical coefficients, correspondingly in N/m^5 и N/m^8; A_{x2} – test object amplitude, mm; c – spring stiffness.

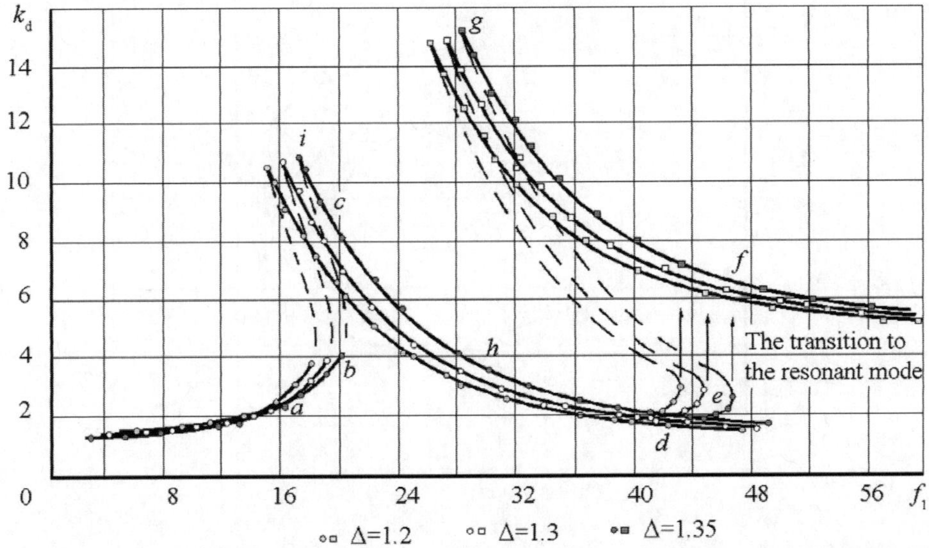

Figure 2. Experimental amplitude frequency characteristics of the test stand with parametric excitation (P_0=0.6 MPa, e=1.3 mm, m=42 kg): point i – the main resonant mode (f_1/f_2=1); point g – the second resonant mode (f_1/f_2=2).

Figure 3 shows the parameters of the second resonant mode existence.

Ranges ab are obtained at negative value of the root in the formula (5), bc – at positive value. The dots indicate the experimental values of the parameters c_{sh} and Δ, where the stable resonance process occurs. The lower ab ranges are in good agreement with the experimental data, and the top rangers are not limited to (bc), i.e. the resonance occurs around the whole value field c_{sh} and Δ above the curve ab. In fact, it is not so: experiments demonstrate the limitations of the above area.

The points c are characterized by the fact that the preloading to a size h in them has a limiting value. At the same time the radical expression in the formula (5) vanishes:

$$1-\frac{2\alpha^2 k_2 f^3 A_{x2}}{m(a_1 k_2 e f^2)}=0 \qquad (6)$$

If draw the horizontal axis through the points c and expand areas ac around them 180 °, we obtain the limiting areas, approximately corresponding to the experiments results.

Left fields are limited minimally by acceptable preliminary shell preload, taking into account the equation (2):

$$h_{min}=d-0.5d_0+e. \qquad (7)$$

Qualitatively, amplitude frequency characteristics of the main and second resonances (Figure 2) do not differ, but the amplification of the oscillations in the second resonance mode is much higher, so the system dynamic response factor reaches k_d=15…16. The quality factor of the system increases and this is of theoretical interest.

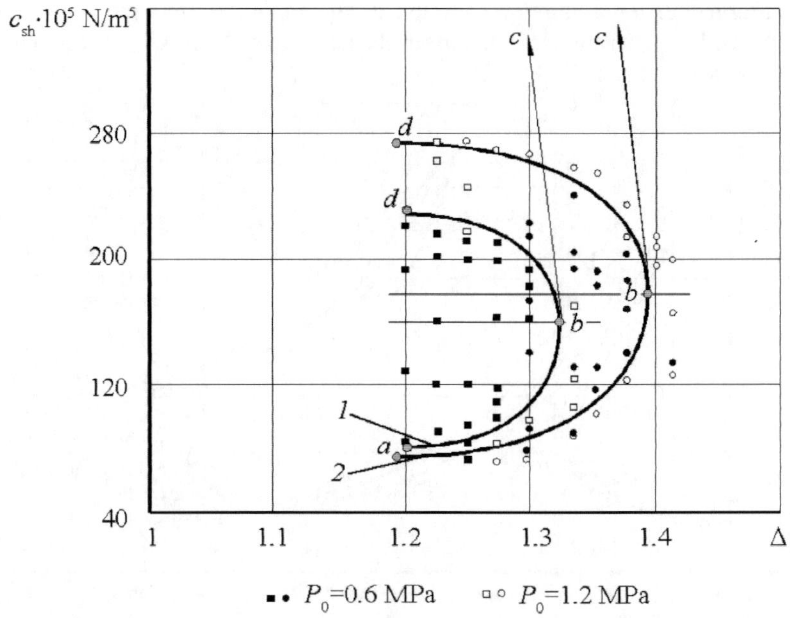

Figure 3. Parameters value range c_{sh} and Δ (d_0=20 mm, b=45 mm), where the second resonant mode is implemented *1* – e=1.22 mm; *2* – e=1.67 mm.

The most informative content about the energy exchange in oscillatory systems has the phase conditions [11, 23]. Figure 4 shows the experimental oscillograms of the second resonance.

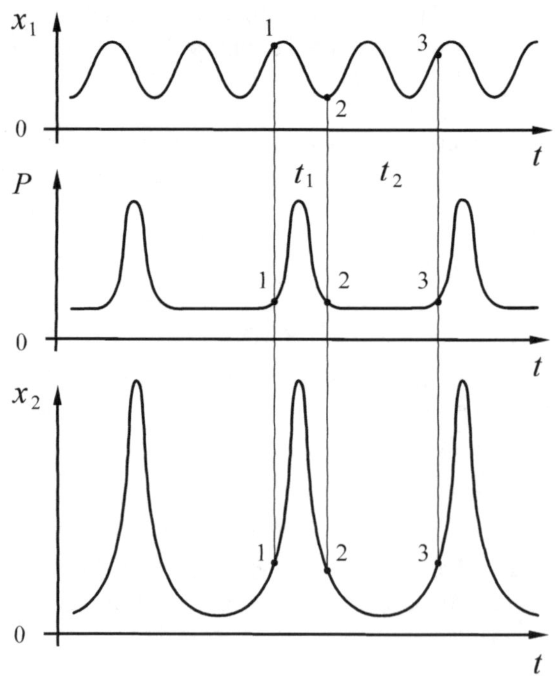

Figure 4. The vibration oscillograms in the second resonance: x_1 и x_2 – the plunger oscillations of generator and hydraulic cylinder, P – pressure in its cavity and shell.

The pressure oscillogram shows that the dynamic interaction in the system is in the section 1-2 – the pressure reaches the maximum value. There is a pause in the 2-3 section – the pressure value approaches to zero and this is explained by the fact that the hydraulic cylinder plunger, while moving upwards, «pulls» out of the shell almost all of the liquid which was accumulated in it at P_0. There is no interaction in the pause.

The signal section x_1, at the moment of the interaction pulse (points 1 and 2), corresponds to the rotation angle of the generator eccentric ~210°...235°. This is much larger than at the first resonance – 135°...150° (Figure 4 does not demonstrate). Obviously the interaction angles and the phase shift between x_1 and x_2 are linked to energy exchange. However, unlike linear systems, the measurement of the phase shift on the displacement of the oscillation peaks is impossible here, since the phase conditions varies also within the period of oscillation («fast phase»).

The original method was used to processing the oscillograms. The oscillation in a nonlinear system can be represented by a vector rotating with variable frequency. By means of sequential constructions it is possible to establish a change in the «fast phase» depending on rotation angle and? the vector of the input signal is x_1.

Figure 5 shows its variation and the average integral value of ψ_{ai} at the second resonance at Δ=1.2 mm and P_0=0.6 MPa.

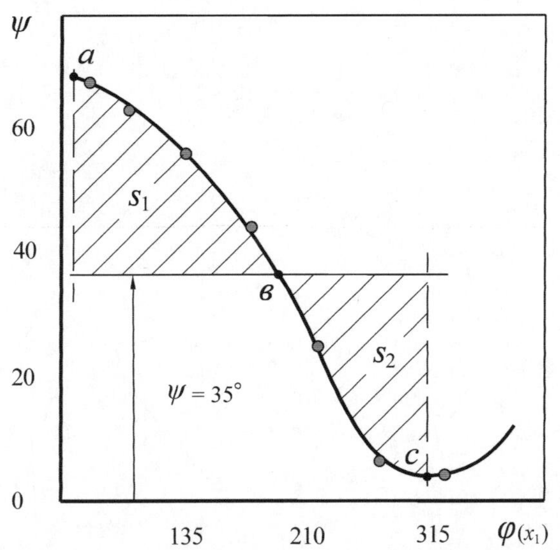

Figure 5. The diagram of the definition for the average integral value of the phase shift.

There is energy exchange in the system at the dynamic interaction, so the phase shift is determined on the section between the dashed lines (Figure 5) corresponding to p. 1 and 2 on the oscillogram $P(t)$ (Figure 4). The average integral value of the phase was determined by the equality of the dashed areas s_1 and s_2.

Thus, several oscillograms were processed at different frequencies ω_1. At the points of the maximum amplitude A_2 of the second resonance, the phase error between x_1 and x_2 is 39 °, whereas in the first resonance, ψ_{ai} was 74 °.

The vector diagrams (Figure 5) correspond to the points at the beginning of the interaction (p. *a*, Figure) at average integral value of the phase (p. *в*) and at the end of the interaction (p. *c*).

For linear system, the energy balance on one period is defined as

$$K+P+E_a = E_0 + E_{in}, \tag{8}$$

where E_0 – energy level of the system in resonance, E_0=K+P=const.

The balance of reactive energies (kinetic K and potential P) and the balance of dissipative energy E_d and the input energy E_{in} are saved. The elastic and inertial force vectors as well as driving force and resistance forces are directed opposite to one another and the phase shift is equal to 90°. Here the energy exchange is significantly different. The interaction between the generator and hydraulic cylinder starts at ψ=64°, and the output from it corresponds to 8° (Figure 6).

All values in the variation interval ψ are less than 90°, therefore it can be seen on the vector diagrams that the driving force F not only compensates the resistance force, but adds its own part – the component $F\ Cos\psi$ to the reactive forces $\bar{\Phi}$ and $\bar{F}_y$.This component is directed along the x_2 axis and performs positive work. This means that during the interaction period (p. 1 and 2 in Figure 4), the generator introduces into the system additional energy ΔE_0.

As a comparison, in the first resonance, the interaction in the system begins at ψ=105°, and the output from it at 45°, therefore at the beginning of the interaction the reactive component F Cosψ is directed opposite to the x_2 axis and performs negative work, braking the oscillations of the hydraulic cylinder (Figure 6). The average integral value of the «fast phase» is less than the phase shift in the linear system (74° against 90°), so the amplitude A_2 in our case is larger than for a linear system with equivalent dissipative energy scattered during a period, but less than at the same system in the resonance regime at ψ=35°. At point 1 (Figure 4), the hydraulic cylinder plug enters with the energy:

$$E_{01}=K_1+P_1, \tag{9}$$

and from point 2 comes out with increased energy:

$$E_{02}=K_2+P_2>E_{01}, \tag{10}$$

however:

$$\Delta E_0=E_{02}-E_{01}. \tag{11}$$

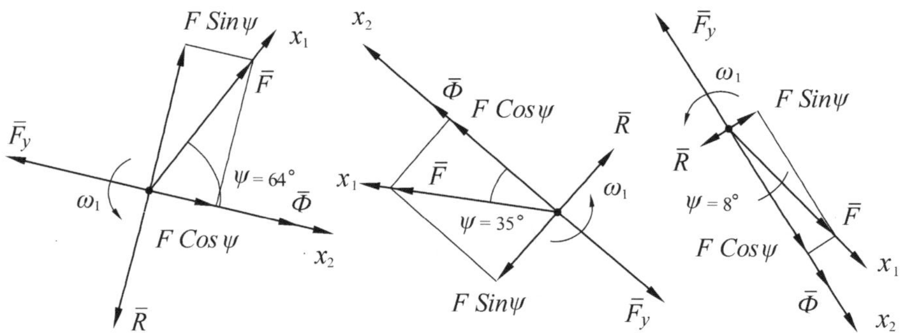

Figure 6. Vector diagrams at parametric resonance:
$\bar{F}$ – vector of the driving force is equivalent to the kinematic excitation,
$\bar{\Phi}$, $\bar{R}$ и $\bar{F}_y$ – vectors of inertial forces, viscous drag and elastic force.

As follows from the energy exchange, the generator inputs energy that compensates the dissipative energy of frictional forces while another quotient of the input energy is accumulated in the form of reactive energy of the actuator. The non-return of this energy is due to the «detachment» of the hydraulic cylinder plunger in the pause, when the reactive energy ΔE_0 is converted to the active E_a in the section of 2-3 (Figure 4).

The second resonance effect is of great practical value. This resonance effect allows the significant expansion of the frequency range of the implemented oscillations spectrum with the amplitude and frequency modulation at a variable parameter h. Figure 2 shows the amplitude-frequency characteristics of the oscillation circuit, obtained by continuous sweep of the excitation frequency f_1 in the range of 8...56 Hz. At first the system passes through the first resonance with slow amplitude A_{x2} increasing at a-b range and its following leap to the value at the point c. However, transition to point c is a transition process shown in timed sweep oscillations (Figure 7).

Further, the amplitude is reduced in accordance with the gain slope, passing through the minimum at the point *d*; but at the point *b* there is the second leap to the amplitude value A_{x2} at the point *f*. During *e-f* transition the oscillation frequency of the test platform instantly reduces in half as the generator f_1 basic frequency is captured by frequency f_2 at their ratio $f_1/f_2=\omega_1/\omega_2=2$. Therefore, at the range of *e-f* (Figure 6) the transition process is much shorter. At the point *f* there are two possible options for further displacement of the excitation frequency f_1 in the following direction:

- frequency increases with decrease of amplitude A_{x2} with simultaneous increase of f_2;
- strengthening of the amplitude to the maximum value A_{x2} of dynamic coefficient k_d with decrease of oscillations frequency of test platform f_2.

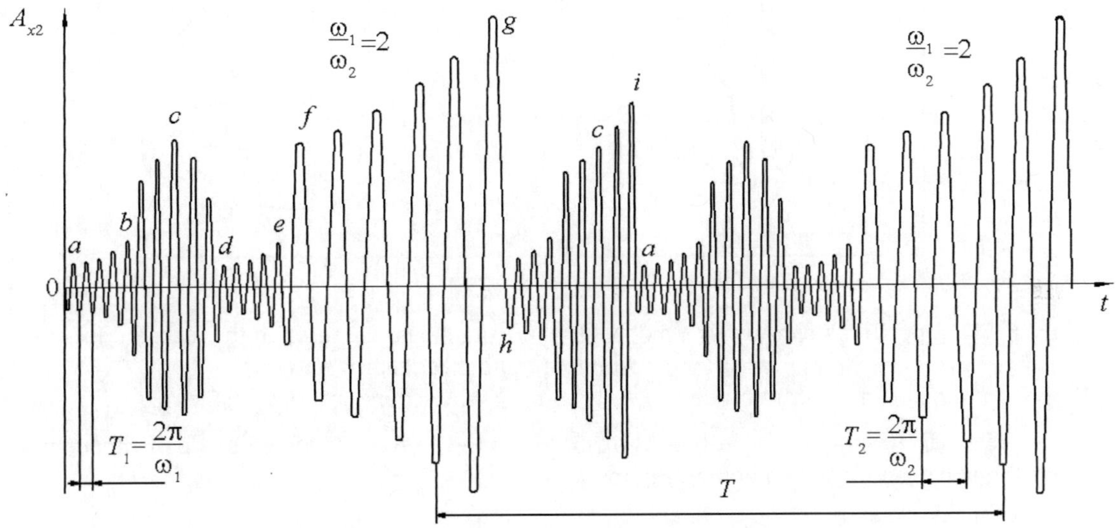

Figure 7. Timed sweep oscillations with the amplitude frequency modulation at stepped transitions in the first and second resonances: *T* – the period of the initial modeling function.

Both designs can be implemented by control system of pump *8* working volume with servo motor No.1 (Figure 1). These designs will change the shape of the modulation and, accordingly, the spectral structure of oscillations. Then the backward displacement of the excitation frequency f_1 from the point *f* to *g*, the frequency f_2 will be reduced in accordance with condition [21]:

$$c+c_{sh}f^2. \tag{12}$$

The amplitude will grow until it reaches the stability limit of the oscillation circuit in the frequency capture mode. System transition to the point *h* is accompanied by instantaneous frequency change f_2 to f_1 and a sudden fall of the amplitude and the dynamic factor. Further, displacement of the excitation frequency f_1 downwards leads to the resonance increase to stability limit of the oscillation circuit at the point *i*. Behind it, the system returns to its original state – point *a* and then there is a repeat cycle.

Graphic illustration of such oscillation process with amplitude and frequency modulation gives a picture of "mobile" line spectrum (Figure 8).

Thus timed sweep oscillations of test stand platform within the frequency range ~8...56 Hz, covering both the resonant modes, demonstrate the feasibility of implementing the amplitude-frequency modulated oscillations, with the possibility to perform control in three ways:

- changing the frequency sweep range of the excitation frequency f_1, which changes the shape of the amplitude-modulation function, and therefore, the harmonic structure of the oscillations;
- changing the position of the reversal points of the sweep frequency excitation on the AFC system;

- random changing of the parameters a_1 and a_2 (1) of the generator oscillator signal by changing the shell preload *h* due to the servo motor No. 2, managed by the generator of occasional signals through the valve *9* (Figure 1).

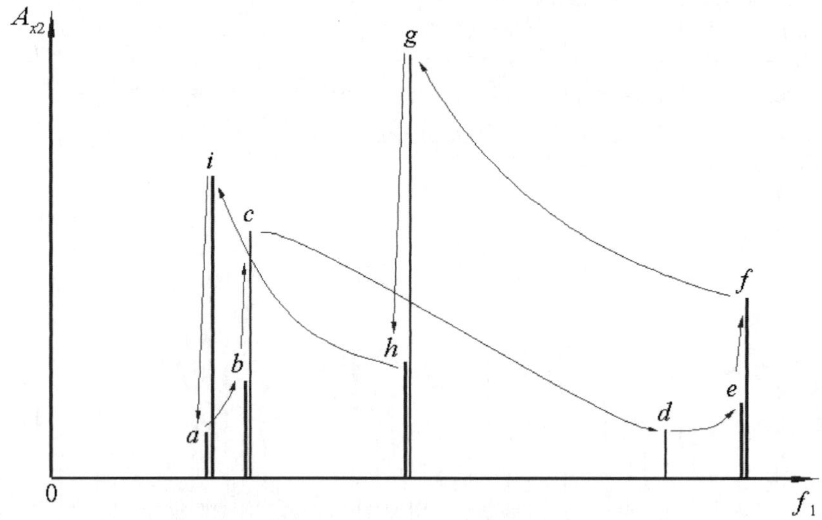

Figure 8. «Mobile» line spectrum as a model for the system passage through the resonant modes with the variables amplitude A_{x2} and frequencies f_1 to f_2:
point *i* – the main resonant mode (f_1/f_2=1); point *g* – the second resonant mode (f_1/f_2=2).

One of the modulation function designs corresponds to the above considered mode of timed sweep excitation frequency and is shown below (Figure 9*a*).

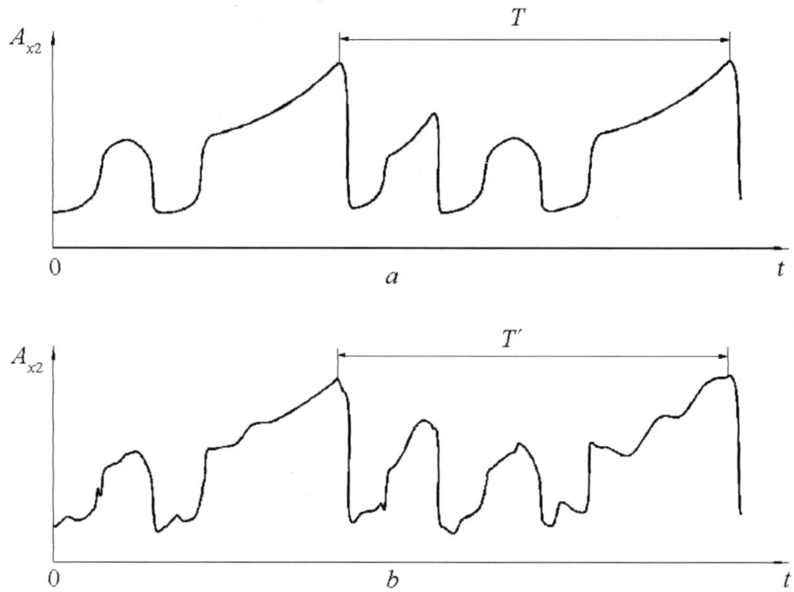

Figure 9. Modulating functions: *a* – original; *b* – fluctuating;
T – original function period; *T′* – the approximate period of the original function.

The elastic shell regulator of oscillations generator (Figure 1) is connected through the transmission with rotary hydro-cylinder shaft, the latter is connected to the hydraulic system via the electrically operated reversing on-off valve *9*. If we use the signals produced by the occasional signal generator [1] as controllers, then due to the randomly changing supplied volumes $-\Delta w$ and $+\Delta w$, the oscillations amplitude of the test platform A_{x2} will receive unpredictable disturbances that lead to fluctuations in the original modulating function (Figure 9, *b*).

Obviously, if rotation angles φ_1 and controller of preload *h* change randomly (Figure 1), the oscillations spectrum of the test platform will be continuous and unpredictable and fluctuate randomly within frequency range from 8…10 to 60 Hz or any other specified range.

4. Summary

The performed research demonstrates that the changing narrowband random vibration mode is one of the most effective methods for vibration testing, including the broad-band vibration tests, further, a stand design allows to implement a large number of test modes. In addition to practical significance of the second resonance, it is also interesting from the theoretical point of view, since despite the external similarity with the parametric one, the energy processes in the dynamical system considered are essentially different.

References

[1] Information on http://www.itestmeasure.com/pdf/prosigNoiseAndVibrationHandbook.pdf
[2] Information on https://www.fs.vsb.cz/export/sites/fs/330/.content/files/VIBDI_skriptaEN.pdf
[3] Harris C., Piersol A. *Harris' shock and vibration handbook* (McGraw-Hill Professional: New York)
[4] Czichos H. *Handbook of Technical Diagnostics. Fundamentals and Application to Structures and Systems* (Springer Verlag Berlin Heidelberg: Berlin)
[5] Lockhart B. *Mechanical Engineering Reference Manual* (NY Research Press: New York)
[6] Ahirrao N., Bhosle S., Nehete D. 2018 *Procedia Manufacturing* **20** 434–439
[7] Forental V., Forental M., Nazarov F. 2015 *Procedia Engineering* **129** 695–701
[8] Lei L., Haihong H., Fu Zh., Zhifeng L. 2017 *Journal of Cleaner Production* **165** 1407–1419
[9] Mehn-bol D, Strel'nikov A. 2009 *Jerdenet-Progress* 29–33
[10] Seimert M., Gühmann C. 2018 Measurement **108** 201–206
[11] Petrovas A., Rinkevičienė R. *Automatic Control Theory I, II: A Laboratory Manual* (Technika: Vilnius)
[12] Nakra B. *Theory And Applications Of Automatic Controls, third ed* (New Age International: New Delhi)
[13] Nizhegorodov A., Gavrilin A., Moyzes B. 2016 *J. Phys.: Conf. Ser.* **671** 012037 doi: 10.1088/1742-6596/671/1/012037
[14] Nizhegorodov A., Gavrilin A., Moyzes B. 2015 *Key Engineering Materials* **685** 320–325 doi: 10.4028/www.scientific.net/KEM.685.320
[15] Hongyuan Q., Jianming Y., Stephen B., Jinghan Zh. 2017 *Journal of Sound and Vibration* **406** 74–88
[16] Wei Zhao, Yi-Min Zhang 2018 *Mechanical Systems and Signal Processing* **113** 77–89
[17] Tandon N., Nakra B. *Vibration and Noise Engineering* (I.K. International Publishing House Pvt. Ltd.: New Delhi)
[18] Edward H. S. *Mechanical engineer's reference book* (Butterworth-Heinemann: Oxford)
[19] Steinwolf A. 2012 Random Vibration Testing Beyond PSD Limitations Sound&vibration **40** 12–21.
[20] Deng N., Ostadan F. 2008 analysis *The 14 th World Conference on Earthquake Engineering* 04-02-0024
[21] Nizhegorodov A., Gavrilin A., Moyzes B., Ditenberg I., Zharkevich O., Zhetessova G., Muravyov O., Bets M. 2016 *Vibroengineering* **18** 3734–3742
[22] Information on http://www.labview.ru/solutions/172/item3228.
[23] Adams M. *Rotating machinery vibration: from analysis to troubleshooting* (, CRC Press, Taylor & Francis Group: Boca Raton)
[24] Halit E. *Acceleration, Vibration, and Shock Measurement* (Taylor & Francis Group, CRC Press

LLC: Abingdon)
[25] Plotnikova I., Efremova O., Tchaikovskaya O. 2017 *J. Ponte* **73**(4) 158–162 doi: 10.21506/j.ponte.2017.4.45
[26] Plotnikova I., Redko L., Chicherina N., Efremova O., Serebryakov S. 2018 *IOP Conference Series: Materials Science and Engineering* **457**(1) 012021 doi: 10.1088/1757-899X/457/1/012021
[27] Wan D., Bao Y., Tian Y., Liu Z., Qiu Y. 2016 *Key Engineering Materials* **680** 3–8
[28] Goinaraghi F., Kuo B. *Automatic Control Systems tenth ed.* (, McGraw-Hill Education: New York City)
[29] Odnokopylov I. G. *et al* 2016 *IOP Conf. Ser.: Mater. Sci. Eng.* **189(1)** 012010 doi**:** 10.1088/1757-899X/189/1/012010
[30] Gavrilin A., Chuprin E., Moyzes B., Halabuzar E. 2014 *Proceedings of 2014 International Conference on Mechanical Engineering, Automation and Control Systems* 6986947 doi: 10.1109/MEACS.2014.6986947

IOP Conf. Series: Materials Science and Engineering **516** (2019) 012032 doi:10.1088/1757-899X/516/1/012032

Fluxgate sensor modeling

P Baranov, A Kolomeytsev and I Zatonov

Tomsk Polytechnic University, Tomsk, Russia
E-mail: bpf@tpu.ru

Abstract. The paper presents a finite element analysis to build the fluxgate model using the COMSOL Multiphysics environment. COMSOL Multiphysics is a powerful interactive environment for modeling and solving all kinds of scientific and engineering problems. Recommendations are given to the optimization of the element geometry and settings of materials and the finite element mesh. Specific configure settings are described for the built-in Physics Interfaces and the solver.

1. Introduction

Welding of steel pipes and structures is often accompanied by the magnetic blowout effect induced by residual magnetism. This phenomenon leads to the instability of the welding process, metal scattering, and the formation of such defects in a welded joint as pores, faulty fusion and penetration, slag inclusions, and others. Sometimes, welding is impossible due to the blow-out or electrode sticking [1–5]. The main cause of the pipeline residual magnetism is the use of magnetic-field flow detectors for inspecting mechanical conditions of pipelines.

The additional contributing factors to the pipeline residual magnetism are geomagnetic field, elastic mechanical stresses, processing pipe magnetism, and extremely close transmission cables. As the pipe magnetism does not allow to obtain a good quality welded joint, their demagnetization prior to welding is a necessary technological procedure. Since a complete demagnetization cannot be achieved, welding is allowed therefore to be performed at a weak residual magnetism which does not substantially affect the welding process.

The residual pipe magnetism is measured with magnetometers comprising a magnetic field sensor and an electronics module. A magnetic field sensor is the main element of any magnetometer and intended for the transformation of the magnetic flux density in electric signal, usually, the voltage. In the design of magnetic field sensors, various physical effects such, for example, as Hall effect, Gauss effect, Suhl effect and others are used. Fluxgate sensors and magnetometers are the most preferable devices for measuring residual pipe magnetism. Also fluxgate sensors are widely used in space and geophysical explorations [6–9], navigation, attitude and stabilizing control systems together with gyro and accelerometer [10-–12], magnetic resonance imaging, brain imaging etc.

However, a mathematical calculation of the fluxgate, and evaluation of its sensitivity is rather a complicated problem because a fluxgate is a non-linear transducer comprising a ferro-magnetic core [13–14]. It is advisable to involve computer simulation facilities in the fluxgate design. The aim of this paper is to present the finite element analysis to build the fluxgate model using the COMSOL Multiphysics environment.

2. Fluxgate geometry

A differential fluxgate configuration is the most widespread configuration used in the magnetic field measurements. A schematic view of the differential fluxgate is given in Figure 1. This schematic will be used to build the fluxgate model.

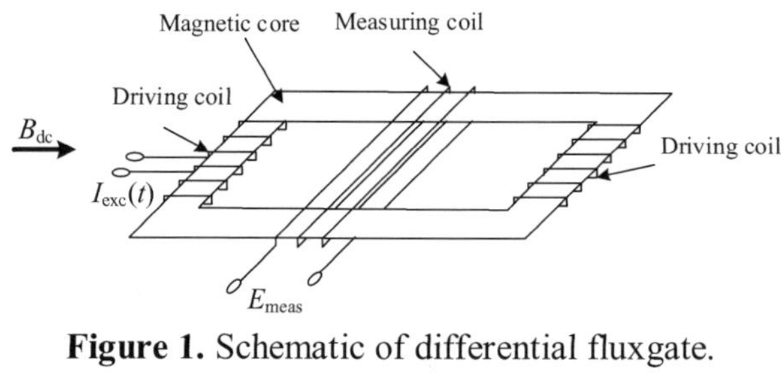

Figure 1. Schematic of differential fluxgate.

The fluxgate modeling in COMSOL Multiphysics begins with the description of its physical shape. In most cases, the main requirement for the fluxgate geometry is to deliver its real dimensions to the computer model as accurately as possible, including necessary assumptions. Thus, sharp edges and corners of diverse model elements must be rounded. This helps to avoid the emergence of extremely tensed areas with a high concentration of magnetic field lines (fringing effect) as illustrated in Figure 2. Owing to this avoidance, a better convergence is achieved when solving the problem of the fluxgate modeling. When modeling non-cylindrical inductance coils, buckling of the conductor must be rounded also. These manipulations are activated with the Fillet operator, which considerably facilitates the formation of turns with the Coil Geometry Analysis setting parameter.

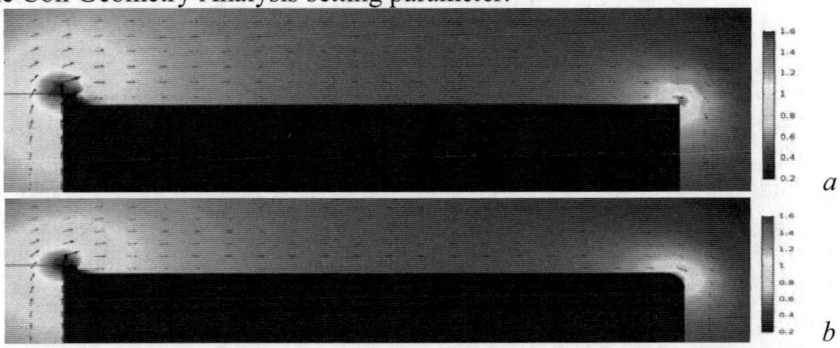

Figure 2. Fillet creation at a corner and magnetic flux density distribution in the fluxgate: *a* – without filleting; *b* – with filleting.

The model of the fluxgate geometry is depicted in Figure 3.

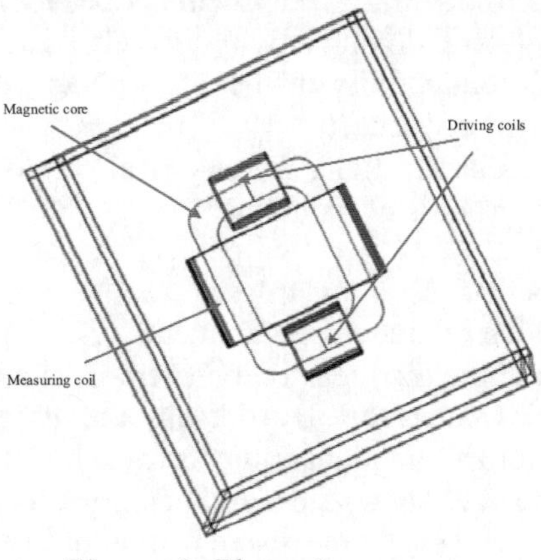

Figure 3. Fluxgate geometry.

3. Built-in materials settings

The material of the core is the essential parameter in the interaction with the magnetic field. Depending on a problem to be solved, the magnetic material can be either soft or hard. The main difference between these materials is the magnetization curve which is the dependence between the magnetic flux density and the field intensity. In COMSOL Multiphysics, HB Curve allows the use of a curve that relates magnetic flux density B and the magnetic field

H. Despite a broad range of built-in materials, COMSOL Multiphysics may not provide the required material, so a user has to modify one of the predefined materials or define a completely new material. The properties of the required material are often known well in advance. Also worth highlighting is the fact that the creation of a new material is accompanied by the indication of a great number of magnetization curve points. The accuracy of solutions will depend on this indication as well as repeatability of ferro-magnetic core properties at a multiple remagnetization. If a user possesses only the approximate data on the material properties, COMSOL Multiphysics offers the Interpolation commands for the diagram description. A linear interpolation allows a user to rather appropriately describe the magnetization curve. As for a cubic interpolation, it provides the magnetization curve with a greater smoothness and, as a consequence, a uniform flux density with the growing field intensity. The curve extrapolation should be linear.

A computation of magnetic fields implies the presence of the airspace. Usually, a coil is in the air. The predefined air material should be additionally modified. The electrical conductivity of the air should be different from zero. This is rather important for the convergence achievement when solving the problem of the fluxgate modeling, since it will not be tackled until the default value of zero for the electrical conductivity of the air is not changed. A real value of the electrical conductivity of the air is about 10–15 S/m. So, it is suffice to set this value between 10^{-4}–10^{-2} S/m. It is worth noting that with increasing value of the electric field, the electrical conductivity will play an ever greater role in the process of modeling.

4. Predefined physics interface variables

Physical processes which relate to measurements of changes in the magnetic field and the direction of the flux density in materials, can be described by the Magnetic Field physics interface. By default, the Physics Interfaces are provided with the discretization of quadratic forms, which require a significant amount of the random access and read-only memories (not less than 16 GB). Therefore, it is advisable to select the discretization of linear forms.

The dependence between the magnetic flux density and the field intensity in the ferro-magnetic core is described by the Ampère's Law feature. For this, we select the HB Curve in Constitutive Relations settings. The magnetic field is created by the coils in which the current flows. The coil properties, the number of turns and the current direction in the coil are described by the Magnetic Fields interface coil features. For solving the non-linear problem, it is important to use the additional calibration of the vector potential of the magnetic field. In the case of the symmetrical finite element model (FEM), its building can be facilitated by its splitting, for example into eight parts. Then, such features as the Perfect Magnetic Conductor and Magnetic Insulation should be additionally applied for the magnetic field. The Magnetic Insulation feature does not allow the magnetic field lines to pass and directs them parallel to the symmetry plane. On the contrary, the Perfect Magnetic Conductor feature allows both the magnetic flux and lines to pass perfectly normal to the symmetry plane.

5. Meshing

Mashing is another important step in the solution of the problem relating to the fluxgate modeling. Due to the fact that the fluxgate is driven by the variable magnetic field, the skin effect is rather essential for the process of modeling. It must be considered in the FE model as early as at 10 kHz frequency. For this, it is necessary to introduce the additional boundary layers in the mesh as depicted in Figure 4. The addition is provided by a click the Boundary Layers node.

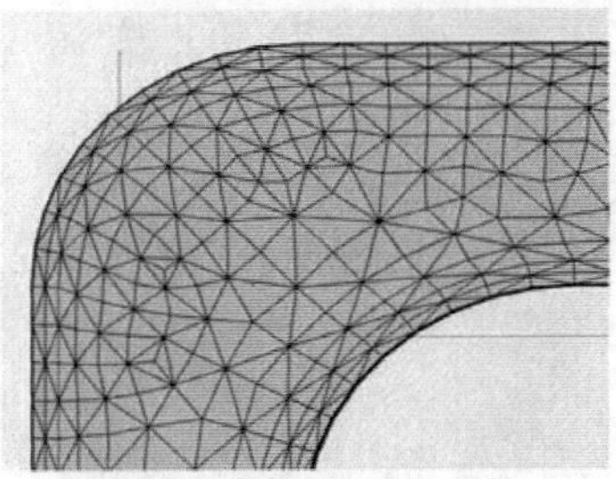

Figure 4. Finite element mesh for ferro-magnetic core. Two boundary layers are added.

For the best symmetry, the FEM should be split into several similar elements. The mesh is then created for one of these element and copied for other ones. This technique allows us to gain the similar size of the finite elements for each side of the symmetry plane. Also, it allows avoiding errors caused by a non-uniform mesh generation inside the element during the automatic meshing. FEM of the fluxgate coils should consider the skin effect and prescribe the mesh accurately enough to generate the coil model without breaks. As a result, the geometry of turns is more precise and the convergence is achieved when solving the problem of the fluxgate modeling. For example, Figure 5 presents the FEM of the fluxgate measuring coil.

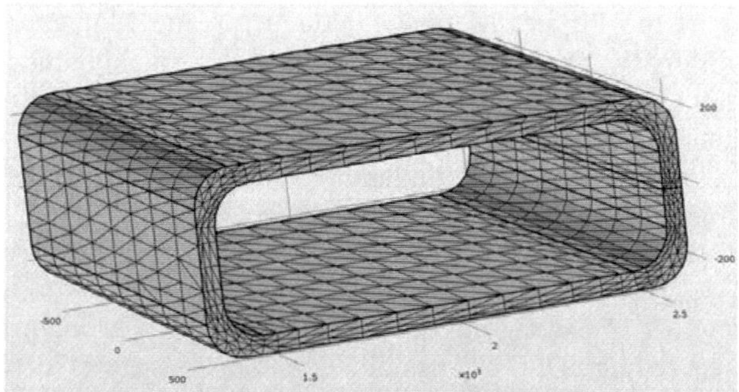

Figure 5. FEM of fluxgate measuring coil.

Figure 6 presents the FEM of the fluxgate with the predefined mesh.

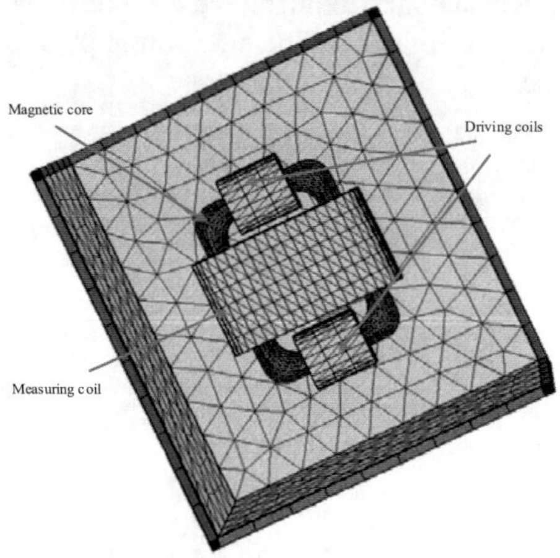

Figure 6. FEM of the fluxgate.

6. Studies and solver settings

COMSOL Multiphysics lets a user to tune solver settings. Our problem is a time-dependent, and its solution includes three steps, namely the Geometry Modeling, and the selection of a Stationary and a Time-Dependent solver. At the first step of the geometry modeling, we analyze both the drive and measuring coils. The second step implies the solution of the stationary problem in the absence of the variable drive field, and the third step includes a modeling of the fluxgate operation at the alternating current of the prescribed amplitude and frequency. In order to improve the convergence of the problem solution, the type of the linear solver should be changed from the iterative to direct. The direct solver is used for modeling the magnetic field. The Jacobian update operator is to be set on Every Iteration. We select the Constant

IOP Conf. Series: Materials Science and Engineering **516** (2019) 012032 doi:10.1088/1757-899X/516/1/012032

(Newton) Iteration for the non-linear method. The maximum number of iterations should not be less than 25. The accuracy class is selected depending on the required accuracy of the study. The Time Stepping option is either selected as BDF Free by default or changed by the Generalized Alpha Strict option, which is more stable.

The direct solver PARDISO is chosen for modeling. The time stepping and the modeling step are selected such that to achieve a satisfactory data display on the studied frequencies, i.e. each signal period should be minimum 100 samples per period.

As a result of calculations, COMSOL Multiphysics builds up the diagram for the flux density distribution. The obtained diagram can be supplemented with arrows which indicate the direction of the magnetic flux density and induced currents. The resulting model of the fluxgate is presented in Figure 7. Figure 8 shows the output signal of the measuring coil obtained during measuring the residual magnetism which equals 10 μT.

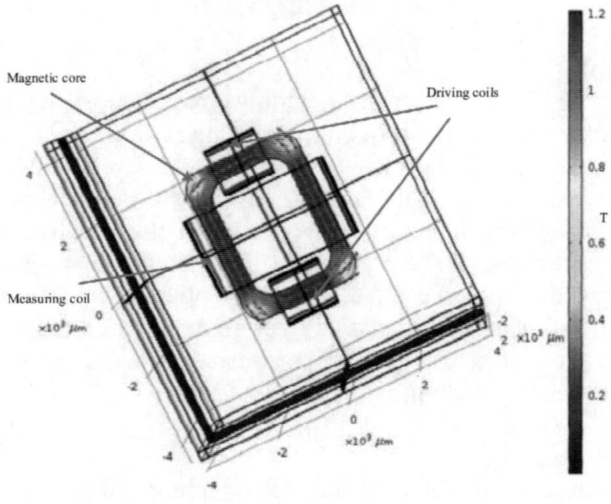

Figure 7. Resulting fluxgate model in COMSOL Multiphysics.

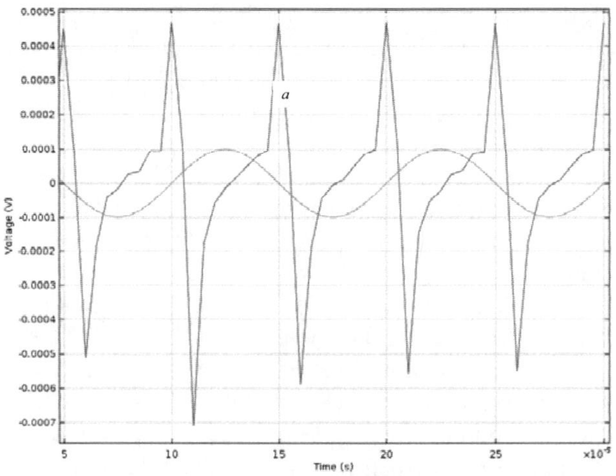

Figure 8. Output signal (a) of the measuring coil obtained during measuring the residual magnetism.

7. Conclusions

A study of non-linear systems, including magnetic systems, is rather complicated from the viewpoint of both modeling settings and the availability of a serious computational power. Nevertheless, the opportunity of modeling magnetic systems in the COMSOL Multiphysics environment allowed us to considerably reduce the labour and time costs and also predict the properties of the object under study. The presented finite element analysis for building the fluxgate model with the COMSOL Multiphysics environment can be recommended for the selection of electrical modes of the fluxgate operation, optimization of the structure and the core material in order to increase the measurement accuracy and upranging.

Acknowledgments

The research is carried out at Tomsk Polytechnic University within the framework of Tomsk Polytechnic University Competitiveness Enhancement Program grant and funded from Russian Science Foundation (RSF), Project № 17-79-10083.

References

[1] Blakeley P. J., Simkin J. 1992 *Welding and Metal Fabrication* **60**
[2] Hernandez J. H. E. *et al* 2009 *Oil and Gas Journal* **107**
[3] Espina-Hernández J. H. *et al* 2010 *Proceedings of the Biennial International Pipeline Conf.* **3**

[4] Cui J. *et al* 2011 *Natural Gas Industry* **31**
[5] Benjaboonyazit T. 2014 *Engineering Journal* **18**
[6] Ripka P. 2000 *Magnetic Sensors and Magnetometers* (Artech house: Boston)
[7] Carr C. 2005 *Ann. Geophys* **23**
[8] Ripka P. 2010 *IEEE Sensors Journal* **10**
[9] Lenz J. E., Edelstein A. S. 2006 *Magnetic Sensors and Their Applications* **6**
[10] Hava Can *et al* 2015 *Journal of Superconductivity and Novel Magnetism* **28**
[11] Nesterenko T. G. *et al* 2015 *XXI IMEKO World Congress 'Measurement in Research and Industry'*
[12] Nesterenko T. G. *et al* 2015 *IOP Conference Series: Materials Science and Engineering* **81**
[13] Butta M. 2009 *Sensors and Actuators* **156**
[14] Baranov P. V. *et al 2018 MATEC Web of Conf.***158**

IOP Conf. Series: Materials Science and Engineering **516** (2019) 012033 doi:10.1088/1757-899X/516/1/012033

Phased arrays of ultrasound emitters controlled by binary signals for acoustic levitation

S Roslyakov, F Emelyanov, N Erzakova, E Sivkov

Tomsk State University, Tomsk, Russia
E-mail: sdy@mail.tsu.ru

Abstract. A phased array of ultrasonic emitters for acoustic levitation has been developed. The array is controlled by a multichannel signal generator. The generator allows you to create 64 channels of independent signals with controlled phase. The possibility of connecting of 4 arrays of emitters, each of the 320 elements. The possibility of providing levitation of foam particles in the field of two oppositely directed arrays was experimentally shown.

1. Introduction

To ensure ultrasonic levitation, it is necessary to control the ultrasound field with high accuracy. Due to the focusing of radiation from a variety of emitters, it is possible to provide high field amplitudes in a given region of space [1–7]. The use of phased arrays of ultrasonic emitters with digital signal generation [1] allows you to control the ultrasonic field and move the levitating particles along a predetermined trajectory. It is advisable to create a system of phased arrays of blocks [2] of the same design. The phase control of each emitter allows you to set the desired field distribution, and varying the values of the amplitudes of the emitters does not significantly affect the final field in the focus area. Such technologies are used for the manipulation of liquids [3]. In works [1–4, 6, 7] for phased arrays, radiators are used at a frequency of 40 kHz of the type MA40S4/S. These radiators are matched with air and allow you to create a pressure of sufficient amplitude for levitation, even water droplets. In papers [5] and [6], only two signal sources with independently controlled phases are used for single axis movement. In [5], the possibility of levitating an object with dimensions greater than the length of the wave is shown, and the required spatial distribution of the field is set using a vibrating aluminum plate attached to two ultrasonic transducers. Field focusing can be achieved by placing emitters at different depths [6], without the need to use a complex signal control system for each emitter.

In this paper, we propose a system of a 64 channel amplifier with a digital control by a microcontroller for a system of phased arrays, irradiating the region of ultrasonic levitation of particles. The ultrasonic signal is generated by the binary outputs of the microcontroller, in fact, a 64-channel 1-bit DAC is implemented. Binary signals pass through a band-pass filter, which allows you to get almost sinusoidal signals that are fed to amplifiers TDA7297 (32 pieces of dual-channel chips are used). Using a 64-channel signal generator, it is proposed to control 4 flat gratings of radiators, each of which consists of 320 elements. This scheme allows you to control the frequency and phase of the emitted signal, which should be sufficient to maintain levitation.

2. System overview

A multichannel digital-to-analog converter (ADC) requires the transmission of a large data stream, which is not available to low-performance microcontrollers. To control 64 channels of 8-bit ADCs with a frequency of 100 kHz, a stream of 6.4 MB per second is required. Given that the signal phase is of greater importance in field control, the amplitude of the signal can be set constant. It is proposed to use a 1-bit DAC, or in fact, binary outputs of a microcontroller for setting the signal of a phased array of ultrasonic emitters. It is supposed to use the MA40S4/S ultrasound emitters with a resonant frequency of 40 kHz, which will limit the spectrum of filtered and amplified signal generated by a single-bit DAC. The binary signal from the controller is fed to the low-pass filter and then to the amplifier based on the TDA7297 chip. The amplified signal is fed to the ultrasonic emitters MA40S4/S.

Figure 1 shows a photograph of the developed unit for generating and amplifying 64 ultrasonic signals with controlled phase based on a STM32F407 microcontroller and TDA7297 amplifiers. The radiators of the amplifiers are fixed on the aluminum frame, which ensures sufficient cooling during operation.

Figure 1. Photo of 64-channel digitally controlled generator with amplifiers.

The microcontroller is controlled in real time via a USB interface from a personal computer. The program provides continuous field control even when receiving data and control signals, which allows you to smoothly move levitating particles. Up to 64 independent ultrasonic emitters can be connected to the signal outputs of the board.

We consider the possibility of controlling particles in three-dimensional space, which is ensured by the use of 4 flat phased arrays consisting of 320 elements (20×16) with independent control of 16 channels (column), placed in rows of 20 elements. In this case, 20 elements placed in column will be controlled by one signal and will be in phase. Arrays are placed in pairs opposite each other. The phase difference between the counter lattices provides the movement of particles horizontally, and the variation of the focusing height of all the lattices ensures the movement of the levitating particle along the vertical axis. Figure 2 shows a photograph of a 64-channel generator connected to two phased arrays of emitters.

Figure 2. 64-channel generator connected to two phased ultrasonic arrays.

Figure 3*a* shows the signal of one of the channels after passing through the low-pass filter and the amplifier. As a result of amplification, the amplitude of the signal increases and minor distortions appear. However, after emission, the signal becomes harmonic, since the MA40S4 / S emitters have a narrow bandwidth (Figure 3*b*). The signal in Figure 3*b* is measured with the help of the ultrasonic receiver MA40S4 / R.

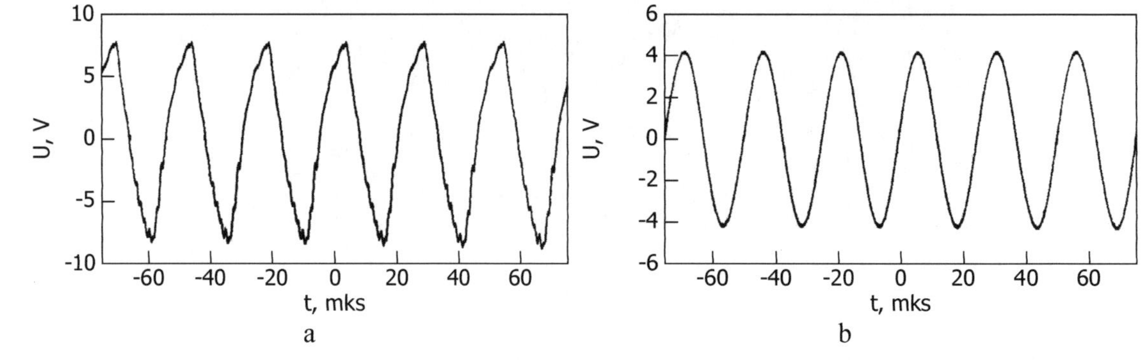

Figure 3. Signal after filtration and amplification (a) and signal after radiation (b).

To test the possibility of levitation of particles in the ultrasound field created by the developed grids, an experiment was conducted on the levitation of foam particles (Figure 4). Two grids were placed at a distance of 16 cm from each other and directed towards (one on top, the other on the bottom), which ensured the formation of standing waves, in the nodes of which the particles levitate. The emitter signals were set to form plane waves with a horizontal phase front. As a result of the interference of counterpropagating plane waves, a standing wave was formed with a step of 4 mm nodes.

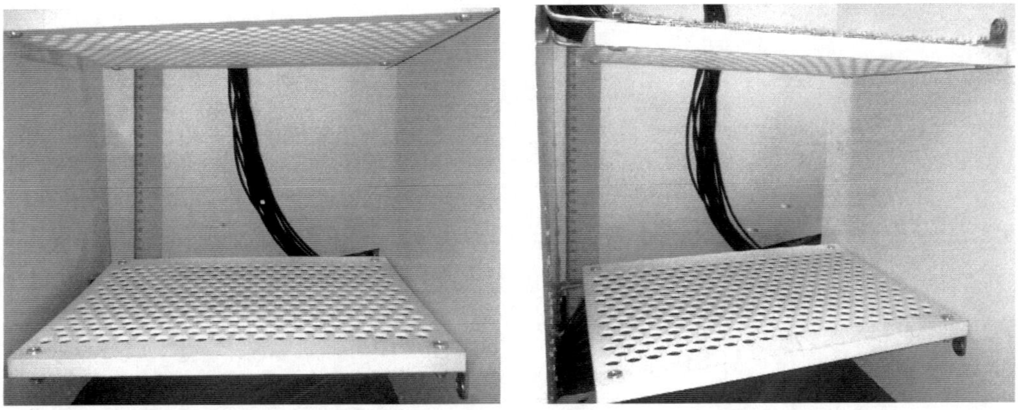

Figure 4. Photographs of experiment on levitation of Styrofoam particles.

3.Conclusions

A control system for phased arrays of radiators for ultrasonic levitation in air has been developed. It is shown that it is enough to use a single-bit digital-analog converter to generate control signals with a predetermined phase. Based on the STM32F407 microcontroller and TDA7297 amplifiers, a 64-channel ultrasound signal generator at a frequency of 40 kHz was created. During the experiments, this generator was connected to two phased arrays of 320 emitters each, which ensured the levitation of foam particles in the air.

Acknowledgments

The scientific research was carried out with the support of the grant of the Russian Science Foundation No. 17-79-20051.

References

[1] Ochiai Y. *et al* 2014 *PLoS ONE* **9(5)** e97590 doi: 10.1371/journal.pone.0097590
[2] Marzo Asier *et al* 2015 *Nature Communications* **6** 8661 doi: 10.1038/ncomms9661.
[3] Andrade M. A. B. *et al* 2018 *Review of Scientific Instruments* **89(12)** 125105 doi:10.1063/1.5063715
[4] Freeman E. *et al* 2018 *Proceedings of the 7th ACM International Symposium on Pervasive Displays - PerDis '18* doi:10.1145/3205873.3205878
[5] Masuda K. *et al* 2018 *IEEE Transactions on Ultrasonics, Ferroelectrics, and Frequency Control* 1–1 doi:10.1109/tuffc.2018.2870069
[6] Sukhanov D. *et al* 2018 *MATEC Web of Conferences* **155** 01017 doi:10.1051/matecconf/201815501017
[7] Watanabe A. *et al* 2018 *Scientific Reports* **8(1)** doi:10.1038/s41598-018-28451-5

IOP Conf. Series: Materials Science and Engineering **516** (2019) 012034 doi:10.1088/1757-899X/516/1/012034

Estimation of phase homogeneity of LiTiZn ferrites from the temperature dependence of the initial permeability near the Curie point

A V Malyshev, A Petrova, A P Surzhikov

Tomsk Polytechnic University, Tomsk, Russia
E-mail: abk9@tpu.ru

Abstract. The estimation of the chemical homogeneity of lithium-substituted ferrite ceramics based on an analysis of the temperature dependences of the initial permeability near the Curie point was made. The transition of ferrite from ferrimagnetic to paramagnetic state at the Curie point is sensitive to the phase and chemical homogeneity of the material. It is shown that the best homogeneity is in ferrite products of the SL-187 brand compared with samples of ferrite ceramics, sintered in the laboratory conditions.

1. Introduction

Lithium-substituted ferrites have a wide range of applications in low-power microwave devices (phase shifters, switches, etc.). In addition, they have a low cost compared to other ferrites.

As is known, almost all the electromagnetic properties of ferrites are structurally sensitive, i.e. dependent on the content and volume distribution of defects in the microstructure.

It is known that ferrite ceramics obtained by classical ceramic technology has a number of structural defects and side phase inclusions [1, 2]. In a number of experimental works [3–6], the possibility of assessing the ferrite ceramics quality and its phase homogeneity according to the degree of decrease in the initial permeability on the temperature dependence near the Curie point is indicated.

Earlier, we proposed a method based on the mathematical processing of the experimental temperature dependences of μ_i. It is shown that this method allows controlling the defects level affecting the processes of magnetization reversal in ferrites [7]. However, it was found that this method is not effective enough to estimate the chemical homogeneity of the magnetic phase of the initial powders and finished products. The transition of ferrites from the ferrimagnetic to the paramagnetic state is sensitive to the chemical homogeneity of the material. A characteristic of the transition is the rate of decline of the initial permeability on a temperature scale.

It is known that the Curie point T_c of the material is highly sensitive to both chemical composition and the presence of controlled and uncontrolled impurities in the material. Consequently, any deviation of the chemical composition from the nominal level leads to a change in T_c, causing the blur of the curve $\mu_i(T)$ in the vicinity of T_c. Most clearly, this heterogeneity can be revealed by differentiating the $\mu_i(T)$ curve during the transition from the Hopkinson peak to the paramagnetic state of the material.

Therefore, to assess the chemical homogeneity of LiTiZn ferrites obtained under laboratory conditions, as well as industrial products of the SL-187 brand, in this work was used measurements of the initial permeability near the Curie point.

2. Experimental techniques

The synthesis of microwave ferrite was carried out on ceramic technology at the factory "Ferrite-Domen Co.", St. Petersburg. Industrial ferrite powder synthesized by oxide technology from a mechanical mixture of oxides and carbonates was used for the manufacture of samples.

The binder additive in the form of 10 % aqueous solution of polyvinyl alcohol (PVA) in an amount of 12 % by weight of the powder was introduced into the powder before pressing. Then the mixture was thoroughly mixed, sequentially rubbed through sieves and kept for 24 hours in the air at room temperature for more complete penetration of the binder solution into the powder particles. Compaction of samples in the form of disks and toroids was carried out using one-sided cold pressing using a PGr-10 hydraulic press. The pressure was 200 MPa. Exposure was three minutes.

Samples were sintered in laboratory conditions in the air in a resistance furnace with an integrated thermostat. Ferrite samples were sintered in 2 regimes: $T_S = 1010$ °C, $t_S = 2$ h; $T_S = 1100$ °C, $t_S = 1$ h, where T_S is the sintering temperature, t_S is isothermal holding time. After sintering in the laboratory conditions, the samples had the form of toroids with dimensions: the outer diameter $D = 18.5$ mm, the inner diameter, $d = 14$ mm, thickness $h = 2$ mm. Toroids from the examined ferrite were used as the cores of the inductor. Industrial ferrite products of the SL-187 grade were used for comparative studies. These ferrite products were made under the production conditions at the factory "Ferrite-Domen Co." at $T_S = 960$ °C; $t_S = 8$ h. The dimensions of products: the outer diameter $D = 20.5$ mm, the inner diameter $d = 10.9$ mm, thickness $h = 3.5$ mm.

The chemical formula: $Li_{0.649}Fe_{1.598}Ti_{0.5}Zn_{0.2}Mn_{0.051}Bi_{0.002}O_4$.

The phase composition of the samples was determined using an ARL X'TRA X-ray diffractometer (Switzerland) with a Peltier Si(Li) semiconductor detector and Cuk_α radiation. XRD patterns were measured for $2\theta = (10 - 80)°$ with a scan rate of 0.02 °/s. The phase compositions of the examined samples were determined using the PDF-4 powder data base of the International Centre for Diffraction Data (ICDD). The XRD patterns were processed by the full-profile Rietveld analysis using the Powder Cell 2.5 software.

The morphology of the sintered samples grains was investigated using a Hitachi TM-3000 scanning electron microscope (SEM). The average grain size was calculated using the intercept method.

The Curie point T_c was determined from the temperature dependence of μ_i by drawing a tangent to the curve of the site of a rapid drop in μ_i values. The intersection of this tangent with the temperature axis determined the value of T_c. Also, this temperature was determined by the results of the mathematical processing of the temperature dependence of the initial permeability.

Measurement of inductance L of an inductor with a core of ferrite toroid was carried out by the bridge method [7].

3. Result and discussion

Figure 1 presents the X-ray diffraction pattern of samples sintered in laboratory conditions.

As can be seen from the Figure 1 obtained data confirmed the formation of single phase homogeneous cubic spinel structure.

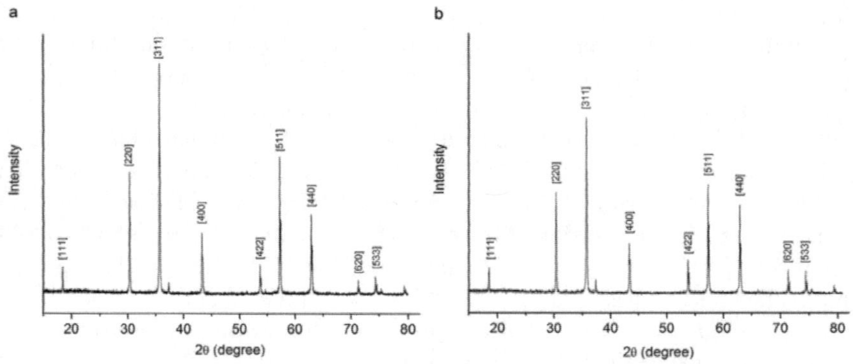

Figure 1. The X-ray diffraction patterns of samples sintered in the laboratory conditions at T_s = 1010 °C, t_s = 2 h (a) and T_s = 1100 °C, t_s = 1 h (b).

Figure 2 presents a microstructure of the sintered ferrite ceramics. According to the data presented in Figure 2 sintered ferrite ceramics have well-formed grains. However, the grains are strongly agglomerated and have intragranular porosity. The average grain size showed similar values (2.5 ± 0.3) μm for both types of laboratory samples.

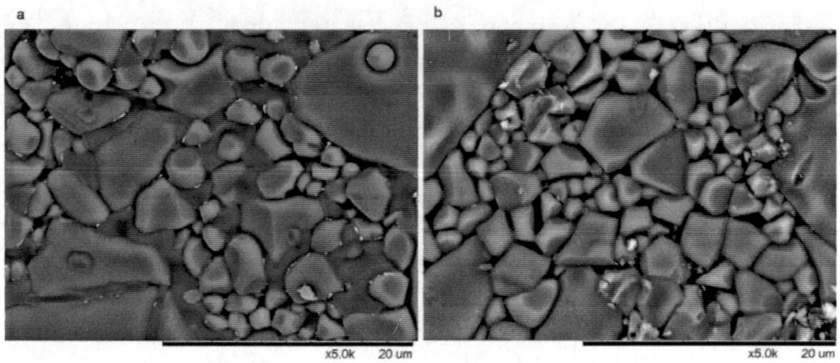

Figure 2. SEM micrographs of samples sintered in the laboratory conditions at T_s = 1010 °C, t_s = 2 h (a) and T_s = 1100 °C, t_s = 1 h (b).

At the same time, it was found the difference between the microstructure of laboratory samples and the microstructure of ferrite products of the SL-187 brand. Such materials are characterized by higher porosity and an average grain size of 50-100 microns (Figure 3).

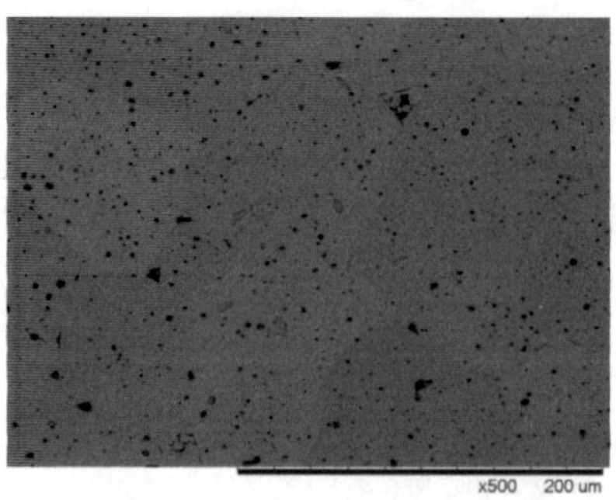

Figure 3. SEM micrographs of industrial ferrite product SL-187.

Magnetic hysteresis loops are a traditional method for magnetic materials control. Comparing the parameters of the hysteresis loop allows estimating the irregularity and the presence of non-magnetic inclusions of ferrite ceramics. Figure 4 shows magnetic hysteresis loops obtained by the traditional oscillographic method.

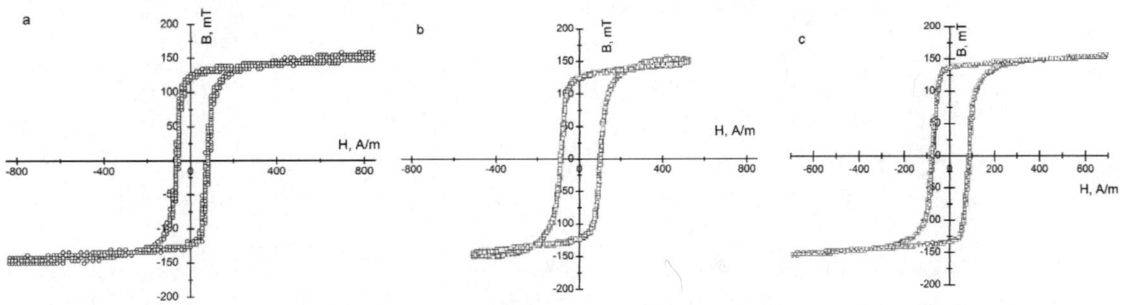

Figure 4. Magnetic hysteresis loops of samples sintered in the laboratory conditions at T_s = 1010 °C, t_s = 2 h (a). T_s = 1100 °C, t_s = 1 h (b), and industrial ferrite product SL-187 (c).

From a comparison of the parameters of the hysteresis loops, it can be seen that similar parameters of the hysteresis loop are characteristic of laboratory samples, while for an industrial product there are lower values of the coercive force (Table 1).

Table 1. Hysteresis loop parameters of ferrite ceramic samples

Samples type	B_s (mT)	B_r (mT)	H_c (A/m)
T_s=1010 °C, t_s=2 h	170	127	95
T_s=1100 °C, t_s=1 h	168	129	98
Industrial product SL-187	160	143	81

From Table 1 it can be seen that, according to the value of the coercive force H_c, industrial products are characterized by the smallest defectiveness due to smaller internal stresses affecting the magnetic reversal of ferrite magnetic domains according to the Neel theory [2].

Figure 5 shows the temperature dependences of the initial permeability near the Curie point. A sharp decrease in the initial permeability near the Curie point suggests the formation of a single-phase and chemically homogeneous structure of ferrite ceramics. There are sharp decreases of initial permeability near the Curie point. This behavior assumes the formation of a single-phase and chemically homogeneous ferrite structure.

According to the temperature dependences of the initial permeability in our work [7], the parameters of the phenomenological expression and the defects level were calculated. It was shown that the defects level in industrial products is lower compared with samples sintered in laboratory conditions. For samples manufactured under laboratory conditions, the optimal defects level was obtained for the sintering regime T_S = 1010 °C, t_s = 2 h.

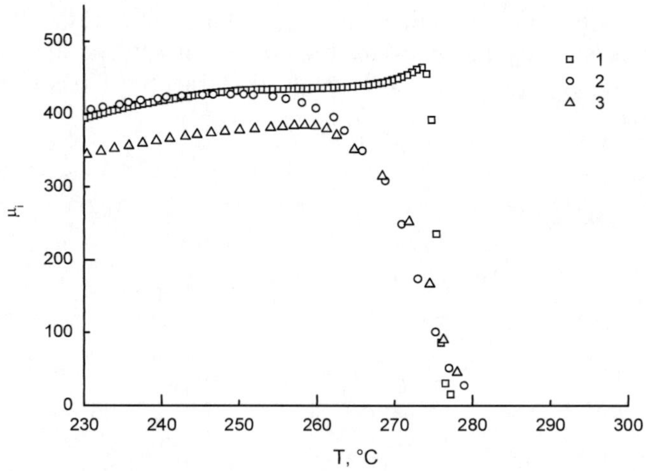

Figure 5. Comparison of the temperature dependencies of initial permeability for the samples sintered in industrial (1) and laboratory (2) conditions (2 – at 1010 °C for 2 h; 3 – at 1100 °C for 1 h) in the vicinity of the Curie point.

Figure 5 shows temperature dependencies of initial permeability of industrial products and the samples sintered in laboratory conditions. From a comparison of obtained experimental data, it follows that the greatest homogeneity is characteristic of the industrial products and slightly worse in the samples sintered in laboratory conditions.

4. Conclusion

In this work, the analysis of the experimental data showed that the transition of ferrite from ferrimagnetic to paramagnetic state at the Curie point is sensitive to phase and chemical homogeneity (homogeneity) of the material.

It is shown that the rate of decrease of the temperature dependence of initial permeability in the vicinity of the Curie point can be a quantitative characteristic of homogeneity [1]

Acknowledgements

This research was supported by the Ministry of Education and Science of the Russian Federation in part of the “Science” program, Project 3.4937.2017.

References

[1] Verma A., Chatterjee R. 2006 *J. Magn. Magn. Mater.* **306** 313–320
[2] Smith J. and Wijn H. P. J. 1959 *Ferrites: Physical Properties of Ferromagnetic Oxides in Relation to their Technical Applications* (Phillips Technical Library: Eidnhoven)
[3] Khan Z. H. *et al* 2013 *J. of Alloys and Comp.* **548** 2018–215
[4] Parvin R. *et al* 2016 *J. Magn. Magn. Mater.* **401** 760–769
[5] Ghodake J. S. *et al* 2015 *J. Magn. Magn. Mater* **378** 436–439
[6] Hu J. *et al* 2007 *Physica B: Cond. Mat.* **400** 119–123
[7] Malyshev A. V. *et al* 2018 *J. Magn. Magn. Mater.* **456** 186–193

IOP Conf. Series: Materials Science and Engineering **516** (2019) 012035 doi:10.1088/1757-899X/516/1/012035

Optical system for monitoring ultrasonic waveguide

A Kuzovova

Tomsk State University, Tomsk, Russia
E-mail: sdy@mail.tsu.ru

Abstract. An optical system for visualization of ultrasonic waveguide aimed to estimate its vibrations is proposed. The optical method does not affect the observed fluctuations, is non-contact and allows you to take measurements quickly. An analytical calculation of the optical system was carried out. According to the results of numerical modeling, an optical system was assembled, which is a microscope. Using the developed microscope, the tip of the ultrasonic waveguide was visualized.

1. Introduction
Ultrasonic vibrations are used in flaw detection, medicine, tomography and remote sensing [1–4]. The problem of measuring the amplitude of ultrasonic oscillations, in particular for ultrasonic waveguides, is relevant. The use of ultrasonic vibrations allows us to accelerate the course of physicochemical processes, both in technological processes and in biological media. The most important parameter characterizing the operation of the ultrasonic oscillatory system is the amplitude of oscillations. Depending on the process under consideration, the amplitude of oscillations is chosen to be maximum in a certain narrow frequency range, and it should be taken into account that if the amplitude is too high it will lead to the destruction of the oscillatory system. Currently, both contact and contactless methods are used to measure the amplitude of oscillations. In the case of contact methods, the measuring sensor is in direct contact with the measured surface. Among contactless methods, optical methods are the most common.

An optical method for imaging ultrasonic vibrations based on a microscope, digital video camera and stroboscopic illumination is proposed. Strobe lighting with a frequency of tens of kilohertz is carried out using high-speed infrared LEDs.

2. Numerical modeling
For the analytical calculation of the geometric dimensions of the optical system, a number of approximations were adopted. First, we consider that the following approximation is valid: the lenses are thin, that is, any beam passing through the optical center of such a lens does not experience refraction and does not change the direction of propagation. The second is that the lens system is considered centered. This means that the main optical axes of all lenses are the same. Consider an optical system consisting of a combination of 4 convex and concave lenses: three convex and one concave (Figure 1). The concave lens allows reducing the size of the optical system.

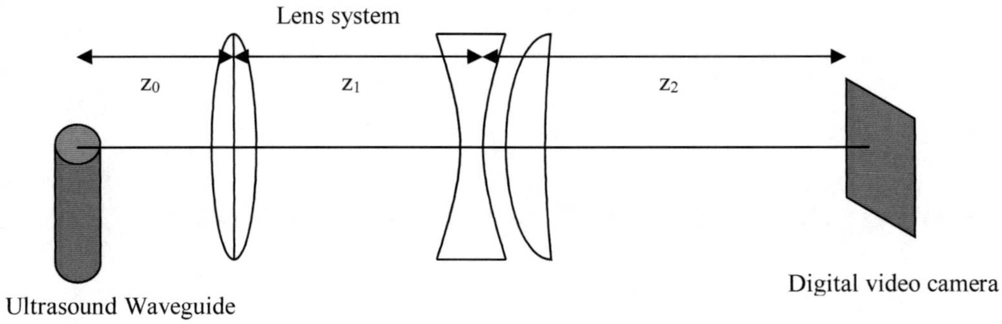

Figure 1. Optical system layout.

A numerical simulation of the lens system was carried out to verify the possibility of upscaling the image. We consider the image of the object in Figure 2, which is located at distance a from the lens which is equal to the focal length of the lens F_1 i.e. $F_1 = a$. This test image has elements that would be easy to recognize after passing throw the optical system.

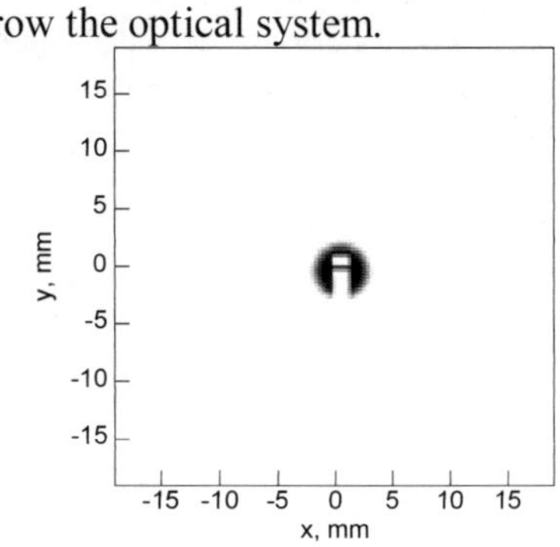

Figure 2. Source image of a source.

Let's produce calculation of the optical field passing through the lens system and falling on the matrix of a digital video camera using the representation of the field in the spectrum of plane waves. The wave field created by a source we denote by function $M(x,y)$. A 2D Fourier transform allow us to go into the spectrum of plane waves.

$$\widetilde{M}(k_x,k_y)=\frac{1}{(2\pi)^2}\int\int_{-\infty}^{+\infty} M(x,y)e^{-ik_x x-ik_y y}dxdy \tag{1}$$

The process of wave propagation in the area from a source to a plane-convex collecting lens will be described by the multiplier $e^{iz_0 k_z}$ that is derived from a solution of the Helmholtz equation. Spectrum of the field passed in space distance z_0, we will present in the following form: $\widetilde{U}_0(k_x,k_y)=\widetilde{M}(k_x,k_y)e^{iz_0 k_z}$, where $k_z=\sqrt{k^2-k_x{}^2-k_y{}^2}$, k – wave number. Taking the inverse Fourier transform of the field spectrum, we obtain the field itself at the input of the first lens:

$$U_0(x,y)=\int\int_{-\infty}^{+\infty}\widetilde{U}_0(k_x,k_y)e^{ik_x x+ik_y y}dk_x dk_y \tag{2}$$

Accounting of the influence of the first and second lenses on the field is described by a multiplier: $L_1(y,x)=e^{ik\left[F_1-\sqrt{F_1{}^2+X^2+Y^2}\right]}e^{ik\left[F_2-\sqrt{F_2{}^2+X+Y^2}\right]}$. The field at the output of the first two lenses can be represented as follows: $U_1(x,y)=U_0(x,y)L_1(y,x)$ in the approximation of the phase screen for a flat lens. Taking the Fourier transform of the field at the output of the first system of two lenses, we obtain the spectrum of the field past the lens system:

$$\widetilde{U}_1(k_x,k_y)=\frac{1}{(2\pi)^2}\int\int_{-\infty}^{+\infty}U_1(x,y)e^{-ik_x x-ik_y y}dxdy \tag{3}$$

We consider the process of wave propagation in the next section from the second flat-convex lens to the flat-concave lens in the form of a spatial spectrum multiplier: $e^{iz_1k_z}$. The spectrum of the field passed in the distance z_1 we will present in the following form: $\widetilde{U}_2(k_x,k_y)=\widetilde{U}_1(k_x,k_y)e^{iz_1k_z}$. Taking the Fourier transform of the field spectrum, we obtain the field itself at the output of a flat-concave lens:

$$U_2(x,y)=\int\int_{-\infty}^{+\infty}\widetilde{U}_2(k_x,k_y)e^{ik_x x+ik_y y}dk_x dk_y \tag{4}$$

Taking into account the influence of concave and convex lenses on the field described by the phase factor $L_2(y,x)=e^{ik\left[-F_3+\sqrt{F_3^2+X^2+Y^2}\right]}e^{ik\left[F_4-\sqrt{F_4^2+X^2+Y^2}\right]}$. The field at the output from a concave and convex lens is represented as follows: $U_3(x,y)=U_2(x,y)L_2(y,x)$ .Taking, the Fourier transform from of the field at the output of a concave and convex lenses we obtain the spectrum of the field, which passed the lens system.

$$\widetilde{U}_3(k_x,k_y)=\frac{1}{(2\pi)^2}\int\int_{-\infty}^{+\infty}U_3(x,y)e^{-ik_x x-ik_y y}dxdy \tag{5}$$

Then the travel of wave from output of lens system to video camera is characterized by multiplier of spatial spectrum $e^{iz_3k_z}$. Spectrum of the field passed in space distance z_3 we present in the following form: $\widetilde{U}_4(k_x,k_y)=\widetilde{U}_3(k_x,k_y)e^{iz_3k_z}$. Taking the inverse Fourier transform from the field spectrum, we get the field itself at the video camera matrix aperture:

$$U_4(x,y)=\int\int_{-\infty}^{+\infty}\widetilde{U}_4(k_x,k_y)e^{ik_x x+ik_y y}dk_x dk_y \tag{6}$$

An image of a given test source after passing through such a lens system will be enlarged (Figure 3).

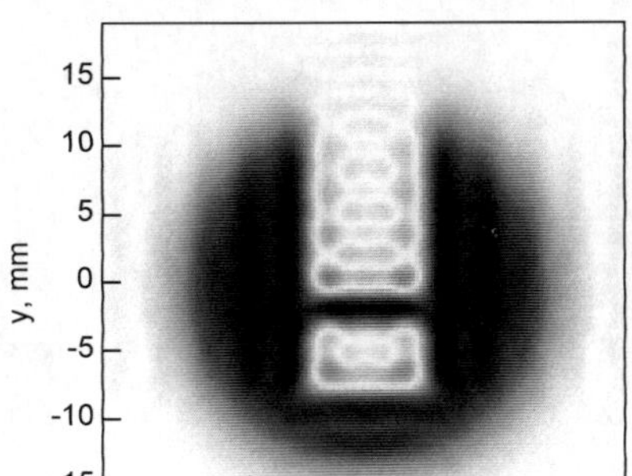

Figure 3. Image of the source after passing through the lens system (for comparison, the original image is shown in the center).

The scaling factor of the considered optical system will be calculated by the following formula:

$$k=\frac{F_2}{F_1}\frac{F_4}{F_3} \tag{7}$$

IOP Conf. Series: Materials Science and Engineering **516** (2019) 012035 doi:10.1088/1757-899X/516/1/012035

where - F_1, F_2, F_3, F_4 are the focal lengths of the lenses. Formula (7) is derived by geometry optics approximation and corresponds to the numerical results presented on Figure 3. According to (7) for lens system with $F_1 = 90$ mm, $F_2 = 250$ mm, $F_3 = -90$ mm, $F_4 = 200$ mm.

On the basis of analytical calculations, the optical system of four lenses Figure 4 was proposed.

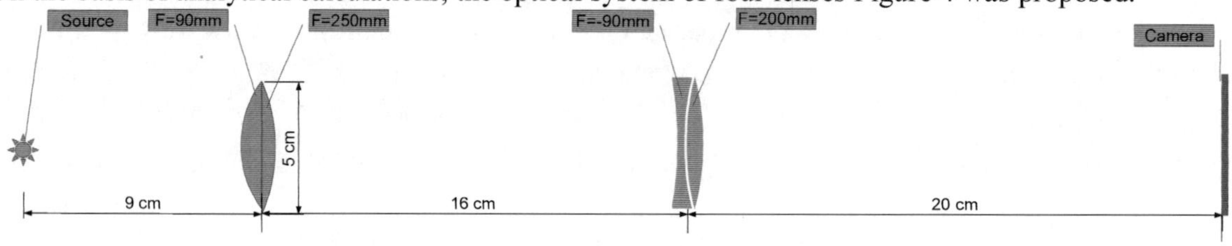

Figure 4. Scheme of the proposed optical system.

3. Experimental results

The principle of operation is based on one of the main optical properties of light - the refraction of light rays when passing the boundary of media with different densities. On the basis of the carried out analytical calculations, two blocks of lenses were assembled. The first block of lenses consists of two flat convex lenses with a diameter of 50 mm with focal lengths F_1 = 90mm and F_2 =250mm. The first lens is focused on the source at distance z_0 =90mm (the focal length of the lens $F_1 = z_0$). The lenses are placed closely to reduce the geometric dimensions of the optical system, but as shown by the results of analytical calculation they could be separated. On distance $z_1 = F_2 - F_3$=160mm placed the second block of lenses with a diameter of 50mm. This block consists of a flat-concave lens with focal length $F_3 = -90$ mm and flat-convex lens with focal length $F_4 = 200$ mm, which are placed close to each other. The flat-concave lens was taken from the idea that such a lens has a focus behind the lens, which allows us to reduce the geometrical dimensions of the entire system. On distance $z_2 = F_4$ =200 mm the matrix of the video camera (with the resolution 640x480 pixels) is placed. The image of the source is formed on the matrix of video camera. According to (7) a 6.17 upscale is provided. According to analytical calculations, an optical system was manufactured, which, in essence, is a microscope (Figure 5).

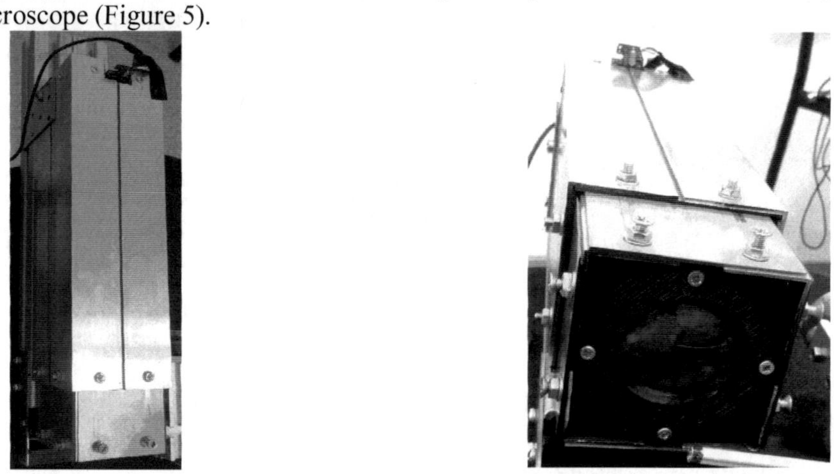

Figure 5. Photos of the microscope design containing: movable lens unit.

A microscope is a construction of two blocks: the first block is a lens block; the second block is the video camera matrix located in box of a larger size. A block with lenses is slid into the block of camera matrix; this allows changing the distance from the camera to the blocks of lenses and, thus, changing the magnification of the microscope.

IOP Conf. Series: Materials Science and Engineering **516** (2019) 012035 doi:10.1088/1757-899X/516/1/012035

With the help of the developed optical system, the ultrasonic waveguide was monitored. The installation consists of a developed microscope, a two-coordinate positioning system (a two-axis support), an ultrasonic wave guide, a stroboscopic illumination, a personal computer Figure 6.

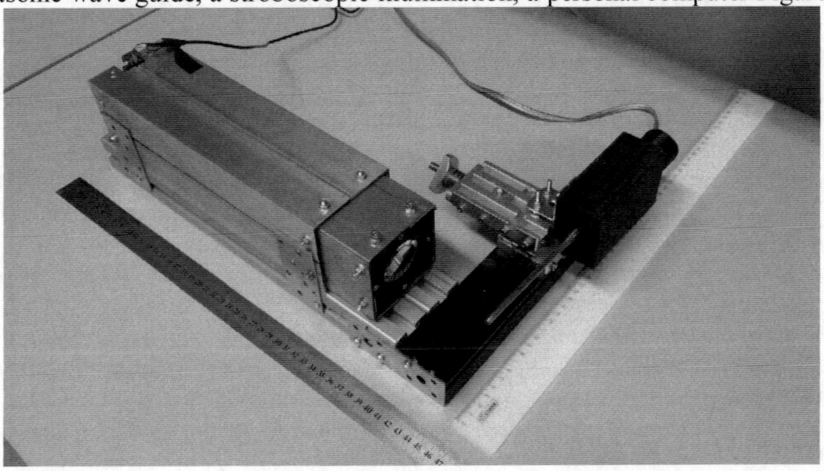

Figure 6. Photograph of the installation for monitoring oscillations of the tip of the ultrasonic waveguide.

Monitoring the oscillations of the tip of the ultrasonic waveguide is carried out as follows. The ultrasonic waveguide is placed in a holder placed on a two-axis support so that its tip is in the focus of the microscope, and is connected to the control unit. The microscope is located on the platform and is connected via a USB interface to a personal computer. The tip of the ultrasonic waveguide is illuminated with stroboscopic illumination with a frequency close to the resonant frequency of the ultrasonic waveguide carried out using high-speed infrared LEDs. The light reflected from the surface of the tip of the ultrasonic waveguide falls on the lens system of the microscope and forms the image on the digital matrix of the video camera. This allows us to observe the vibrations of the ultrasonic waveguide on the screen of a personal computer. Figure 7 shows the image of ultrasonic waveguide boundary.

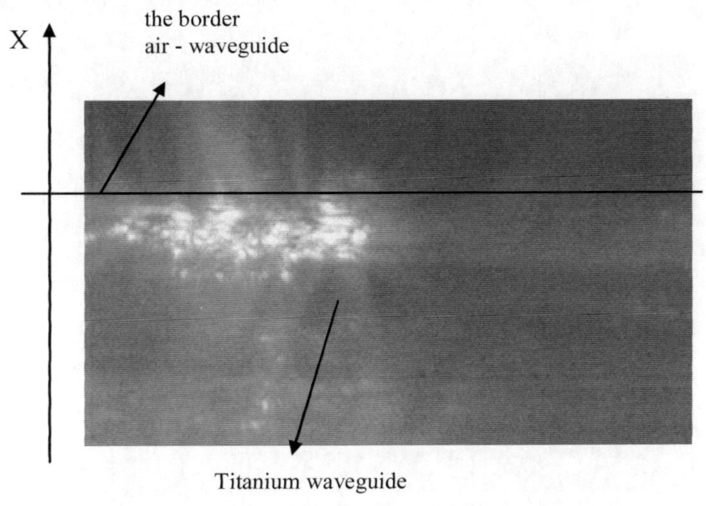

Figure 7. Image of ultrasonic waveguide tip border.

IOP Conf. Series: Materials Science and Engineering **516** (2019) 012035 doi:10.1088/1757-899X/516/1/012035

Thus, the developed optical system allows us to observe the boundary of the tip of the ultrasonic waveguide and monitor the vibrations of its tip. Observation of ruler image showed that applied camera allows to achieve spatial resolution of 1.25 mkm per pixel.

4. Conclusion

A numerical simulation of the optical system representing of the microscope was carried out. The calculated magnification factor of the optical system calculated from the simulation results coincided with analytical estimates based on the approximation of geometric optics. A microscope was assembled, which allows visualization of the air-wave guide boundary with an analytically predicted increase. The possibility of obtaining images with a resolution of 1.25 μm per pixel, which is sufficient for observing ultrasonic vibrations of waveguides, was experimentally shown.

Acknowledgments

The work was supported by the Ministry of Education and Science of Russia within the framework of the project "Creation of a high-tech complex of ultrasound surgery" (Unique identifier of the project RFMEF157517X0163).

References

[1] Makarov L.O. 1983 *Acoustic measurements in the processes of ultrasonic technology* (Mechanical Engineering: Mocsow)

[2] Leonov G.V., Khmelev V.N.,. Savin I.I, Abramenko D.S. 2005 *Intrnational Workshops and Tutorials on Electron Devices and Materials EDM'2005* 64–67

[3] Leonov G.V., Khmelev V.N., Savin I.I., Abramenko D. S. 2006 *Method of measuring the amplitude of oscillations.* The patent of the Russian Federation №2292530

[4] Khmelev V.N., Abramenko D.S., Barsukov R.V., Genna D.V. 2010 *Electronic Journal "Technical Acoustics"* **2(13)** 1–13

IOP Conf. Series: Materials Science and Engineering **516** (2019) 012036 doi:10.1088/1757-899X/516/1/012036

The analysis of optimal heat discard parameters in the combined cooling system of radio-electronic equipment

A Rumyantsev[1], A Lebedev[2], A Murashov[3], A Timoshenko[1], K Muratov[4]

[1]OJSC Academician A.L. Mints Radiotechnical Institute, Moscow, Russia
[2]Yaroslavl State Technical University, Yaroslavl, Russia
[3]Yaroslavl Higher Military School of Air Defense, Yaroslavl, Russia
[4]Tyumen Industrial University, Tyumen, Russia
E-mail: YVikulova@oaorti.ru

Abstract. The paper presents the mathematical model of the combined cooling system and calculation methods of optimal heat discard parameters of radio-electronic equipment as compared to standard cooling methods in high-power radiolocation stations. The optimization problem is solved to determine the control actions: the flow rate of cooling air in the air cooling system and the coolant in the liquid cooling system. The total cooling costs are considered as the main function, they consist of the costs of the cooling air and coolant circulation and the costs of cooling the heat carriers themselves. Restrictions are the conditions of air and liquid cooling thermal balance and maintaining the set temperature in the space where the radio-electronic equipment operates. The optimization problem in a limited area is solved by the bitwise search method. Air and circulating fluid cooling is carried out in the devices with free cooling which are located outside the radio-electronic system. The parameters of thermal modes of operation of these devices are given in the form of tables which are processed by means of regression analysis methods to solve the optimization problem. The air cooling system uses heat pipes with radiators which ensure efficient removal and dissipation of heat. Heat removal from radiators occurs in the turbulent mode, as the analysis shows. In conclusion a comparative analysis of different types of cooling systems energy consumption is carried out, depending on the required heat power and outside air temperature.

1. Introduction

Forced air cooling is the most common method of electronic equipment cooling; this is due to low energy consumption, simplicity of design and high reliability in operation. However, the constant power increase of modern electronic equipment leads to the need for the increase of the heat flux rate required for cooling, which can be achieved by using liquid cooling systems or combined systems [1–7]. Despite the fact that the described above systems do exist, the problem of effective cooling is not completely solved [8–14]; this is explained by low organization efficiency of the cooling medium interactions with the elements of the device, the inconsistency of interaction of the subsystems in different operating modes and lack of reliable methods of calculation and simulation.

2. Research Method

The authors of this paper introduce the new design of the combined air-liquid cooling system (Figure 1), which provides an opportunity for effective cooling at different temperatures. The effectiveness of the system is confirmed by a series of experimental theoretical studies carried out on a manufactured pilot plant. The effect of temperature and heating power on the power consumption has been analyzed during the experiments. The following source data have been used in the calculation of the cooling system, where Q [(kW] is total heat radiation power; C_2 [J/(kg°K)] is liquid specific heat; ρ_2 [kg/cm^3] is liquid density; Δt_2[°C] is temperature difference of the liquid on the refrigerating unit condenser (RU); C_1 [J /(kg·K)] is air specific heat; ρ_1 [kg/cm^3] is air density; t_{out} [°C] is ambient temperature; t_{in} [°C] is air temperature in the area, where the electronic equipment is installed. Moreover, the radiators operational characteristics are presented in the paper.

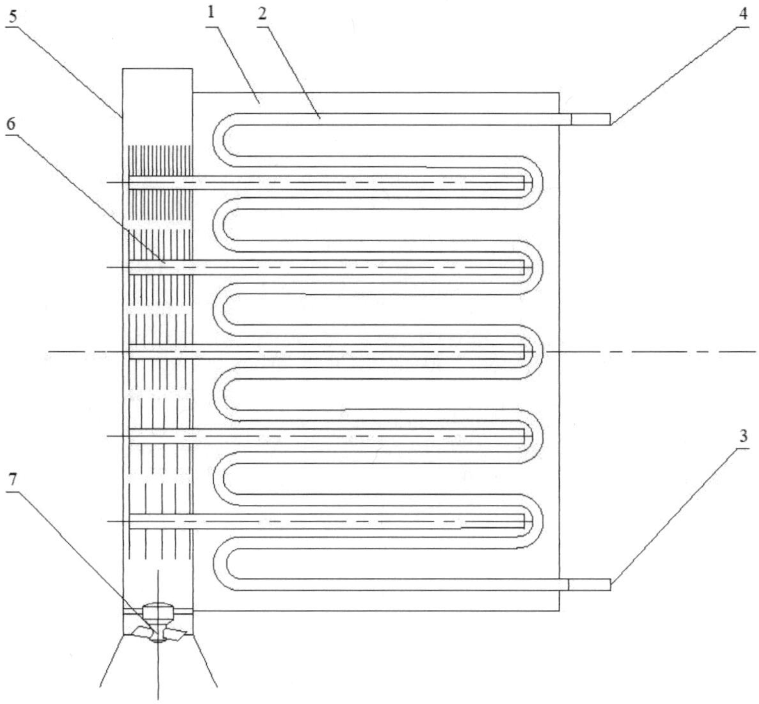

Figure 1. The general scheme of the combined cooling system of the radio-electronic unit, where 1 is frame; 2 are channels for the coolant; 3 and 4 are unions; 5 is air section shell; 6 are air unit rods with fins; 7 is fan.

3. Results and Discussion

The total expenditures on cooling consist of the power supplied for the coolant and air circulation and the power consumed absorbed by the liquid cooling system and the air-cooling system. Consequently, this can be expressed in the following way:

$$N_S = N_{cl} + N_{ca} + N_1 + N_2. \qquad (1)$$

Power expenditure for air and coolant circulation is represented by the Equation [2]:

$$N_{ca} = \frac{10^{-3} P_a V_1}{\eta_1}[kW],\ N_{cl} = \frac{10^{-3} P_2 V_2}{\eta_2}[kW], \qquad (2)$$

where P_1 is the hydrodynamic thrust of the fan which varies in the range of 60÷100 Pa; η_1 is the mechanical performance index of the fan which is measured in the range of 0.5÷0.8; P_2 is the hydrodynamic thrust of the pump, which varies in the range of 600÷800 kPa; η_2 is the mechanical performance index of the pump which is measured in the range of 0.7÷0.9.

The power consumption of external cooling systems depends on ambient temperature and cooling capacity, while the compressors have variable performance. Experimental data concerning power

IOP Conf. Series: Materials Science and Engineering **516** (2019) 012036 doi:10.1088/1757-899X/516/1/012036

consumption of air and liquid cooling systems (depending on ambient temperature and cooling capacity) are presented in Tables 1 and 2. The power values at temperatures from -20 to 5 °C are 0.050 [kW] and are not given in the tables.

Imagine the correspondence $N_i = N_i(Q,t)$ in the form of surfaces of the 3rd order and planes N=0.050, describing the negative temperature conditions:

$$N_1=0.052-0.01Q+(0.02Q^3-0.013Q^2+0.0017Q-0.00014)t+(0.0002Q-$$
$$-0.000035)t^2+(0.0000052Q-0.00000031)t^3; \quad (3)$$
$$N_2=0.045+0.02Q-0.025Q^3+(-0.0011Q-0.00067)t+(0.0002Q-$$
$$-\quad 0.000073)t^2+(0.0000133Q-0.00000532)t^3$$

Table 1. Power consumption at heat power 0.5 [kW].

t_{out}	N_1			
	Q=0.2 [kW]	Q=0.3 [kW]	Q=0.4 [kW]	Q=0.5 [kW]
10	0.052	0.054	0.054	0.058
15	0.054	0.056	0.064	0.068
20	0.056	0.068	0.082	0.094
25	0.058	0.084	0.110	0.150
30	0.062	0.100	0.140	0.170

Table 2. Power consumption at heat power 0.7 [kW].

t_{out}	N_2			
	Q=0.4 [kW]	Q=0.5 [kW]	Q=0.6 [kW]	Q=0.7 [kW]
10	0.054	0.058	0.060	0.062
15	0.058	0.060	0.070	0.074
20	0.060	0.070	0.088	0.106
25	0.064	0.086	0.120	0.150
30	0.070	0.110	0.170	0.220

The graphic presentation of the data $N_i = N_i(Q,t)$ is displayed in Figures 2 and 3.

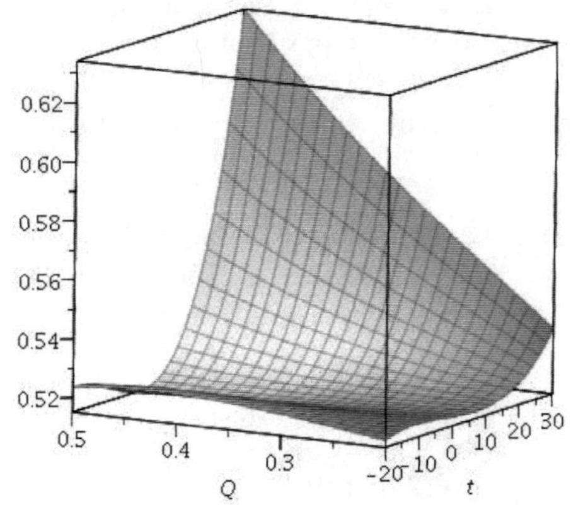

Figure 2. Graphic representation of correspondence $N_1 = N_1(Q, t)$.

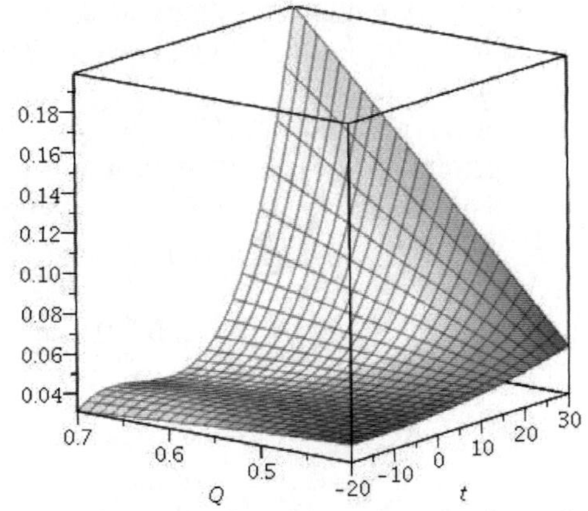

Figure 3. Graphic representation of correspondence $N_2 = N_2(Q, t)$.

We will search for the minimum power consumption defined by Equation (1) in a limited area. One of the equations included in the constraint system is the heat balance equation which is presented in the following way:

$$Q = Q_1 + Q_2 \, , \tag{4}$$

where Q_1 and Q_2 are the power of thermal radiation discarded by air and liquid. Here is the following:

$$Q_2 = C_2 \cdot \rho_2 \cdot V_2 \cdot \Delta t_2. \tag{5}$$

The magnitude Q_1 is calculated by a well-known method [3]. The air velocity in the air-cooling system is determined by the Equation $w=V_1/S_k$, where S_k is the radiator(s) heat pipes free area. The following characteristics are used in the calculations, where L is length [m]; S_{cros}, S_{long}, $S_{cros.r}$ are cross-sectional area of the air-duct and the longitudinal and cross-sectional area of the radiator [m^2]; Z is number of radiator fins.

The area of effective cross section is defined by the Equation: $S_k = S_{long} - S_{cros.r}$. Now the amount of heat discarded by convection is $Q_l = P_{conv} = Z\,\lambda_m m S_{long} w$, where λ_m is thermal conduction coefficient of the radiator metal [W/(m^2°K]; $m = (\frac{\alpha_c}{\lambda_m} \cdot L)^{1/2}$, $\alpha_c = \frac{Nu\lambda_a}{L}$ [W/(m2°K)], λ_a is thermal conduction coefficient of air [W/(м^2°K)], $Nu = 0{,}032Re^{0{,}8}$ is the Nusselt number, which is connected with the Reynolds criterion determined by the expression $Re = d \cdot w/v$, v is air kinematic viscosity [m^2/s], d is distance between the radiator fins.

Having calculated Q_1 we then count the air temperature differential on the condenser $\Delta t_1 = Q_1(C_1\rho_1V_1)^{-1}$. The average air temperature in the area is defined by the Equation $\bar{t}_a = t_{a.in} + \Delta t_a/2$, where $t_{a.in}$ is air temperature on entering the area. Hence, the condition limiting the problem space is presented as follows:

$$\bar{t}_a = t_{out}. \tag{6}$$

The optimization problem (1) with the constraints (4) and (6) solved by the gradient descent method [4] in the Maxima system at the given power of heat emission: Q=0.7 kW. Therein it was taken into account that the magnitude Q_1 is actually a function of the air consumption V_1. The calculations included the values of the model parameters: C_2.=4187; ρ_l.=1000 kg/m^3; Δt_l = 5°C; C_1. = 1005 J/[kg°K]; ρ_1.= 1.26 kg/m^3; t_{out}.= 20°C; t_{conf}=22°C; P_2. = 60 kPa; η_{out} = 0.7; P_1 = 80 Pa; η_a. = 0.7; Z. = 10; λ_m=2.44·10^2 [W/(m^2°C)]; λ_1 = 0.026 [W/(m°K)]; S_{cros}= 0.0285 m^2; L= 0.397 m; $S_{cros.r}$ = 0.0093 m^2; S_{long} = 0.12 m^2; d=0.015 m.

Figure 4 displays the dependencies $Q_1 = Q_1(V_1)$ at three values of the parameter S_{cros}. The upper curve corresponds to S_{cros} =0.0195 m^2, the middle S_{cros} =0.0295 m^2 and the lower S_{cros} =0.0395 m^2. Figure

5 shows the lines of equal quantities of the discarded heat at V_1, V_2 different for three values of Q. The upper graph corresponds to Q=900 W, the intermediate Q=800 W, the lower Q=700 W.

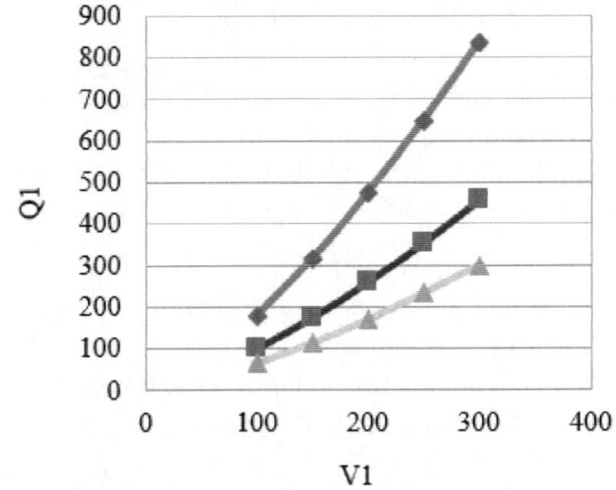

Figure 4. Rated dependencies $Q_1 = Q_1(V_1)$.

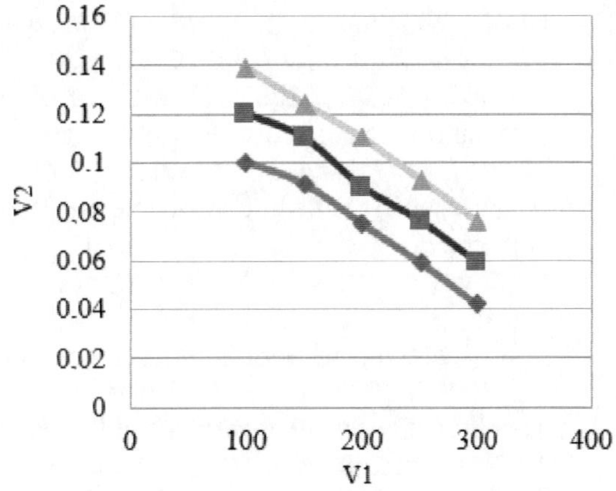

Figure 5. Lines of equal amounts of discarded heat.

Figures 6 and 7 reflect the dependences $N_s = N_s(V_1, V_2)$ at t=30°C (Figure 6), t=10°C (Figure 7). The range of values Ns=0 which lie in the horizontal plane correspond to those values V_1, V_2 which are incompatible with the heat balance condition.

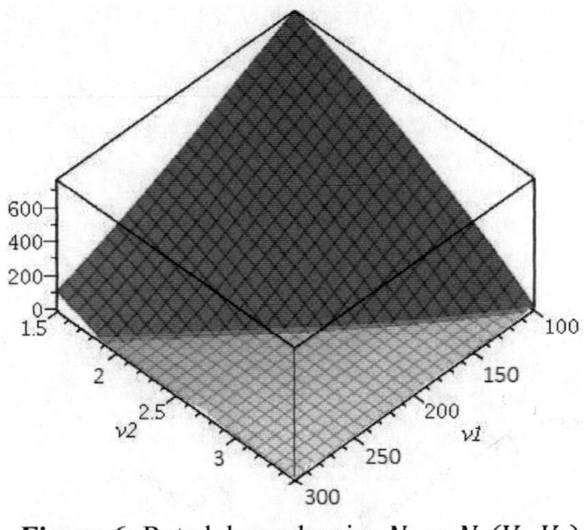

Figure 6. Rated dependencies $N_s = N_s(V_1, V_2)$ at t=30°C.

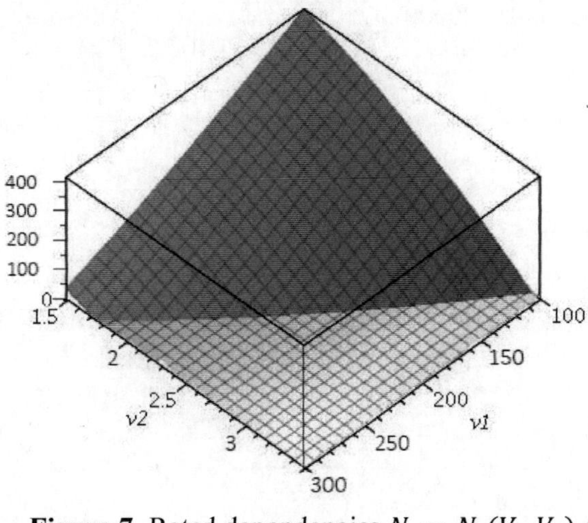

Figure 7. Rated dependencies $N_s = N_s(V_1, V_2)$ at t=10°C.

In consequence of solving the optimization problem (1) in a limited area (6) the values of the control parameters V_1, V_2, were obtained which ensure the minimum power spent on cooling the radio-electronic unit *Ns*:

$$t = -10^0C,\ V_1 = 150\ m^3/h,\ V_2 = 7.66\cdot 10^{-3}\ m^3/h,\ N_s = 70.62\ W$$
$$t = 10^0C,\ V_1 = 250\ m^3/h,\ V_2 = 7.63\cdot 10^{-2}\ m^3/h,\ N_s = 63.53\ W$$
$$t = 30^0C,\ V_1 = 250\ m^3/h,\ V_2 = 1.11\ m^3/h,\ N_s = 214.12\ W$$

4. Conclusion

A mathematical model of optimal heat discard in the air-liquid cooling system of radio-electronic equipment has been developed. The model makes it possible to define the parameters of air and liquid cooling systems for radio-electronic equipment with heat emission power 0.8 kW.

IOP Conf. Series: Materials Science and Engineering **516** (2019) 012036 doi:10.1088/1757-899X/516/1/012036

References

[1] *Refrigerating systems and heat pumps. Safety and environmental requirements* (2014) GOST (State standart) **2** EN 378-2-2014 (Standards Publishing House: Moscow)
[2] Kordon M. Ya. *et al* 2005 *Heat Engineering: Study Guide* (PSU: Penza) (in Russian)
[3] Boev S *et al* 2019 *AEBMR* **47** 581–584
[4] Lyashkov V. I. 2005 *Theoretical Foundations of Thermal Engineering: Study Guide* (Mashinostroenie: Moscow) (in Russian)
[5] *HSPICE® Reference Manual: MOSFET Models. Version D-2010.12* (2010) URL: http://www.ele.uri.edu/courses/ele448/HspiceRef/hspice_mosmod.pdf
[6] Naberezhnykh A I *et al* 2012 *Proc of 1th Int Conf* (2012) **1** 321–327 URL: http://www.spsl.nsc.ru/FullText/konfe/october2012vol1.pdf (in Russian)
[7] Yazan K. 2008 *Patent DE* WO2008/025630 URL: https://patentscope.wipo.int/search/en/detail.jsf?docId=WO2008025630&redirectedID=true
[8] Ruy S *et al* 2007 *Proc of IEEE Int Conf ICPE'07* 59–61 doi: 10.1109/ICPE.2007.4692348
[9] Zanchini E. *and* Lazzari S. 2009 *Proc of IEEE Int Conf ICREPQ* **1(7)** 668–675 doi: 10.24084/repqj07.469
[10] Bazhenov A. A. *and* Yarovikov V. I. 2007 *Measurement Techniques* **50(12)** 1282–1290
[11] Pons J. L. *et al* 2004 *Sensors and Actuators A: Physical* **110(1–3)** 336 – 343
[12] Boev S. F. *et al* 2018 *IOP Conf. Ser.: Mater. Sci. Eng.* **457** 012003 doi: 10.1088/1757-899X/457/1/012003
[13] Boev S. F. *et al* 2018 *IOP Conf. Ser.: Mater. Sci. Eng.* **457** 012010 doi: 10.1088/1757-899X/457/1/012010
[14] Plotnikova I. *et al* 2016 MATEC Web of Conferences **79** 01019 doi: 10.1051/matecconf/20167901019

IOP Conf. Series: Materials Science and Engineering **516** (2019) 012037 doi:10.1088/1757-899X/516/1/012037

Adaptive neuro-fuzzy classifier for evaluating the technology effectiveness based on the modified Wang and Mendel fuzzy neural production MIMO-network

V I Syryamkin, S V Gorbachev, M V Shikhman

National Research Tomsk State University, Tomsk, Russia
E-mail: egs@sibmail.com

Abstract. The paper describes practical results gained during synthesis of the classification model for estimating technology effectiveness when solving multi-criteria problems of expert data analysis (Foresight) aimed to identify technological breakthroughs and strategic perspectives of scientific, technological and innovative development.

1. Introduction

Foresight research is becoming a popular tool for solving multi-criteria tasks of expert data analysis that allow identifying technological breakthroughs and strategic perspectives of scientific, technological, and innovative development [1]. As a system of methods, Foresight is constantly developing and improving; for the last 20 years, an extensive experience of its practical application has been accumulated [2]. The analysis of publications proves the effectiveness of the combination of qualitative and quantitative methods of expert data processing [3]. Moreover, with the constant complication of the system of methods, it is important to develop intellectual approaches that can provide scientifically valid transparent conclusions for Foresight research [4].

The purpose of this study is to develop a modified MIMO-structure of the adaptive fuzzy Wang and Mendel neural production network based on the fuzzy decision tree and adaptive algorithms of neuro-fuzzy inference when solving multi-criteria expert analysis problems (foresight).

2. Initialization of the modified Wang and Mendel fuzzy neural production MIMO-network

Let us have a data set D describing multidimensional feature objects (innovative technologies), where in each line there are m numerical values for the attributes $x_1, x_2, \ldots, x_m$ and one class label C_i of the set $C = \{C_1, C_2\}$.

Let x_{kj} be the value of the k-th input variable of the j-th observation; C_j is the value of output variable of the j-th observation; K is the number of observations; m is the number of input variables.

Rule i: IF x_1 is A_{1i} AND x_2 is A_{2i} AND ... AND x_m is A_{mi} , THEN $y_1 = r_i(1)$ AND $y_2 = r_i(2)$, where A_{ki} is the linguistic term by which the k-th input variable is estimated.

Figure 1 shows the structural-functional diagram of the constructed adaptive neuro-fuzzy classifier with a logic output that implements the MIMO- mapping. The module solves the problem of evaluating the effectiveness of innovative technologies as multidimensional feature objects using the fuzzy decision tree (stage 1) and fine-tuning of the parameters of membership functions and conclusion rules by hybrid learning algorithms (step 2). It consists of seven layers with the following basic functions of the nodes.

IOP Conf. Series: Materials Science and Engineering **516** (2019) 012037 doi:10.1088/1757-899X/516/1/012037

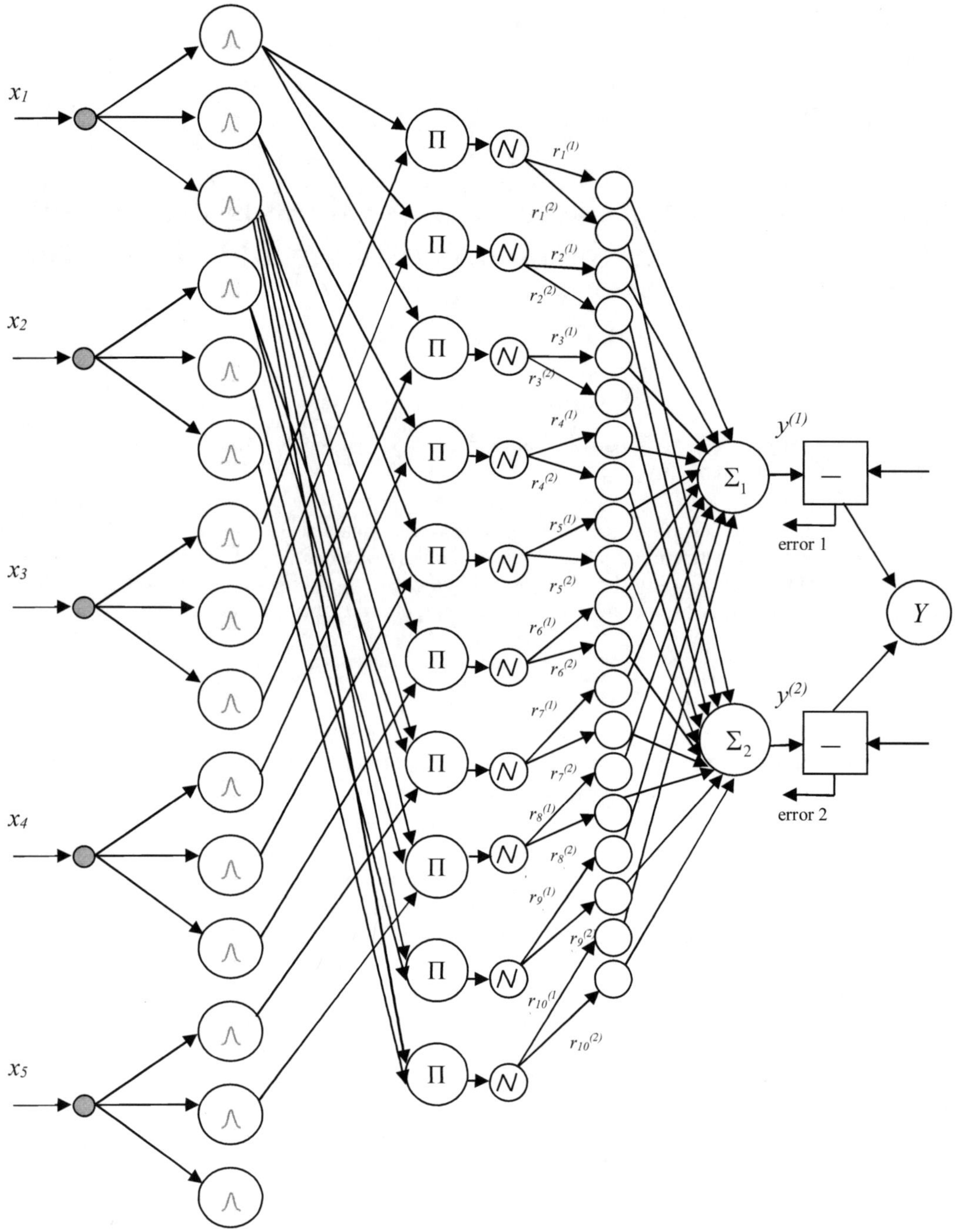

Figure 1. Structural-functional diagram of the adaptive neuro-fuzzy classifier based on the modified Wang and Mendel fuzzy neural production network with MIMO-structure.

Layer 1 (input).

Indicators evaluated by experts as innovative technologies, for example, the degree of the technology influence on global problems arising within the framework of the most pressing global crises (environmental, food, geopolitics, and energy) are considered as inputs.

The nodes in the *1*-st layer accurately convey the input attribute values into the *2*-nd layer.

Layer 2 (fuzzification) is characterized by the conversion of numerical attribute values into linguistic terms in order to reduce the information and present it in a form understandable for experts. Each *k*-th input variable is evaluated by linguistic term A_{ki}.

Layer 3 (aggregation) determines the degree of truth conditions for each of the fuzzy rules (branches of the tree); it is a non-adaptive layer of *AND*-neurons that simulate logical conjunction *AND* by multiplication of membership functions:

$$w_{ij} = \prod_k \mu_{ki}\left(x_{kj}\right) = \prod_k \frac{1}{1 + \left|\frac{x_{kj} - c_{ki}}{\sigma_{ki}}\right|^{2b_{ki}}}, i = 1,\dots,n,$$

where i is the rule number, the corresponding branch of the decision tree; in this case $i=1,\dots,10$; n is the number of fuzzy rules.

The multiplication is calculated over all variables x_k presented at the i-th tree branch and included in the i-th rule.

w_{ij} can be interpreted as the weight or the degree of truth of the i-th rule (tree branch) condition in the j-th example.

The nodes of this layer are non-adaptive, they expect the relative degree (weight) for the implementation of fuzzy rules among all the rules in the j-th example according to the formula:

$$\overline{w_{ij}} = \frac{w_{ij}}{\sum_{i=1}^{n} w_{ij}}.$$

Layer 5 (activation) has 2 leaves at each branching point of the tree (rules) in accordance with the number of classes. On the leaves of the tree, it determines the degree of truth of the rule conclusions (the contribution of each rule) for each class using the weighting parameters r_i^1 and r_i^2 that are initialized, for example, by the method of the nearest neighbor search.

Note that the initialization of the parameters of membership functions (centers and widths) of the rule conditions and conclusions is provisional and they are to be fine-tuned on the next phase.

For each class, adaptive nodes in layer 5 calculate the degree of truth of rule conclusions with the propagation of the j-th signal according to the formula:

$$y_{ij}^{(1)} = \overline{w_{ij}} r_i^{(1)},\ y_{ij}^{(2)} = \overline{w_{ij}} r_i^{(2)},\ i = 1,\dots,n.$$

Layer 6 (defuzzification) combines the degree of truth of the conclusions of each rule to obtain the membership functions of the output variable. For each class, non-adaptive nodes in this layer generate the output variable values $y_j^{(1)}$ and $y_j^{(2)}$, which represent the degree of truth allocating the j-th signal to the class as the sum of the truth degrees of all rules:

$$y_j^{(1)} = \sum_{i=1}^{n} y_{ij}^{(1)},\ y_j^{(2)} = \sum_{i=1}^{n} y_{ij}^{(2)}.$$

Thus, for the j-th example, neurons of the *6*-th layer implement the function approximation, which can be written as follows:

$$y_j = \frac{\sum_{i=1}^{n}\left(r_i \prod_k \mu_{ki}\left(x_{kj}\right)\right)}{\sum_{i=1}^{n}\prod_k \mu_{ki}\left(x_{kj}\right)} = \frac{\sum_{i=1}^{n} r_i \prod_k \frac{1}{1 + \left|\frac{x_{kj} - c_{ki}}{\sigma_{ki}}\right|^{2b_{ki}}}}{\sum_{i=1}^{n}\prod_k \frac{1}{1 + \left|\frac{x_{kj} - c_{ki}}{\sigma_{ki}}\right|^{2b_{ki}}}}. \quad (1)$$

After fine-tuning parameters (step 2), *layer 7 (output)* calculates the total degree of truth Y_j of the predicted class l_j of the j-th signal according to the method of the right modal value (based on the maximum degree of truth):

$$\text{the network output } Y_j = \max\left(y_j^{(1)}, y_j^{(2)}\right),$$

$$\text{the predicted class } C_j = \arg\max_{l=1,2}\left(y_j^{(l)}\right).$$

The accuracy of fuzzy inference is achieved by collaboration of rules, which belong to the same class.

3. Adaptive tuning of parameters of the neuro-fuzzy classifier

The constructed network of fuzzy inference with forward signal propagation (Figure 2) has five input variables, each of which was fuzzified into three linguistic terms. Branches of the fuzzy decision tree that connect the layers 2 and 3 form the rule base. It is obvious that each node in layer 3 corresponds to exactly one rule. Thus, the rule base includes 10 rules.

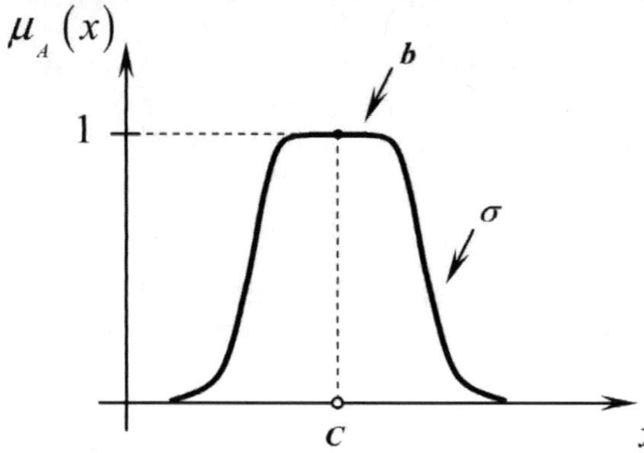

Figure 2. Generalized bell-shaped membership function.

We describe a hybrid-learning algorithm for center and width parameters of the bell-shaped membership functions and rule conclusions. In the hybrid algorithm, the parameters to be adapted are divided into two groups: linear parameters r_i of the 5-th layer and the parameters c_{ki} and d_{ki} of nonlinear membership functions in the 2-nd layer. Correction of the parameters is performed in two stages.

Stage 1.

When fixing specific values of the membership function parameters, we calculate linear parameters r_i by solving a system of linear equations (the value obtained in the initialization are in the first cycle). With the known values of membership functions, formula (1) in the j-th example is a linear combination of the parameters r_i with constant coefficients $\overline{w_{ij}}$:

$$y_j = \sum_{i=1}^{n} \overline{w_{ij}} r_i, \tag{2}$$

where $\overline{w_{ij}} = \dfrac{w_{ij}}{\sum_{i=1}^{n} w_{ij}} = \dfrac{\prod_k \mu_{ki}(x_{kj})}{\sum_{i=1}^{n}\prod_k \mu_{ki}(x_{kj})} = const$ is the normalized degree of truth of conditions at the i-th rule, $i=1,\dots,n$.

With K training examples ($X(j)$, d_j), $j=1, \dots K$) and by substitution of the output signal with the expected value of d_j from (2), we obtain the system of K linear equations:

$$\begin{cases} \overline{w_{11}}r_1 + ... + \overline{w_{n1}}r_1 = d_1 \\ ... \\ \overline{w_{1K}}r_1 + ... + \overline{w_{nK}}r_1 = d_K \end{cases}$$

which can be conveniently written in the matrix form:

$$\begin{bmatrix} \overline{w_{11}} & \overline{w_{21}} & ... & \overline{w_{n1}} \\ \overline{w_{12}} & \overline{w_{22}} & ... & \overline{w_{n2}} \\ ... & & & \\ \overline{w_{1K}} & \overline{w_{2K}} & ... & \overline{w_{nK}} \end{bmatrix} \begin{bmatrix} r_1 \\ r_2 \\ ... \\ r_n \end{bmatrix} = \begin{bmatrix} d_1 \\ d_2 \\ ... \\ d_K \end{bmatrix}.$$

In the abbreviated form: $w \cdot r = d$.

The dimension of matrix W is equal to $K{\cdot}n$. With the help of pseudoinverse of W matrix, the solution can be obtained in one step:

$$r = W^{+}d,$$

where W^{+} is pseudoinverse of matrix W.

Pseudoinverse of matrix W is done using the singular value decomposition *(SVD)* with the following reduction of its dimension.

Recursive Greville and Fadeev algorithms belong to universal methods for finding the matrix pseudo-return. In this paper, we give the Greville algorithm for the pseudo-inversion of matrices.

4. The Greville's algorithm for the pseudo-inversion of matrices

Let us imagine that there is matrix $W \in R^{K \times n}$ and w_k is its *k*-th column, *k=1, ..., n.*

Let W_k be a matrix composed of the first k columns of the matrix W:

$$W_k = [w_1 \; w_2 \; ... \; w_k].$$

In case $k = 1$: $W_1 = w_1$.

In case $k = 2,..., n$: $W_k = [W_{k-1} \; w_1]$.

$W_n = W$.

Matrix $W^{+} \in R^{n \times K}$ can be computed using the recursive algorithm.

1. Initialization

$$W_1^{+} = \begin{cases} 0, \text{ if } w_1 \\ \dfrac{w_1^T}{\|w_1\|^2}, \text{ otherwise.} \end{cases}$$

2. In a cycle for $k = 2,...,n$

$$W_k^{+} = \begin{bmatrix} W_{k-1}^{+}(I - w_k f_k) \\ f_k \end{bmatrix}.$$

where I is the *1*-st matrix of order K,

$$f_k = \begin{cases} \dfrac{c_k^T}{\|c_k\|^2}, \; c_k = \left(i - W_{k-1}W_{k-1}^{+}\right)w_k, \; c_k \neq 0 \\ \dfrac{w_k^T \left(W_{k-1}^{+}\right)^T W_{k-1}^{+}}{1 + \left\|W_{k-1}^{+}w_k\right\|^2}, \; c_k = 0 \end{cases}$$

Obtained on the last step, the matrix W_n^{+} is the desired pseudo-inversion matrix.

Stage 2.

After fixing the values of the linear parameters r_i, the actual outputs y_j of the network for each j-th sample, $j=1,\dots, K$ are calculated. We use a linear relationship for this calculation:

$$\begin{bmatrix} y_1 \\ y_2 \\ \dots \\ y_K \end{bmatrix} = Wr$$

Next, we calculate the error vector $\varepsilon = y - d$.

The training is based on minimizing the differentiable target function according to the method of least squares specified with the use of the Euclidean norm as a mean square deviation E [5]:

$$E = \frac{1}{2}\left(y_j - d_j\right)^2,$$

where y_j is the output value obtained as a result of fuzzy inference;
d_j is output (reference) value from the training sample.

The necessary condition for minimization of the error is that its derivatives in the parameters c_{ki}, σ_{ki}, and b_{ki} of generalized bell-shaped membership functions of the k-th input variable $\left(k \in [1,m]\right)$ on the i-th branch (the rule) are equal to zero.

The error signals are sent through the connected network in the direction to the input of the network (back propagation of error) up to the 2-nd layer, where components of the target function gradient may be calculated. After the formation of the gradient vector, the parameters are specified using one of the gradient methods of learning, for example, the method of steepest descent [6]:

$$c_{ki}(t+1) = c_{ki}(t) - \eta_c \frac{\partial E(t)}{\partial c_{ki}},$$

$$\sigma_{ki}(t+1) = \sigma_{ki}(t) - \eta_\sigma \frac{\partial E(t)}{\partial \sigma_{ki}},$$

$$b_{ki}(t+1) = b_{ki}(t) - \eta_b \frac{\partial E(t)}{\partial b_{ki}}.$$

After refinement of non-linear parameters, the process of adaptation of linear parameters (the 1-st stage) and non-linear parameters (the 2-nd stage) is restarted. This cycle is repeated until all process parameters are stabilized.

Formulas for calculating the target function gradient for one learning example *(x,d)* take the following form:

$$\frac{\partial E(t)}{\partial c_{ki}} = \left(y(x) - d\right)\sum_{i=1}^{n} r_i \frac{\partial \overline{w_i}}{\partial c_{ki}},$$

$$\frac{\partial E(t)}{\partial \sigma_{ki}} = \left(y(x) - d\right)\sum_{i=1}^{n} r_i \frac{\partial \overline{w_i}}{\partial \sigma_{ki}},$$

$$\frac{\partial E(t)}{\partial b_{ki}} = \left(y(x) - d\right)\sum_{i=1}^{n} r_i \frac{\partial \overline{w_i}}{\partial b_{ki}}.$$

where the partial derivatives to the parameters of the *bell-shaped membership function* are calculated by the following formula:

$$\frac{\partial \overline{w_h}}{\partial c_{ki}} = \frac{\delta_{hi} m(x_k) - l(x_k)}{\left[m(x_k)\right]^2} \prod_{s,s\neq k} \mu_{si}(x_s) \frac{\left[\frac{2b_{ki}}{\sigma_{ki}}\left(\frac{x_k - c_{ki}}{\sigma_{ki}}\right)^{2b_{ki}-1}\right]}{\left[1+\left(\frac{x_k - c_{ki}}{\sigma_{ki}}\right)^{2b_{ki}}\right]^2},$$

$$\frac{\partial \overline{w_h}}{\partial \sigma_{ki}} = \frac{\delta_{hi} m(x_k) - l(x_k)}{\left[m(x_k)\right]^2} \prod_{s,s\neq k} \mu_{si}(x_s) \frac{\left[\frac{2b_{ki}}{\sigma_{ki}}\left(\frac{x_k - c_{ki}}{\sigma_{ki}}\right)^{2b_{ki}}\right]}{\left[1+\left(\frac{x_k - c_{ki}}{\sigma_{ki}}\right)^{2b_{ki}}\right]^2},$$

$$\frac{\partial \overline{w_h}}{\partial b_{ki}} = \frac{\delta_{hi} m(x_k) - l(x_k)}{\left[m(x_k)\right]^2} \prod_{s,s\neq k} \mu_{si}(x_s) \frac{\left[-2\left(\frac{x_k - c_{ki}}{\sigma_{ki}}\right)^{2b_{ki}} \ln\left(\frac{x_k - c_{ki}}{\sigma_{ki}}\right)\right]}{\left[1+\left(\frac{x_k - c_{ki}}{\sigma_{ki}}\right)^{2b_{ki}}\right]^2}$$

for $h=1,2,..,n$, (n is a number of tree branches (rules)),
δ_{ki} denotes the Kronecker delta:

$$\delta_{ki} = \begin{cases} 1, & k = i \\ 0, & k \neq i \end{cases}.$$

$l(x_k) = \prod_s \mu_{si}(x_s)$, $m(x_k) = \prod_s \mu_{si}(x_s)$ for all variables x_s occurring on the i-th branch of the decision tree and included into the i-th rule.

5. The calculation of the target function gradient relative to the parameters of the Gaussian membership function

Note that the final formulas for the calculation of the target function gradient relative to the parameters of the membership function depend on the used definition of the function error on the network output and on the type of the membership function. For example, when using the *Gaussian function*:

$$\mu(x) = \exp\left[-\left(\frac{x-c}{\sigma}\right)^2\right].$$

The appropriate formula of the target function gradient for a single example (x, d) from the training samples takes the following form:

$$\frac{\partial E}{\partial c_{ki}} = (y-d)(r_i - y)\overline{w_i}\frac{2(x_k - c_{ki})}{(\sigma_{ki})^2},$$

$$\frac{\partial E}{\partial \sigma_{ki}} = (y-d)(r_i - y)\overline{w_i}\frac{2(x_k - c_{ki})^2}{(\sigma_{ki})^3},$$

where $\overline{w_i} = \frac{\prod_k \mu_{ki}(x_{kj})}{\sum_{i=1}^{n}\left[\prod_k \mu_{ki}(x_{kj})\right]}$.

In the practical implementation of the hybrid learning method of the network, the dominant factor in adaptation is the first stage when the weights r_i are chosen using pseudo-inversion [7]. To balance its influence, the second stage (selection of non-linear parameters by gradient descent method) is repeated in each cycle.

Table 1 presents the average ratios of RMS errors in experiments on test data in 10-multiple cross-validation, the standard deviation of errors is presented in parentheses.

Table 1. The root means square error classification before and after the parametric settings of the neuro-fuzzy classifier.

No	Test set	Fuzzy decision tree		Modified Wang and Mendel fuzzy neural production MIMO- network	
		Average ratio of RMS errors	Standard deviation	Average ratio of RMS errors	Standard deviation
1	Breast	1.49	(0.00)	0.29	(0.0)
2	Credit	7.81	(0.00)	2.18	(0.48)
3	Cylinder	6.16	(4.51)	1.43	(3.18)
4	Diabetes	21.84	(1.27)	4.19	(1.15)
5	Gamma	21.13	(0.70)	2.53	(3.41)
6	Glass	39.13	(0.00)	5.85	(0.43)
7	Haberman	26.67	(0.00)	3.27	(1.47)
8	Heart	14.44	(5.60)	2.89	(0.87)
9	Ionosphere	3.99	(7.35)	0.53	(0.00)
10	Iris	8.00	(2.67)	3.48	(2.56)
11	Liver	36.76	(4.65)	4.13	(3.16)
12	Segmentation	12.38	(2.33)	0.14	(0.07)
13	Spam	28.98	(0.00)	2.51	(0.18)
14	Steel	20.78	(0.93)	4.17	(0.24)
15	Vehicle	25.37	(1.80)	2.95	(1.64)
16	Wine	5.00	(6.43)	0.18	(3.82)

6. Conclusion

In conclusion, we note that the practical importance of the modified Wang and Mendel fuzzy neural production MIMO-network is determined by the great potential opportunities of its application in intellectual systems for analysis and control of multi-criteria decision-making [8] in the case of processing multidimensional semi-expert data describing promising technologies and scientific and technical (innovation) objects.

The results of the study are listed below.

1. Instead of classical fuzzy production network, we can use neuro-fuzzy decision tree structure that automatically defines the base of fuzzy rules for solving problems of complex object classification in a multidimensional semi-space of features using expert data as a neuro-fuzzy system. We can also apply the inverse cycles of iteration in hybrid method of back propagation of error for adjusting the parameters (centers and widths of membership functions and the weight parameters of the rule conclusions in the leaves of the tree).
2. Synthesized knowledge base may be interpreted as a decomposition of the space of influencing factors on the field with blurred boundaries, within which the response function takes a fuzzy value. The number of such fuzzy regions is equal to the number of rules.
3. The described neural network method of adapting parameters of a decision tree (a fuzzy hierarchical system) based on the error backpropagation from leaves to the root node improves

classification accuracy of the fuzzy decision tree saving the interpretation without changing the tree structure.

4. In contrast to the known heuristic methods for the synthesis of neuro-fuzzy networks, which lead to the construction of a parametrically complex and cumbersome model with a large number of neurons, the method proposed increases the speed of the classical algorithm because of the optimized network and reduced the number of calculations.

5. From the point of view of the expert methods, we have received a new intellectual logically transparent foresight tool for conducting foresight studies

Acknowledgments
The paper was supported by “The Tomsk State University competitiveness improvement programme” under grant (No 8.2.24.2018) and by the Russian Foundation for Basic Research (grant No 16-29-04388).

The authors are grateful to Tatiana B. Rumyantseva from Tomsk State University for English language editing.

References

[1] Sokolov A. V. 2007 *Foresight* **1(1)** 8–15
[2] Georghiou I., Cassingera Harper J., Keenan M., Miles I. 2018 *The Handbook of Technology Foresight* (Edward Elgar Publishing: England)
[3] Morozova N. V. 2014 *Bulletin of the Chuvash University* **3** 178–182
[4] Gorbachev S. V., Syryamkin V. I., Syryamkin M. V., Vaganova E. V. 2017 *Innovation 2017* 66–70
[5] Kruglov V. V., Long M. I., Golunov R Y 2001 *Fuzzy logic and artificial neural networks* (FIZMATLIT: Moscow)
[6] Ezhov A. A., Shumsky S. A. *1998 Neurocomputing and its application in economics and business: teaching aid* (MEPhI: Moscow)
[7] Gorbachev S. V. 2015 *International Conference on Cognitive Computing and Information Processing* 341–346
[8] Abramova T. V., Vaganova E. V., Gorbachev S. V., Gribovsky M. V., Syryamkin V I, Syryamkin M. V., Yakubovskaya T. V. 2012 *The cognitive system of monitoring and forecast of scientific-technological development of the state* (Publishing house of Tomsk state University: Tomsk)

IOP Conf. Series: Materials Science and Engineering **516** (2019) 012038 doi:10.1088/1757-899X/516/1/012038

The roll pressure dependence of the roll forming mill on the basic profile parameters

I Manzhurin, B Bakhyt, N Yelena, S Yelena

Karaganda State Technical University, Karaganda, Kazakhstan
E-mail: elena.sidorina.78@mail.ru

Abstract. In the article results of the roll pressure dependence of the roll forming mill on the basic profile parameters are given. The pressure roll forming mill has different values with different conditions of curved profile production. It depends on the thickness of the original billet, the bend's radius and also it depends on other parameters: the width of the bendable elements, rolls' diameters, the hem's angles, the settings of the mill, the material properties, etc. It is difficult to solve such a complicated problem - taking into account the influence all factors with using the general theory of plasticity. The issue is complicated due to the influence of these factors, most of which are statistical in nature, different in magnitude and direction. It determines the pressure on the rolls as a stochastic variable.

1. Introduction

Stochastic value, in the pressure on the rolls' case, can be fully described if the law of its distribution is known, in other words, the relationship that establishes between the possible values of the variable and the corresponding probabilities. The law of distribution can be established on the basis of the statistical material analysis. The source [1–10] shows that the pressure on the rolls follows a log-normal distribution law. This makes it possible to investigate the dependence of the pressure on the rolls on the profiling parameters by the mathematical processing of experimental data methods, in particular, correlation and regression analysis, mathematical experiment planning. For this purpose, more than 500 pressure values were analyzed with profiling angles and channels from strip t_0= 1 ÷ 5 mm thick from carbon and low-alloy steels along single routes with bending angles ↦=0-15˚; ↦=0-30˚; α=0-45˚; α=0-60˚; α=0-75˚; α=0-90˚ with relative bending radius r/t_0=0.5÷8, the bend flange's width B=50÷350 мм, at different mill settings s/t_0 (where r – the bending radius, mm; s – the gap between the rolls, mm).

2. Materials and Methods

To establish the dependence of the pressure on the mill rolls, based on the statistical (not functional) nature of the pressure dependence on the profiling parameters, it is advisable to use methods that correspond to the nature of the studying process. Therefore, the correlation-regression analysis is the most effective method. The data needed to plot an empirical regression line of the dependence $P=f(r/t_0)$ is given in Table 1.

In Table 1, the relative midpoints of the intervals are calculated by the equation $y'=\frac{(y_{cp}-c)}{i}$, the pressure P values by intervals according to the formula $y_x = c + [\Sigma(m_x y')\cdot i]/\Sigma m_x$, the c= 27 kN; y_x=P; $x=r/t_0$; i=6 кN; Σm_x - the amount included in the interval.

IOP Conf. Series: Materials Science and Engineering **516** (2019) 012038 doi:10.1088/1757-899X/516/1/012038

Table 1. Correlation table of P dependence on r/t_0.

№	$x = r/t_0$ / $y = P$, кN	y'	0÷2 1	2÷4 3	4÷6 5	6÷8 7	8÷10 9	10÷12 11	$\sum m_y$
1	2	3	4	5	6	7	8	9	10
1	0 ÷ 6	-4	54	45	23	7	14	3	146
2	6 ÷ 12	-3	129	31	25	2	-	-	181
3	12 ÷ 18	-2	86	5	13	2	1	-	107
4	18÷ 24	-1	48	1	-	-	-	-	49
5	24 ÷ 30	0	16	2	1	-	-	-	19
6	30 ÷ 36	1	12	-	-	-	-	-	12
7	36÷ 42	2	7	-	-	-	-	-	7
8	42 ÷ 48	3	2	-	-	-	-	-	2
9	48 ÷54	4	2	-	-	-	-	-	2
10	54 ÷ 60	5	1	-	-	-	-	-	1
	Σm_x		351	84	62	11	15	3	526
	$\sum(m_x y')$		-760	-284	-193	-38	-58	-13	
	$y' = \Sigma(my')/\Sigma y'$		-2.16	-3.38	-3.11	-3.45	-3.87	-4.0	
	y_x		14.1	6.72	8.34	6.3	3.78	3.0	

Note. In all tables, y'-relative midpoint; my-sum of pressure values in the interval

Figure 1 shows the experimental regression line. From the analysis of the experimental regression line, it follows that with a relative bending radius increasing to $r/t_0 \approx 5 \div 6$, the pressure decreases, and then the P decreasing slows down. A function of this type is well described by an equation of the logarithmic type.

$$y = 13.186 - 5.605 \ln(x) \tag{1}$$

(reliability coefficient of approximation R^2=0.8715), along which the calculated regression line was plot Figure 1.

The closeness the dependence indicator of y on x with nonlinear regression is the correlation ratio

$$\eta = S(y_x)/S(y) \tag{2}$$

where S – the standard deviation value of the corresponding dispersion estimate. Where $S^2(y_x) = \frac{1}{n}(\Sigma m_x y_x^2) - y_x^2$ – dispersion estimate y_x is relative to total average dispersion estimate y, but $S^2(y_x) = \frac{1}{n}\Sigma m_y (y - y_x)^2$ – dispersion estimate y is relative to total average dispersion estimate y_x.

The correlation ratio is a general the closure connection indicator, not taking into account the communication form. Therefore, in the non-linear correlation case , a regression coefficient is introduced.

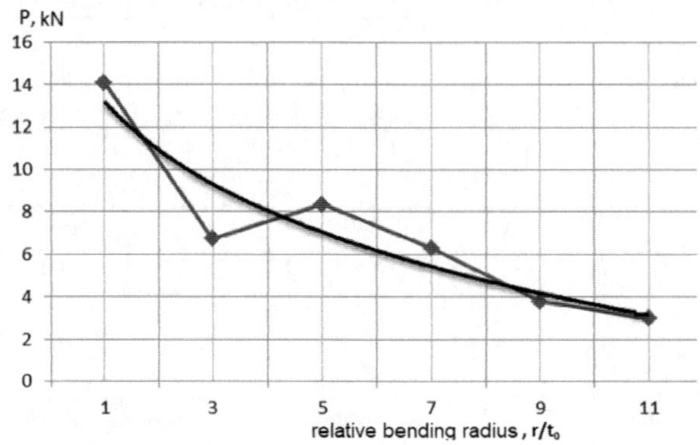

Figure 1. Experimental and calculated regression lines of pressure dependence on relative bending radius.

$$\eta' = \frac{s(y_p)}{s(y)} \tag{3}$$

where $S^2(y_p) = \frac{1}{n}(\Sigma m_x y_p^2) - y^2$ – the estimate of the calculated variances $y_{p'}$ obtained from equation (1) with relative to total average *y*.

The correlation ratio η=0.39 indicates the influence of the relative bending radius on the magnitude of the force on the mill rolls.

Similarly to table 1, for the dependence of pressure P on the width of the flange to be bent a correlation table was created.

The values of pressure plotted on the graph Figure 2 from the correlation and the empirical regression line which is plotted from them shows the dependence of P on the width of the shelf being folded can be approximated by a quadratic function — a second order parabola $y=14.34-4.5376x+1.01167x^2$, accuracy coefficient of approximation for which the calculated regression line is constructed.

$$R^2=0.9761 \tag{4}$$

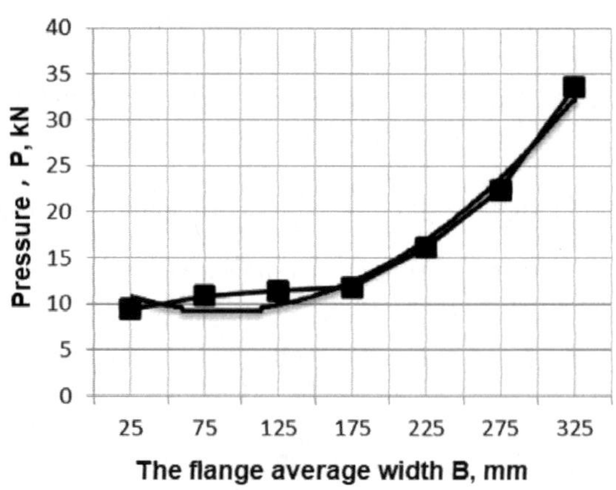

Figure 2. Experimental and calculated regression lines of the pressure dependence on the width of the flange.

The coefficients in equation (4) are calculated by the least squares method and the calculated regression line Figure 2 with the regression coefficient is $\eta'=0.81$; the correlation relation is $\eta_B=0.36$.

The pressure dependence on the angle α^o of the bend is represented by a correlation table similar to Table 1 and in Figure 3.

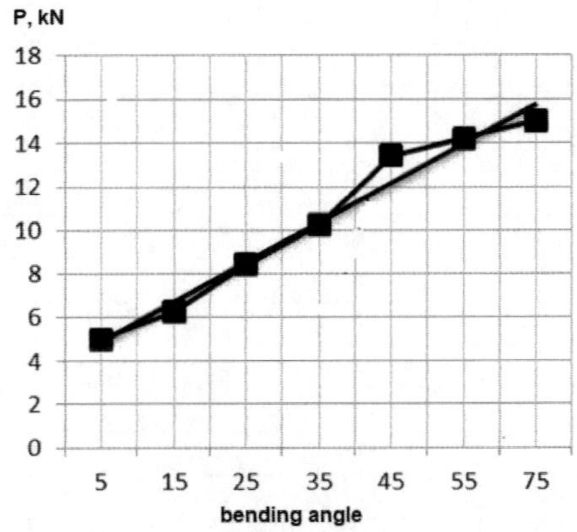

Figure 3. Experimental and calculated regression lines of pressure dependence on bending angle, α°.

The estimated regression line is plotted on the equation

$$y = 3.0743 + 1.8186x \tag{5}$$

Correlation coefficient $r_\alpha=0.91$, approximation reliability coefficient $R^2=0.9739$.

The efforts in the mill stands with a gap between the rolls *s* of equal strip $t_0(s=t_0)$ thickness are not significant. As the initial parameters for the strength calculations, they cannot be taken, because the degree of possible roll forming mill overloading is very high. The reason for such overloads is the change in the gap between the rolls. This is due to the fact that sometimes accidentally, most often deliberately, with non-compliance with the geometry of the profile, they change the compression of the rolls, while the gap is practically not controlled, and the efforts of overloads can exceed the efforts of the normal process several times.

The dependence of pressure on the gap between the rolls is represented by the correlation Table 2 and in Figure 4.

The estimated regression line is based on the equation

$$y = 32.328 - 12.76\ln(x),\ x = \frac{s}{\tau_0} \tag{6}$$

The correlation ratio $\eta_{s/t}=0.55$ indicates a fairly significant dependence of the force on the rolls on the mill setting, and the regression coefficient $\eta' = 0.88$ indicates a good approximation of the dependence of P on s/t_0, the equation described by equation (6), (reliability coefficient of approximation $R_2 = 0.9583$).

IOP Conf. Series: Materials Science and Engineering **516** (2019) 012038 doi:10.1088/1757-899X/516/1/012038

Table 2. Correlation table dependencies *P* from s/t_0.

№	$x=s/t_0$ / $y=P$ [кН]	y'	**0.7**	**0.8**	**0.9**	**1.0**	**1.1**	**1.2**	**1.3**	$\sum m_y$
1	2	3	4	5	6	7	8	9	10	11
1	0 ÷ 6	-4	-	-	6	134	-	2	4	146
2	6 ÷ 12	-3	-	-	11	145	11	8	6	181
3	12 ÷ 18	-2	-	3	6	94	3	1	-	107
4	18÷ 24	-1	-	8	8	31	1	1	-	49
5	24 ÷ 30	0	6	4	-	6	1	1	1	19
6	30 ÷ 36	1	5	1	1	3	3	-	-	12
7	36÷ 42	2	1	1	-	5	-	-	-	7
8	42 ÷ 48	3	-	1	1	-	-	-	-	2
9	48 ÷54	4	-	1	1	-	-	-	-	2
10	54 ÷ 60	5	1	-	-	-	-	-	-	1
	$\sum m_x$		13	19	36	418	16	13	11	526
	$\sum(m_x y')$		12	-4	-67	-1177	-40	-35	-34	
	$\hat{y}'=\sum(m_x y')/\sum m_x$		0.92	-0.21	-1.86	-2.80	-2.50	-2.7	-3.09	
	$\hat{y}_x$		32.52	25.74	15.84	10.20	12.0	10.8	8.46	

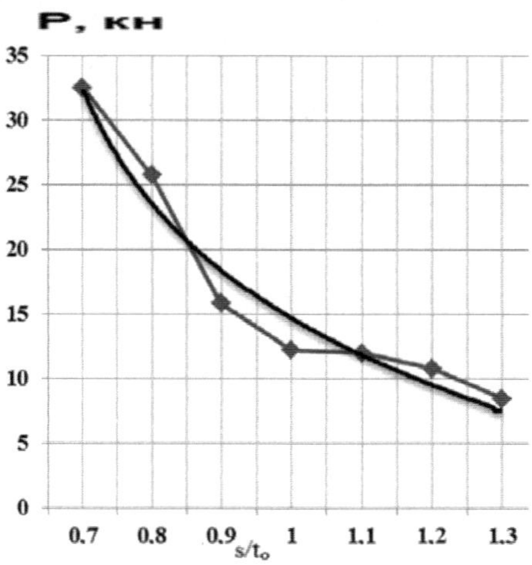

Figure 4. Experimental and calculated regression lines of pressure dependence on mill settings.

Based on the pair correlation equations (1,4,5,6) the general dependence of pressure on profiling parameters can be represented by the equation:

$$P=\left(\frac{s}{t_0}\right)b_1\cdot\left(\frac{r}{t_0}\right)b_2\cdot\alpha b_3\cdot Bb_4 \qquad (7)$$

After logarithm of both parts we get $y=b_0+b_1x_1+b_2x_2+b_3x_3+b_4x_4$

where y =lnP; b_0,b_1,b_2,b_3,b_4— equation parameters; x_1,x_2,x_3,x_4 – logarithms s/t_0, r/t_0, α, B.

To simplify the calculation of the coefficients in (7) on the basis of experimental data using the least squares method using matrices, the variables are coded using the formulas:

IOP Conf. Series: Materials Science and Engineering **516** (2019) 012038 doi:10.1088/1757-899X/516/1/012038

$$X_1 = (\ln x_1 - \ln x_{01})/\ln(\Delta x_{01});$$
$$X_2 = (\ln x_2 - \ln x_{02})/\ln(\Delta x_{02});$$
$$X_3 = (\ln x_3 - \ln x_{03})/\ln(\Delta x_{03}); \quad (8)$$
$$X_4 = (\ln x_4 - \ln x_{04})/\ln(\Delta x_{04}).$$

Before creation a planning matrix, the area in which the experiment will be performed must be specified, other words, boundary values for each factor $x_i\ min$ (lower level) and $x_i\ max$ (upper level) described by the process technology. The midpoint $x_{i\ avr}$ between the boundary values is the planning center or the middle (main) level. For the convenience of matrix theory using in the calculation of coefficients in (7) by the least squares method, the levels are coded in such a way that the upper corresponds to: +1, the middle: 0, the lower: -1. Often a X_i is called an encoded variable. The results of coding are shown in Table 3.

Table 3. Coding of variable factors.

Level	$\frac{s}{t_0}$	$\frac{r}{t_0}$	α°	B	X_1	X_2	X_3	X_4
Upper	1.2	3	90	150	1	1	1	1
Average	1.0	2	60	100	0	0	0	0
Lower	0.8	1	30	50	-1	-1	-1	-1

The planning matrix for two-quarters of the replicas from the complete factorial experiment $N=2^4$ with generating ratios $x_4=x_1x_2x_3$ and $x_4= - x_1x_2x_3$ determining contrasts $x_1x_2x_3x_4 = -1$ and respectively is presented in the Table 4.

Table 4. Planning matrix.

№	$\frac{s}{t_0}$	$\frac{r}{t_0}$	α°	B	Code designation					Output	
					X0	X_1	X_2	X_3	X_4	P,κN	y=lnP
1	0.8	3	30	150	1	-1	1	-1	1	21.5	3.0681
2	1	2	60	100	1	0	0	0	0	11.3	2.4224
3	1.2	1	30	150	1	1	-1	-1	1	5.0	1.6094
4	1	2	60	100	1	0	0	0	0	11.5	2.4424
5	0.8	1	90	150	1	-1	-1	1	1	52.0	3.9513
6	1.2	1	30	50	1	1	-1	-1	-1	3.1	1.1314
7	1.2	3	90	150	1	1	1	1	1	4.7	1.5476
8	1	2	60	100	1	0	0	0	0	10.1	2.3126
9	1	2	60	100	1	0	0	0	0	10.3	2.3322
10	0.8	3	30	50	1	-1	-1	1	-1	15.9	2.7660
11	0.8	1	90	50	1	-1	-1	-1	-1	41.7	3.7542
12	1.2	3	90	50	1	1	1	1	-1	3.3	1.1939

The experiments realization order is randomized according to the table of random numbers.

IOP Conf. Series: Materials Science and Engineering **516** (2019) 012038 doi:10.1088/1757-899X/516/1/012038

The coefficients b_0, b_1, b_2, b_3, b_4 are determined based on the planning matrix and the inverse matrix.

$$(X)^{-1} = \left\| \begin{matrix} 1/12 & 0 & 0 & 0 & 0 \\ 0 & 1/8 & 0 & 0 & 0 \\ 0 & 0 & 1/8 & 0 & 0 \\ 0 & 0 & 0 & 1/8 & 0 \\ 0 & 0 & 0 & 0 & 1/8 \end{matrix} \right\|$$

They are equal to the diagonal terms of the matrix $(X'X)^{-1}$ multiplied by Σy, that is

$$b_0 = \frac{1}{12}\Sigma_1^{12} y_i ; bi = \frac{1}{8}\Sigma_1^{8} x_i y_i ;$$

The selective coefficients for the parameters are determined independently of each other with the lowest possible variances, because the matrix was taken with the mutually orthogonal column vectors. The application of the least squares method provides the minimum estimate of the variance $S^2(y)$. Calculated the coefficients, we obtain an equation reflecting the dependence of the pressure on the rolls on the profiling parameters in coded form

$$y = 2.2776\text{-}1.0071X_1\text{-}0.2338X_2\text{+}0.2340X_3\text{+}0.1664X_4 \tag{9}$$

The adequacy of equation (9) is verified by analysis of variance estimates. The estimate of the variance that defines the inadequacy of the presentation of the results of the experiment with f_1=4 degrees of freedom equal 0.005102, and zero points when f_2=3 the degrees of freedom equal 0.004164. The ratio of these two estimates gives the estimated variances of the Fisher criterion $F_p = 1.22$ *At* a probability of 0.95 and four and three degrees of freedom the table value of the Fisher criterion F_T=*9.1* Comparison $Fp = 1.22 << 9.1 = F_T$ allows to make a conclusion about the obtained mathematical model adequacy of the studied dependence.

After decoding and potentiation equation (9) is given to the form:

$$P = \frac{0.59 \cdot \alpha^{0.42} \cdot B^{0.30}}{(s/t_0)^5 \cdot (s/t_0)^{0.43}}, kN \tag{10}$$

IOP Conf. Series: Materials Science and Engineering **516** (2019) 012038 doi:10.1088/1757-899X/516/1/012038

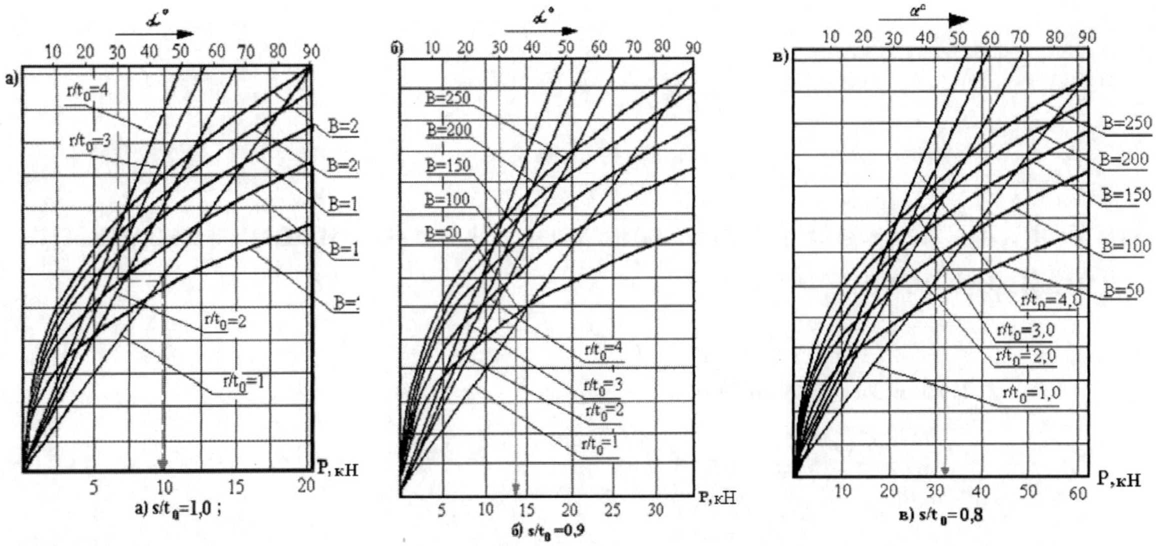

Figure 5. Nomograms for determining the pressure on the rolls on the parameters of the profiling and setting of the mill.

3. Conclusion

In this way, using mathematical methods of experimental data, an adequate pressure dependence equation for the specified parameters was obtained. Also confidence intervals were calculated and nomograms were constructed, according to which, depending on the geometric dimensions of the profile, the profiling route and the mill settings, it is possible to determine the forces occurring in the working rolls of the roll forming mill.

When developing the process of profiling and calculating the calibration of the rolls, you can use the graphs of the pair dependence of pressure on the profiling parameters (Figure 1—3), drawing attention to the mill setting (gap between the rollers). In case of non-compliance with the geometry of the profile is often prescribed preload rolls. At the same time, the gap is practically not controlled, and the force of overloads can exceed the efforts of the normal process several times.

When calculating the pressure from the presented nomograms for design calculations based on the different stiffness of the individual mill sites and the safety factor conditions, the safety factor should be taken 2–3.

References

[1] Bogoyavlenskij K., Ris V., Manzhurin I. 1971 *Tr.LPI Obrabotka metallov davleniem* **(322)** 31 – 34
[2] Manzhurin I., Sidorina Ye., 2012 *ZAO Metallurgizdat* 65 – 68
[3] Adler Yu., 1969 *Vvedenie v planirovanie ehksperimenta* 157 – 161
[4] Adler Yu., Markova E., Yu.V. 1971 *Planirovanie ehksperimenta pri poiske optimal'nyh* 283 – 286
[5] Mitropol'skij A. 1982 *Tekhnika statisticheskih vychislenij* 490
[6] Ventcel E. 1980 *Teoriya veroyatnostej* 571
[7] Nalimov V. Chernova N. 1965 *Statisticheskie metody planirovaniya ehkstremal'nyh ehksperimentov* 340
[8] Nalimov V. 1971 *Teoriya ehksperimenta* 208
[9] Sautin S. 1975 *Planirovanie ehksperimenta v himii i himicheskoj tekhnologii* 48
[10] Kiks Ch. 1967 *Osnovnye principy planirovaniya ehksperimenta* 407

IOP Conf. Series: Materials Science and Engineering **516** (2019) 012039 doi:10.1088/1757-899X/516/1/012039

Increasing the energy efficiency of an autonomous station for unmanned aerial vehicles

R Gimazov[1], S Shidlovskiy[2]

[1]Tomsk Polytechnic University, Tomsk, Russia
[2]Tomsk State University, Tomsk, Russia
E-mail: shidlovskiy@tpu.ru

Abstract. This paper addresses the issue of increasing the energy efficiency of an autonomous station for unmanned aerial vehicles. Low efficiency of modern solar cells is the most important problem of the entire industry of solar energy. The problem of increasing the energy efficiency of installations with solar batteries (photovoltaic installations) for the power supply system of autonomous objects is urgent and connected with the task of maximizing the autonomous operation time. The simulation results showed that the use of extreme regulation increased energy production by 23%, and the use of fuzzy logic significantly increased the speed of reaching the operating point and neutralized voltage fluctuations, which in turn reduced the underproduction of power by an additional 2%.

1. Introduction

The autonomous station of an unmanned aerial vehicle (UAV) performs the functions of a storage, weather station, charger, and communication center. The autonomy of the system is achieved using photovoltaic modules that provide power to the station and charge the unmanned aerial vehicle. Such a system has the possibility of modification in order to increase the energy efficiency of the power supply system. A general view of the station is shown in Figure 1.

The purpose of this work is to increase the energy efficiency of the power supply system of an autonomous UAV station. The tasks of the work include implementation of an effective adaptive charge control algorithm, reorienting photovoltaic modules, and development of an additional block of photocells with a Sun-pointing system.

The well-known and widely used methods for improving the energy efficiency of systems with photovoltaic cells include extreme power control and implementation of the solar tracker. Extreme power control is a method, in which a special charge process algorithm is implemented providing energy extraction at the maximum point of the volt-watt characteristic, which maximizes the energy received from photovoltaic cells. The implementation of the solar tracker is a system for automatic guidance of solar cells on the sun, which increases the amount of incoming energy.

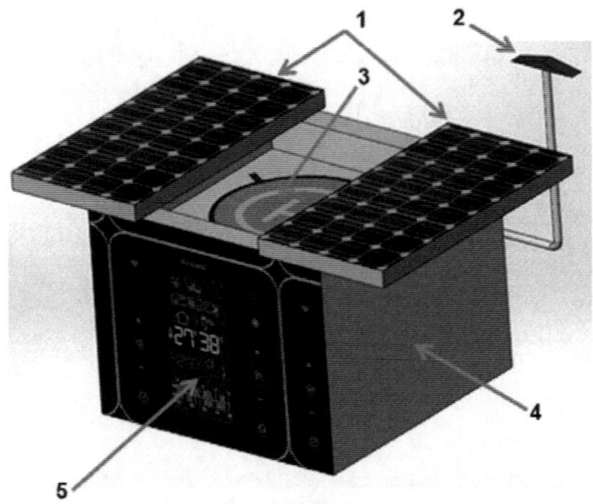

Figure 1. Autonomous UAV station: 1 – movable covers with solar panels; 2 – weather station; 3 – landing field; 4 – frame; 5 – control panel.

Extreme power control or Maximum Power Point Tracking (MPPT) is one of the methods to increase the energy efficiency of photovoltaic modules and wind power plants by obtaining the maximum possible output power of these systems. To implement the MPPT, digital devices are used that analyze the current-voltage characteristic (CVC) to determine the optimal current-voltage pair providing the maximum output power. Figure 2 shows an example of the current-voltage characteristic of a photovoltaic module with the maximum power point indicated on it.

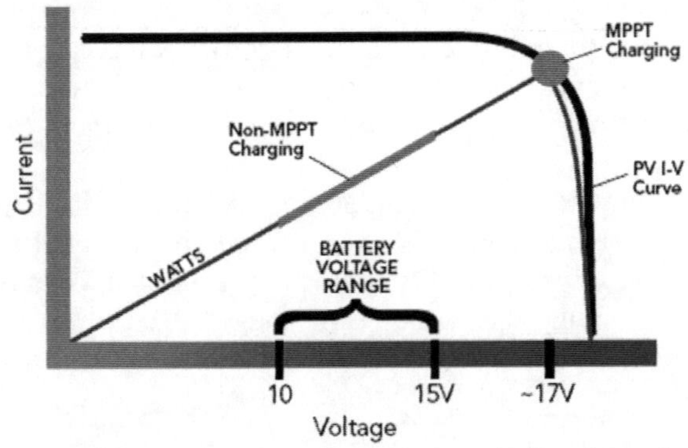

Figure 2. MPPT on the current-voltage characteristic of the photovoltaic module.

The most common MPPT algorithm is the 'perturb and observe' algorithm. In this method, the MPPT device changes the input resistance by a small amount (which results in the voltage changes of the solar installation) and measures the power; if the power increases, the controller continues further adjustments in that direction until power no longer increases. The principle of the 'perturb and observe' algorithm is shown in Figure 3.

The disadvantages of the 'disturb and observe' algorithm include the power fluctuations and the fixed time for hill climbing (the search time for the maximum power point) [1]. Reducing the time for hill climbing leads to an increase in the amplitude of oscillations, which leads to underproduction of power. A decrease in the amplitude of power fluctuations leads to an increase in the search time for the maximum power point. These problems can be solved by using a fairly accurate predictive and adaptive algorithm [2].

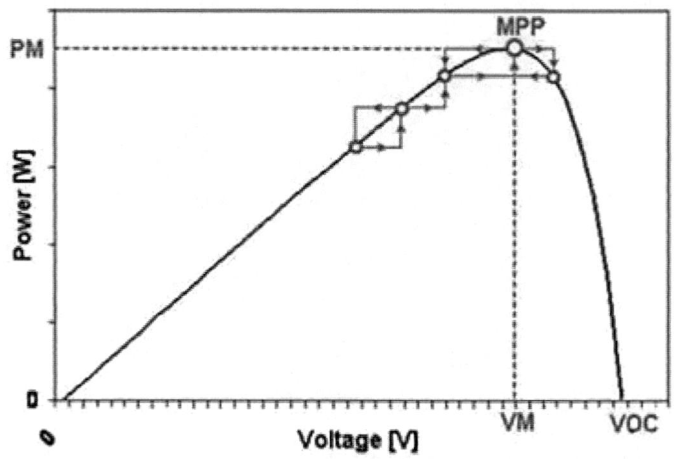

Figure 3. The principle of the maximum power point tracking according to the 'perturb and observe' algorithm.

As an additional measure to increase the energy efficiency for power supply system of the autonomous station, we propose an additional reorientation of photovoltaic modules in order to increase the amount of received light (implementation of a solar tracker). Figure 4 shows the principle of this improvement.

It is known that the maximum of the selected energy is reached when sunbeams are perpendicular relative to the photocell plane, therefore, the concept 3 (in Figure 4) has the best indicators of energy selection.

Adding an extra module of photovoltaic cells allows significant increasing of the amount of energy withdrawn in such a system. However, such a module placed outside the main unit of an autonomous station, may adversely affect the mobility of the plant under development because makes it becomes more cumbersome.

To study the energy efficiency of the charging station, using Matlab / Simulink, we developed a simulation model of the system including models of all the required elements of the system.

The relevance of the presented development is related to the problem of maximizing the autonomous operation time of the UAV. Improving the energy efficiency of the power supply system of the UAV station can significantly increase the autonomy of the presented object.

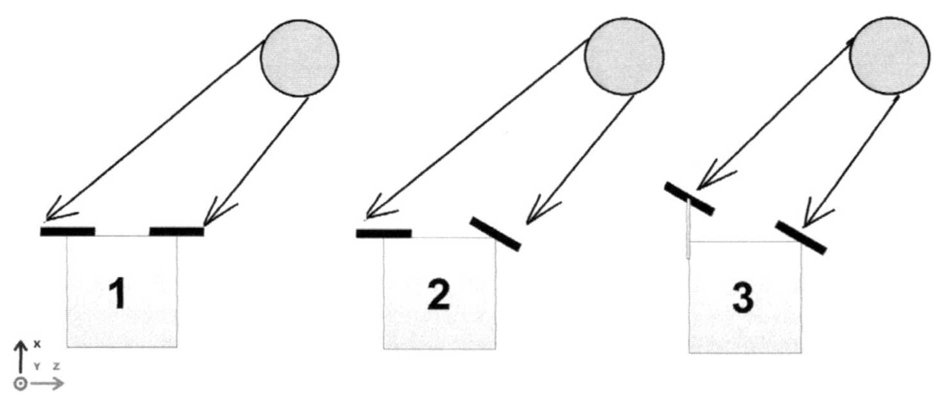

Figure 4. Reorientation of photovoltaic cells (relative movement of the Sun along the *y*-axis): 1 – solar panels are rigidly fixed; 2 – solar panels can change the angle of inclination; 3 - solar panels on additional lifting frames.

IOP Conf. Series: Materials Science and Engineering **516** (2019) 012039 doi:10.1088/1757-899X/516/1/012039

2. Development of the autonomous power supply system

An autonomous object with a solar-powered supply system designed for fast transport must use light and high-strength materials and electronic circuits that consume little energy. The lower is the consumption of electrical energy and the mass of the structure, the more promising is the use of solar panels.

There are several options for building power supply systems based on solar cells [3, 4]. One of the options is to use solar cells as a source of secondary power to recharge batteries. In such a power supply system, power requirements for solar cells are lower in comparison with direct solar power supply. However, in this case, the autonomous system will operate actively only part of the time, and for the rest part of the time, it will recharge its batteries. Another option for using solar cells is direct powering (recharging) of an autonomous object from solar cells. There is also a combined option when solar cells are used as sources of direct and secondary power.

This paper considers the development of a combined power supply system based on solar cells. The idea of building a charger based on a single-board microcontroller for a photovoltaic installation is as follows. A Buck converter is used as a charger and a microcontroller controls its key. When the voltage/current of the photovoltaic module increases, the PWM generator (which is a MPPT controller) increases the pulse frequency, which in turn leads to an increase in the output current.

The functional diagram of the photovoltaic system is shown in Figure 5.

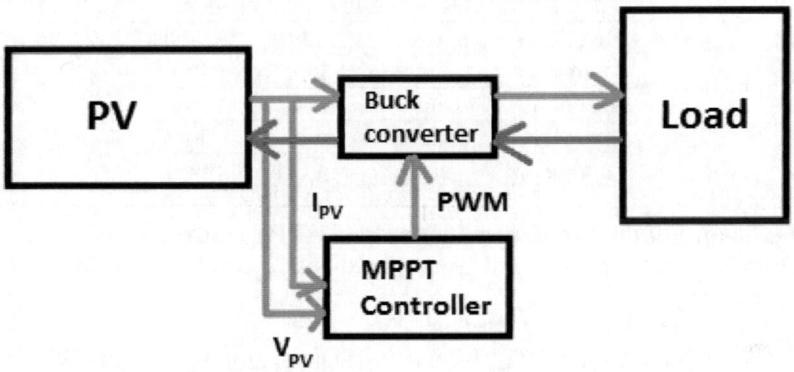

Figure 5. Functional diagram of the photovoltaic system

3. Simulation modeling of the system

The MATLAB/Simulink environment was used to create the autonomous power supply model [5].

In order to evaluate the efficiency of using the MPPT algorithm and the fuzzy regulation-based MPPT algorithm, in addition to the controller itself, we developed a photovoltaic system with a solar battery, a charge controller, a storage battery and a load. The developed model is shown in Figure 6.

The ratio of the change in the power of the solar battery to the change in the voltage of the solar battery (E), and the nature of the change E (dE) are used as input linguistic variables of the fuzzy controller.

$$E(k) = (P(k)–P(k\text{-}1))/(V(k –V(k\text{-}1)) \quad (1)$$

$$dE(k) = E(k)–E(k\text{-}1) \quad (2)$$

The output variable of the fuzzy logic block is the scaling factor of the MPPT algorithm step. The dependence of the output variable value on the input linguistic variables is determined by the set of rules given in the table 1.

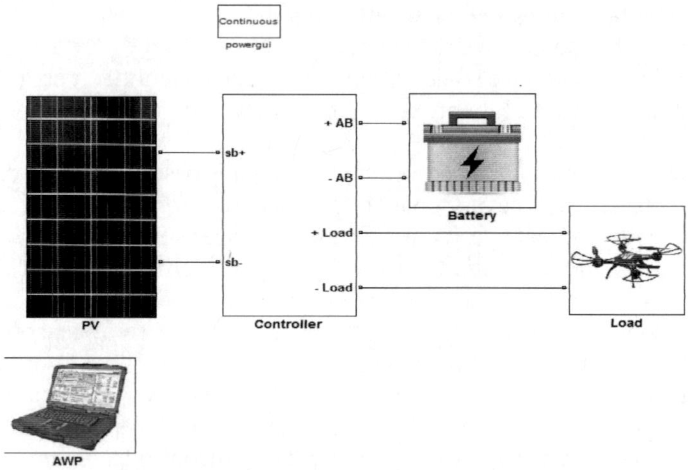

Figure 6. Model of photovoltaic installation.

Table 1. Rules of the fuzzy controller.

E↓ \ dE→	low	middle	high
low	low	low	low
middleL	midL	midL	low
middleH	midH	midL	low
high	high	midH	midL

4. Study of the influence of the fuzzy logic unit on the MPPT controller

To study the effect of a fuzzy logic unit on a solar photovoltaic installation, the following initial model parameters were set:

- Solar battery with a nominal voltage of Un=12V.
- Normal conditions: temperature 25°C, level of solar insolation W = 1000 kW/m^2, angle of rays incidence α = 90° (with the condition that the constant right angle is provided by the solar tracker).
- Consumer load is shutdown.
- The solar battery charges the storage battery with a nominal voltage of 9V.

The simulation results are presented in Figures 7 and 8 (a normal MPPT algorithm is highlighted in green; a fuzzy logic-based algorithm is highlighted in blue).

Figure 7 shoes that the use of the fuzzy logic block in the MPPT controller significantly increased the speed of finding the maximum power point, and also reduced the voltage fluctuations, which in turn neutralized the power fluctuations (Figure 8).

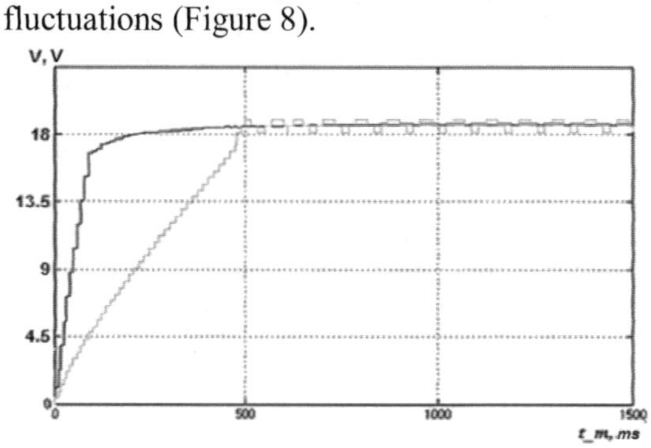

Figure 7. Voltage comparison of the modes of maximum power point tracking: a normal MPPT algorithm is highlighted in green; fuzzy logic-based algorithm is highlighted in blue.

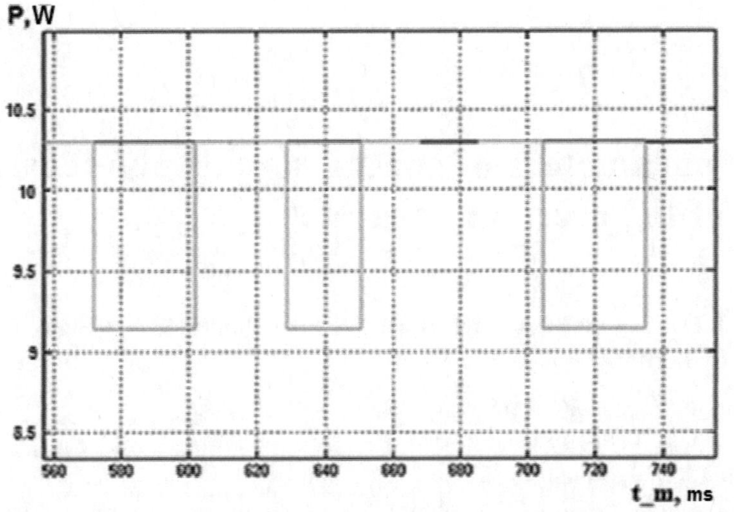

Figure 8. Power comparison of the modes of maximum power point tracking: normal MPPT algorithm is highlighted in green; a fuzzy logic-based algorithm is highlighted in blue.

5. Conclusion

In the course of this work, a power supply system for a solar battery-based autonomous station for an unmanned aerial vehicle has been developed. The effectiveness of the developed system was enhanced by using a solar tracker and MPPT charge controller with an adaptive fuzzy logic-based algorithm. The effectiveness of the MPPT 'disturb and observe' algorithm with adaptation was studied using simulation modeling. As a result of simulation modeling, it was proved that the use of MPPT technology increased energy production by 23%; introduction of a fuzzy logic-based algorithm in the MPPT controller significantly increased the speed of finding the maximum power point, and also neutralized voltage fluctuations, which in turn reduced power underproduction by 2%.

Acknowledgments

The paper was supported by the competitiveness improvement programme of National Research Tomsk State University and grant of the Russian Foundation for Basic Research No 16-29-04388.

The authors wish to thank Tatiana B. Rumyantseva from National Research Tomsk State University for English language editing.

References

[1] Russkin V, Semenov S, Dixon R. 2016 *Bulletin of Tomsk Polytechnic University* **327** (4) 78–87
[2] Shidlovskiy S. 2006 *Automatic control. Reconfigurable structures* (Tomsk State University: Tomsk)
[3] Lovin D. 2007 *How to create android robot by own hands* (Publishing House DMK-Press: Moscow)
[4] Lukutin B., Muravlev I., Plotnikov I. 2015 *Power supply systems with wind and solar powerstations* (Publishing House of Tomsk Polytechnic University: Tomsk)
[5] Kornilov S., Rempe N., Shidlovskiy S. 2016 *Technical Physics* **61**(6) 841–848 doi: 10.1134/S1063784216060116

IOP Conf. Series: Materials Science and Engineering **516** (2019) 012040 doi:10.1088/1757-899X/516/1/012040

Optimizing parameters of speech signals condenser sensor by inner noise of the receiving channel

P A Luchnikov[1], I A Larioshina[2], A P Luchnikov[1], A S Sigov[1], N V Katakhova[1], N I Chernova[1]

[1]Russian Technological University, Moscow, Russia
[2]TUSUR, Tomsk, Russia
E-mail: fisika@mail.ru

Abstract. Noise characteristics of the receiving electroacoustic channel with an electret condenser sensor of speech signal have been considered in the paper. The technique for optimizing the converter' characteristics by inner electrical noise of the signal amplifier as well as the sensor's experimental amplitude-frequency response have been given. It has been found that in order to suppress noise generated by the amplifier it is advantageous to have a Q-factor within $2 \div 4$ and resonance frequency ω_0 in the range of $0.7 \div 0.85\ \omega_2$ in the condenser converter.

1. Introduction

To receive electroacoustic signals from a speech signal object the contact-type sensor converters, or laryngophones are used [1]. When designing a receiving channel of the electroacoustic systems for speech signals the rational function distribution of a given amplitude-frequency characteristic (AFC) formation of the entire receiving channel between the receiving sensor converter and the signal amplifier is of vital importance. With regard to the electroacoustic converter (microphone or laryngophone), the AFC is determined by its design type, for example, condenser sensor, which has the characteristic features of the parameters of the active element in the operating frequency range.

In practice, it is possible to use the concept of the formation of the receiving channel's given AFC when the electroacoustic condenser converter provides the greatest possible given frequency characteristic, and the amplifying channel forms cutoffs at the frequencies ω_1 and ω_2 (Figure 1). Since the mechanical system of electroacoustic condenser converter 1 (Figure 1) does not always provide the given AFC of the channel, the quadrupole 3 adjusting AFC with the characteristic being reverse of the converter's 1 is placed at the output of the linear previous amplifier 2. The quadrupole 3 forms AFC (as a resultant one) with the constant coefficient of transmission of the entire receiving channel in the operating frequency range of $\omega_1 \div \omega_2$.

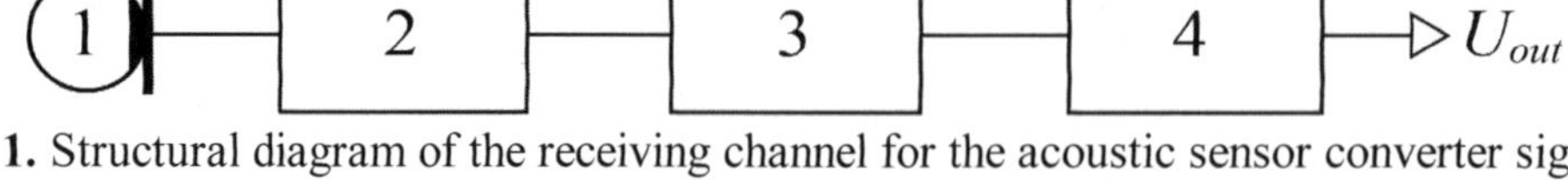

Figure 1. Structural diagram of the receiving channel for the acoustic sensor converter signal gain: 1 – condenser converter of the acoustic signal; 2 – previous linear amplifier; 3 – adjusting quadrupole; 4 – power amplifier; 5 – output signal.

2. Equivalent sensor circuit

For the formation of the channel's uniform AFC let us select the inertial-type electret condenser convertor's (laryngophone's) parameters (Q – factor and resonant frequencyω_0) for maximum protection against the inner noise, having its origin in the speech signal receiving system. To do this, we use its equivalent electric circuit model in the form of a series of R, L, C -circuits (Figure 2a) [1].

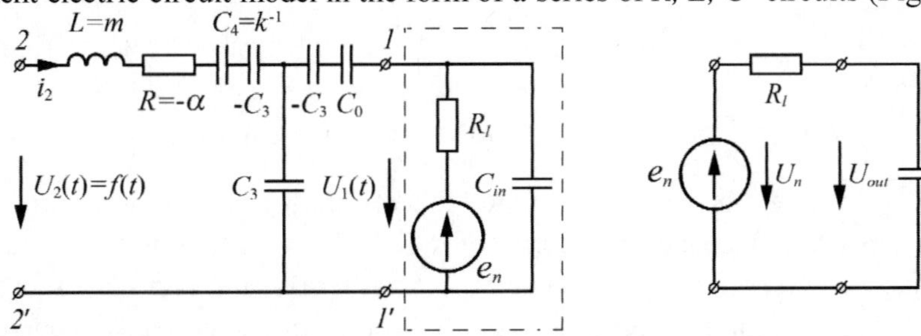

Figure 2. Equivalent circuit of the replacement of electret condenser laryngophone.

3. Amplifier input circuit noise

Input (noise) active resistance R_l of the linear preamplifier can be represented by the equivalent circuit (Figure 2b), which includes the noise-free resistance R_l and the emf source ($e_n = U_n$), connected in series (C_{eq} – equivalent capacitance 1-1′(Figure 2a). The instantaneous value of emf is usually characterized by the mean value on the time by the square of its value, referring to the actual value.

$$\overline{U}_n^2 = U_n^2 = \lim_{T\to\infty} \frac{1}{2T}\int_{-T}^{T} U_n^2(t)dt\ , \tag{1}$$

$$U_n = \lim_{T\to\infty} \sqrt{\frac{1}{2T}\int_{-T}^{T} U_n^2(t)dt}\ . \tag{2}$$

It is known that if the source of the noise possessing spectral power density $N(\omega)$, operates at the linear quadrupole input with the transfer function $K_1(j\omega)$, then the noise power at the output of the quadrupole 3 Figure 1 is equal to:

$$U_{out} = \frac{1}{2\pi}\int_{\omega_1}^{\omega_2} N(\omega)|K_1(j\omega)|^2 d\omega\ , \tag{3}$$

where $\omega_1 \div \omega_2$ is the quadrupole pass band.

Using equation (3) it is possible to calculate the noise level at the output of the amplifier 2 (Figure 1) generated by the e_n source at the bandwidth within its operating frequencies range of $\omega_1 \div \omega_2$. The characteristic of the amplifier is taken as being ideally rectangular with the gain coefficient equal to 1.

When calculating the transmission coefficient $K_1(j\omega)$ of the sensor we consider the input clamps of the equivalent circuit (Figure 2a) as the emf e_n , while the output ones is 1-1′. In this case, the input 2-2′ is short-circuited. For the low frequency region, we can assume that the part of the converter's equivalent circuit, which is parallel to the noise resistance has an absolute capacitive resistance $C = C_0+C_{in}$. Then the expression for the transmission coefficient of the sensor $K_1(j\omega)$ can be written as:

$$K_1(j\omega) = \frac{1/j\omega C}{R_l + 1/j\omega C} = \frac{1}{1 + j\omega R_l C}\ , \tag{4}$$

Then the expression for the noise voltage U_{out}^2 in action can be presented as

$$U_{out}^2 = \frac{4kT_0R_l}{2\pi}\int_{\omega_1}^{\omega_2} \frac{d\omega}{1+\omega^2(R_lC)^2} = \frac{2kT_0}{\pi C}\left[arctg\omega_2R_lC - arctg\omega_1R_lC\right]. \tag{5}$$

From (5) it is quite clear that U^2_{out} depends on the amplifier bandwidth $\omega_1 \div \omega_2$. If $\omega_l = 0$ and $\omega_2 = \infty$, the value U^2_{out} takes its maximum.

$$U^2_{out.m} = \frac{2kT}{\pi C} \cdot \frac{\pi}{2} = \frac{kT_0}{C}. \qquad (6)$$

Let us consider the expression for the normalized effective value of the noise voltage with respect to the square maximum as:

$$U^2_{out} / U^2_{out.m} = m^2 = \frac{2}{\pi}\left[arctg\,\omega_2 R_l C - arctg\,\omega_1 R_l C\right]. \qquad (7)$$

It follows from (7) that at the amplifier output the effective noise level m^2 depends on $\omega_1 R_l C$ nonlinearly. So at the constant frequency ω_1 in terms of $R_l C >> 1/\omega_2$ (when ω_2 is set by the amplifier) the first term in the brackets $arctg(\omega_1 R_l C) \approx \pi/2$ is constant, and the second $arctg(\omega_1 R_l C)$ increases with an increase in R_l. At the same time m^2 decreases. If $R_l C = 1/\omega_1$, which is usually recommended, m^2 is reduced by half, if $R_l C = 5/\omega_1$, m^2 reduces by 5 times. When $R_l C = 10/\omega_1$ m^2 decreases by 9.3.

The graphical function analysis of noise power spectral density at the amplifier output written in the form of

$$S(\omega) = |K_1(j\omega)|^2 \cdot N_0 = \frac{4kT_0 R_l}{1 + \omega^2 (R_l C)^2} \qquad (8)$$

shows that to reduce noise one needs to increase R_l so that the lower frequency limit ω_{low} of the converter would shift to a low frequency region to be several times less than the given ω_1 value for the speech range.

4. Sensor's parameters optimization at noise suppression in the channel

The electret sensor converter is not directly involved in the suppression of the noise generated by the amplifier 2 Figure 1. However, the noise from the preamplifier 2 pass through the adjusting quadrupole 3. The noise level at its output depends on the transmission function $W(j\omega) = 1/K(j\omega)$, which is determined by the transfer function of the condenser converter $K(j\omega)$. When $\omega R_l C_{in} >> 1$ $K(j\omega)$ is written in the form of:

$$K(jw) = K(0) \Big/ \left[(j\eta)^2 + \frac{j\eta}{Q} + 1\right], \qquad (9)$$

where: $K(0) = E_{10} \dfrac{C_0}{C_0 + C_{in}} \cdot \dfrac{1}{\chi}$; $\chi = K - E_{10}^2 \dfrac{C_0 C_{in}}{C_0 + C_{in}}$; $\omega_0 = \sqrt{\chi / m}$; $Q = \sqrt{m\chi}/\alpha$; $\eta = \omega/\omega_0$;

m, ω_0 and χ – are mass, resonance frequency and coefficient of elasticity of the oscillatory system, respectively; α – the total coefficient of friction of the mechanical system; E_{10} is the electric field strength in the air gap between the electret film and the counter electrode.

According to (9), the converter at large R_l is a system of the 2nd order and its transfer function is similar to the transmission function of the adjusting quadrupole assembled on the *RLC*-circuit. Since here, the ω_0 values and Q -factor are relevant to the resonant frequency and quality of the converter electromechanical system we conduct the optimization in terms of suppression of noise, generated by the amplifier.

The noise generated by the amplifier can be taken into account by including noise from the emf source at its input, which is characterized by a certain function of the noise power spectral density $N_a(\omega)$ that for electric circuits on semiconductors in some cases is represented by a polynomial in powers of ω:

$$N_a(\omega) = a_0 + a_1\omega^{-1} + a_2\omega + a_3\omega^2. \qquad (10)$$

To solve the task of optimization of the function $W(j\omega)$, it is necessary to consider the noise power at the output of the system both without the adjusting quadrupole and in its presence. Here, the passage of each noise component should be considered separately.

<u>Noise component</u> $N_a'(\omega) = a_0$

For the output noise power in the absence of an adjusting quadrupole the following expression is true:

$$P_{out} = \frac{1}{2\pi}\int_{\omega_1}^{\omega_2} N_a(\omega)d\omega = \frac{a_0}{2\pi}(\omega_2 - \omega_1), \tag{11}$$

while in its presence it is represented as:

$$P_{out}^* = \frac{1}{2\pi}\int_{\omega_1}^{\omega_2} a_0\left[1 - \left(2 - Q^{-2}\right)\eta^2 + \eta^4\right]d\omega =$$

$$= \frac{a_0}{2\pi}\left[(\omega_2 - \omega_1) - \frac{\omega_2^3 - \omega^3}{\omega_0^2}\cdot\frac{2 - Q^{-2}}{3} + \frac{\omega_2^5 - \omega_1^5}{5\omega_0^4}\right]. \tag{12}$$

Let us consider the expression (8) for at least three different ways of changing the parameters of the converter: a) resonance frequencies ω_0 in a fixed Q-factor; b) parameter χ at m = const, α = const; c) oscillatory system' mass m at χ = const, α = const.

Mode ω_0 = var, Q = const. When $\omega_2 >> \omega_1$ we get

$$\frac{dP_{out}^*}{d\omega_0} = \frac{a_0}{2\pi}\left[\frac{\omega_2^3}{3\omega_0^2}\cdot 2\left(2 - Q^{-2}\right)\frac{4}{5}\frac{\omega_2^5}{\omega_0^5}\right]. \tag{13}$$

From where we have the expression for the relative value of the optimal resonance frequency:

$$\frac{\omega_{r.o}}{\omega_2} = \sqrt{\frac{6}{5}\frac{1}{2 - 1/Q^2}}. \tag{14}$$

In this case the value P_{out}^* is

$$P_{out.o}^* = \frac{a_0}{2\pi}\left[(\omega_2 - \omega_1) - \frac{5}{36}\omega_2\left(2 - Q^{-2}\right)^2\right]. \tag{15}$$

The ratio $P_{out}^* / P_{out} = Q_o^2$ shows the degree of noise suppression at the expense of passing through the adjusting quadrupole.

$$Q_o^2 = \left.\frac{\frac{a_0}{2\pi}(\omega_2 - \omega_1)}{\frac{a_0}{2\pi}\left[(\omega_2 - \omega_1) - \frac{5}{36}\left(2 - Q^2\right)^2\right]}\right|_{\omega_2 >> \omega_1} \approx \left[1 - \frac{5}{36}\left(2 - Q^{-2}\right)^2\right]^{-1}. \tag{16}$$

Similar treatment and numerical calculations have been carried out when using other methods of changing the converter's parameters (*b*) and (*c*) for all components (10) of the noise generated by the amplifier.

The analysis of the optimization results has demonstrated that to suppress noise, generated by the amplifier, in the condenser converter it is reasonable to have Q-factor in the range of 2 ÷ 4, and the resonance frequency ω_0 within the limits of 0.7 ÷ 0.85 ω_2.

5. Experimental results

Samples of laryngophone-type electret capacitance sensor converters, based on the electret film elements 5 have been made experimentally (Figure 3). The electret film 5 is made from fluoropolymer, obtained through the deposition (condensation) from the active gas phase while spraying the initial

IOP Conf. Series: Materials Science and Engineering **516** (2019) 012040 doi:10.1088/1757-899X/516/1/012040

polytetrafluoroethylene in HF-plasma discharge [2, 3] or HF-magnetron [4, 5]. The film thickness was 6 ÷ 8 μm thick.

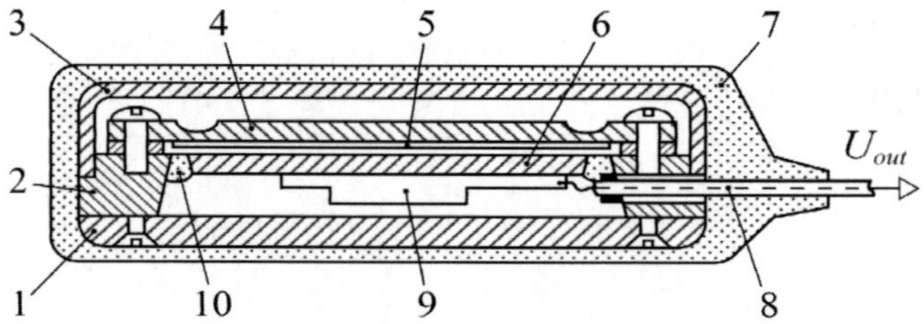

Figure 3. The design of the condenser sensor of speech signals: 1, 2, 3 – parts of sensor housing; 4 – vibrating electrode; 5 – electret film element; 6 – counter electrode; 7 – sealant; 8 – the output of the previous amplifier; 9 – previous amplifier; 10 – counter electrode insulator.

Another way of making the electret element was a planar two-layer structure on a silicon substrate. The first structure layer was made of silicon oxide of 0.7 ÷ 0.9 μm thick, and the outer layer as a protective one was made from F-parylene of 0.2 ÷ 0.5 μm thick. The samples were polarized in corona discharge in strong electric fields, and then were surface charge stabilized at the temperature of 120 °C. The layered electret elements at higher operating temperatures (60 ÷ 100 °C) showed higher stability of the $U(t)$ surface potential.

The electret condenser converter has parameters close to the calculated ones (ω_0 = 2.7 kHz and Q = 2.7...3.4). Typical AFC of the electret speech signals converter is shown in Figure 4. Here: curve 1 – an ideal AFC with due regard for a man's throat vibrations; 2 – AFC of the electret condenser laryngophone; 3 – AFC of the same electromagnetic-type sensor laryngophone.

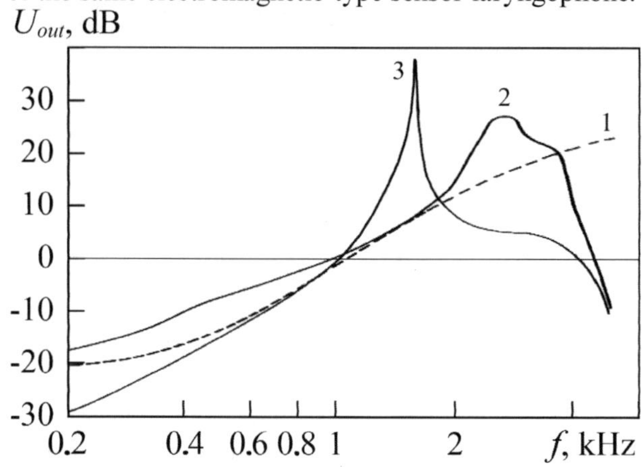

Figure 4. AFC of the condenser sensor of speech signals.

6. Conclusion

The authors have considered noise characteristics of the receiving channel from the electret speech signals condenser converter with the AFC adjusting quadrupole's linear preamplifier placed at the converter's output. The quadrupole characteristic is inverse to the converter's. The parameters of the converter with the greatest protection against the inner noise arising in the electret laryngophone's system of reception and transmission have been analyzed. For effectively suppressing noise generated by the amplifier the sensor condenser converter should have a quality factor Q in the range of 2 ÷ 4, and the self-resonant frequency ω_0 in the region of 0.7 ÷ 0.85 ω_2. It was experimentally found that electret condenser converter of the inertial type (laryngophone) has the optimum frequency response of the channel with high interference immunity at good intelligibility of speech in the range of 100 ÷ 4500 Hz.

Acknowledgments
This work was done at MIREA – Russian Technological University and was supported by the Ministry of Education and Science of the Russian Federation (project № 3.8074.2017/8.9 basic part of the State task).

References

[1] Blank S. F. *et al* 1991 *Communication Technology. Series: FOC.* 1. 16–20. (in Russia)
[2] Buckley D. H. 1981 *Surface effects in adhesion, friction, wear, and lubrication* (ESPC: Amsterdam-Oxford-New York)
[3] Luchnikov P. A. 2018 *IOP Conf. Ser.: Mater. Sci. Eng.* **289** 012037 doi:10.1088/1757-899X/289/1/012037
[4] Belyanin A. F. *et al* 2017 *Russian Technological Journal* **5**(3) 22–40 (in Russia)
[5] Makarova M. *et al* 2018 *Key Eng. Mater.* **781** 8–13 doi:10.4028/www.scientific.net/KEM.781.8

IOP Conf. Series: Materials Science and Engineering **516** (2019) 012041 doi:10.1088/1757-899X/516/1/012041

Merging brain computing interface (BCI) & neural networks for better authentication & recognition

Q Ramzan[1, 2], V I Syryamkin[1]

[1]National Research Tomsk State University, Tomsk, Russia
[2]toggleTechs (a creativity platform), Islamabad, Pakistan
E-mail: qaseemramzan0@gmail.com

Abstract. Brain Computing Interface (BCI) has been proved helpful for the different streams of technology, considering the sensitivity of data in the current era it is required to build new security protocols and authentication models. Just like other fields of technology Brain Computing Interface could also be useful for making the data security better by using BCI as an authentication method without any hard physical inputs. The focus of the issue shifts to 'recognition' of EEG signals pattern and making the authentication model self-learning to increase its efficiency. This leads us to involve Artificial Neural Networks in the authentication system to make it efficient and intelligent.

1. What is a Brain Computer Interface?

What is Brain Computer Interface (BCI)? A brain computer interface (BCI), sometimes called a mind-machine interface (MMI), direct neural interface (DNI), or brain–machine interface (BMI), is a direct communication pathway between an enhanced or wired brain and an external device. BCIs are often directed at researching, mapping, assisting, augmenting, or repairing human cognitive or sensory-motor functions [1, 2].

1.1 Working of BCI

The purpose of a BCI is to detect and quantify features of brain signals that indicate the user's intentions and to translate these features in real time into device commands that accomplish the user's intent. To achieve this, a BCI system consists of 4 sequential components.

1. Signal Acquisition,
2. Feature Extraction,
3. Feature Translation
4. Device Output.

These 4 components are controlled by an operating protocol that defines the onset and timing of operation, the details of signal processing, the nature of the device commands, and the oversight of performance. An effective operating protocol allows a BCI system to be flexible and to serve the specific needs of each user.

Electrical signals from brain activity are detected by recording electrodes located on the scalp, on the cortical surface, or within the brain. The brain signals are amplified and digitized. Pertinent signal characteristics are extracted and then translated into commands that control an output device, such as a spelling program, a motorized wheelchair, or a prosthetic limb. Feedback from the device enables the user to modify the brain signals in order to maintain effective device performance. BCI = brain-computer interface; ECoG = electrocorticography; EEG = electroencephalography [1,3].

IOP Conf. Series: Materials Science and Engineering **516** (2019) 012041 doi:10.1088/1757-899X/516/1/012041

2. What are ANNs (Artificial Neural Networks)?

An Artificial Neural Network (ANN) is an information processing paradigm that is inspired by the biological nervous systems, such as the human brain's information processing mechanism. The key element of this paradigm is the novel structure of the information processing system. It is composed of a large number of highly interconnected processing elements (neurons) working in unison to solve specific problems. NNs, like people, learn by example. A NN is configured for a specific application, such as pattern recognition or data classification, through a learning process. Learning in biological systems involves adjustments to the synaptic connections that exist between the neurons. This is true of NNs as well [4].

2.1 Neuron Model

When a neuron receives excitatory input that is sufficiently large compared with its inhibitory input, it sends a spike of electrical activity down its axon (Figure 1). Learning occurs by changing the effectiveness of the synapses so that the influence of one neuron on another changes [4].

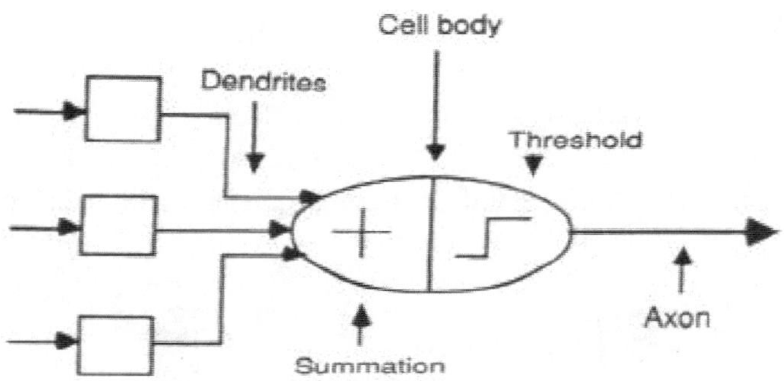

Figure 1. Basic Neuron Model.

We conduct these neural networks by first trying to deduce the essential features of neurons and their interconnections. We then typically program a computer to simulate these features [4].

2.2 Artificial Neuron

An artificial neuron is a device with many inputs and one output (Figure 2). The neuron has two modes of operation; ***the training mode*** and ***the using mode*** [4].

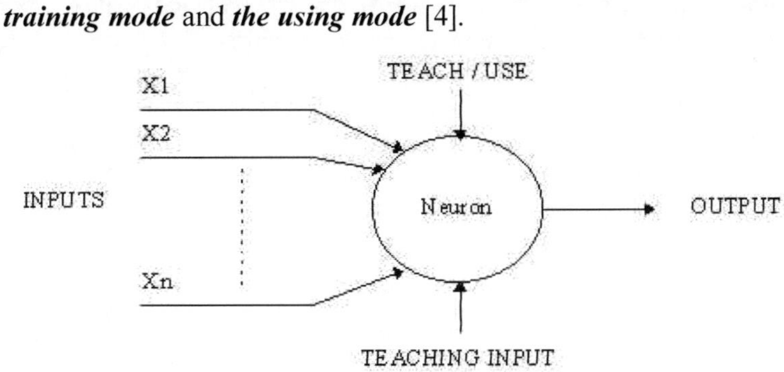

Figure 2. Artificial Neuron Model.

2.2.1 The Training Mode

In the training mode, the neuron can be trained to fire (or not), for particular input patterns [5].

IOP Conf. Series: Materials Science and Engineering **516** (2019) 012041 doi:10.1088/1757-899X/516/1/012041

2.2.2 The Using Mode

In the using mode, when a taught input pattern is detected at the input, its associated output becomes the current output. If the input pattern does not belong in the taught list of input patterns, the firing rule is used to determine whether to fire or not [5].

3. Recognition in Neural Networks

An important application of neural networks is pattern recognition. Pattern recognition can be implemented by using a feed-forward neural network that has been trained accordingly (Figure 3). During training, the network is trained to associate outputs with input patterns. When the network is used, it identifies the input pattern and tries to output the associated output pattern [6].

Figure 3. Recognition Example in Neural Networks.

Suppose a network is trained to recognize the patterns T and H. The associated patterns are all black and all white respectively as shown above [6].

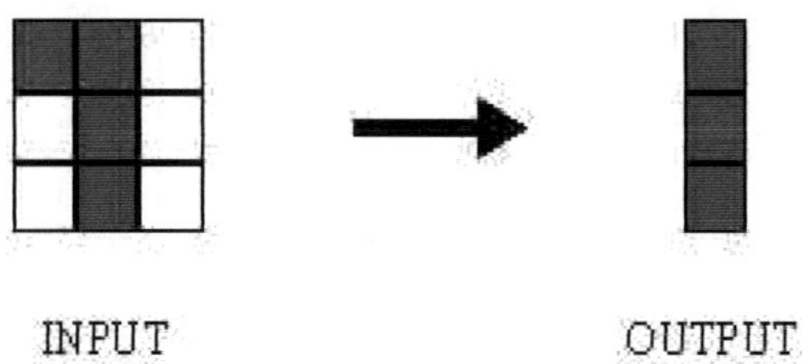

Figure 4. Recognition Example in Neural Networks.

Since the input pattern looks more like a 'T', when the network classifies it, it sees the input closely resembling 'T' and outputs the pattern that represents a 'T' (Figure 4).

4. Complex Neurons

A more sophisticated Neuron is known as the McCulloch and Pitts model (MCP). The difference is that in the MCP model, the inputs are weighted and the effect that each input has at decision making, is dependent on the weight of the particular input. The weight of the input is a number which is multiplied with the input to give the weighted input (Figure 5) [6].

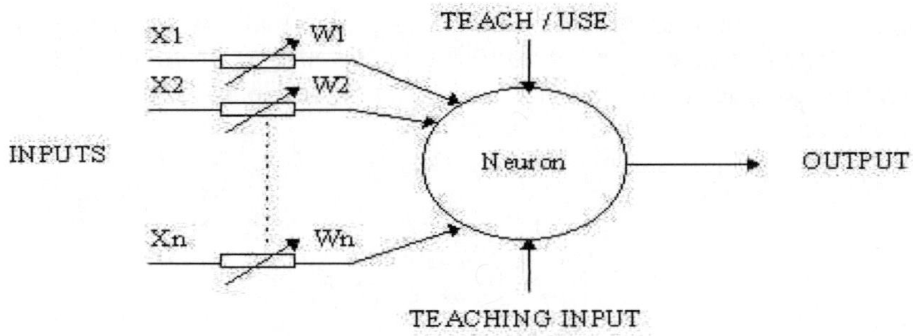

Figure 5. Complex Neuron Model.

5. Combining Brain Computing Interface (BCI) with Neural Networks

Presently ever device is under a threat of security breach and taking into account the sensitivity of personal data and how human lives are half digital makes this issue an important one for humans. It is required to secure our devices with a more secure and innovative protocols instead of the mainstream methods. All the past authentication methods being used by us are breach-able and have a history of either hack attacks or simple security breach due to the weak type of authentication models. All of the authentication models we have so far need an input from the user. This input method is the main loophole which makes it not good enough to protect our devices [1, 8].

If the authentication models will need a physical input from users, then it is highly likely that the user is at risk because the input can be acquired either by trick or by force.

An authentication model based on EEG signals was proposed to overcome this problem and it could be a method of thoughts without any hard physical input. BCI is capable of reading the Brain Signals and then the proposed model in previous paper was developed to authenticate a user based on brain signals and matching the patter with the one stored in a database [1].

5.1 Authentication for BCI with Neural Networks

As we have developed a consensus above that Neural Networks can perform recognition in a very impressive way which is not possible otherwise. So a better thought is to use Neural Networks on the EEG Based Authentication Model and make the Neural Network learn the patterns of user's authentication routine. Here the Neural Network will not only match the pattern to authenticate the users but it will learn gradually the changes in EEG Signals pattern and become more intelligent in a way to recognize the user [6].

The ultimate task will be to train the Neural Network to the point where it will not need to match the EEG Signals with the first every stored pattern but it will be able to recognize the user based on the later learnings which were learnt in result of trainings [6].

5.2 Proposed BCI Authentication Model Combined with Neural Networks

The model shown in Fig. allows the BCI system to send the gathered pattern from input to neural network system which uses different learning techniques to recognize the pattern of signals as well as use this as a learning to improve the previously stored recognition pattern for signals [1].

Headset

EEG headset of any company can be used with the authentication module.

Processing

In processing step all the information gathered from the EEG Headset will be processed for further feature extraction.

Feature Extraction

IOP Conf. Series: Materials Science and Engineering **516** (2019) 012041 doi:10.1088/1757-899X/516/1/012041

Feature extraction is the process of analyzing the digital signals to distinguish pertinent signal characteristics (i.e., signal features related to the person's intent) from extraneous content and representing them in a compact form suitable for translation into output commands. These features should have strong correlations with the user's intent (Figure 6).

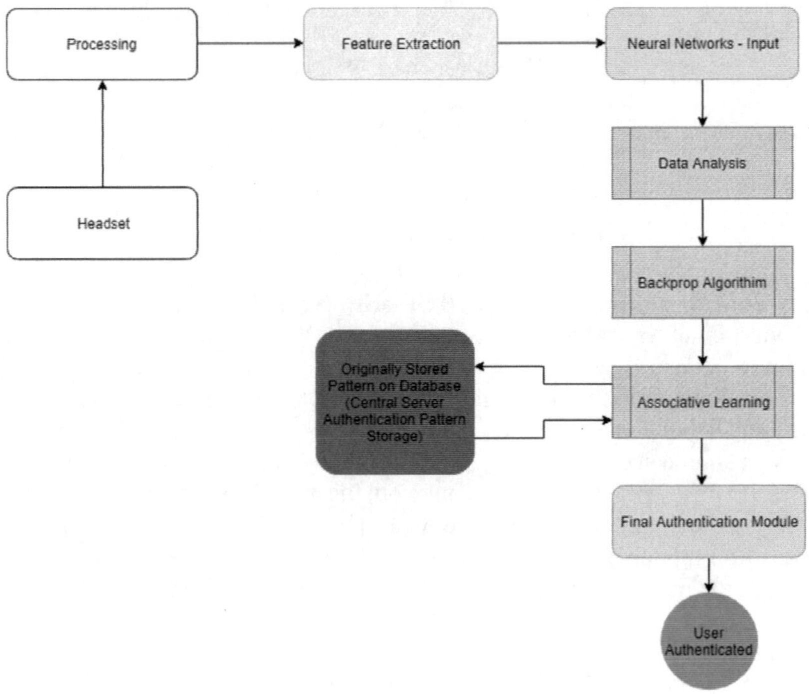

Figure 6. Proposed BCI Authentication Model Combined with Neural Networks.

Neural Networks Input

In this step the meaningful data after feature extraction will be provided to Neural Networks as an input.

Data Analysis

Data Analysis is a step of Neural Networks processing, here the data is analyzed for further processing.

Backprop Algorithm

Backpropagation is a method used in artificial neural networks to calculate a gradient that is needed in the calculation of the weights to be used in the network. It is commonly used to train deep neural networks, a term referring to neural networks with more than one hidden layer.

Associative Learning

An associative neural network (ASNN) is a combination of an ensemble of the feed-forward neural networks and the K-nearest neighbor technique. The introduced network uses correlation between ensemble responses as a measure of distance among the analyzed cases for the nearest neighbor technique and provides an improved prediction by the bias correction of the neural network ensemble both for function approximation and classification. Actually, the proposed method corrects a bias of a global model for a considered data case by analyzing the biases of its nearest neighbors determined in the space of calculated models.

Final Authentication Module

This is a final step of authentication; the final authentication module could either be on server or on the local machine.

IOP Conf. Series: Materials Science and Engineering **516** (2019) 012041 doi:10.1088/1757-899X/516/1/012041

6. Conclusion

This approach specifies the importance of using EEG Based Authentication Model and uncovers the important aspects in creating a recognition system to correctly authenticate a user based on unique patterns found in EEG signals using the above specified algorithm structure. This proposed system uncovers a new way to combine BCI with Neural Networks for the purpose of correctly learning the authentication patterns.

Acknowledgments

The paper is prepared within the framework of the competitiveness improvement programme of National Research Tomsk State University and grant of the Russian Foundation for Basic Research No 16-29-04388. The authors are grateful to Tatiana B. Rumyantseva from Tomsk State University for her assistance in preparing the paper.

References

[1] Ramzan Q. 2018 *MATEC Web of Conferences* **155** 01006 doi: 10.1051/mateconf/201815501006
[2] Melanie Swan. *Future of Brain Computing Interfaces (Block-chaining Your Way into a Cloud Mind)* URL: http://jetpress.org/v26.2/swan.htm
[3] Shumilov V., Syryamkin V., Syryamkin M. 2015 *AIP Conference Proceedings* **1688** 040007 doi: 10.1063/1.4936040
[4] Luke Dormehl URL: https://www.digitaltrends.com/cool-tech/what-is-an-artificial-neural-network
[5] Tino P. URL: https://becominghuman.ai/artificial-neuron-networks-basics-introduction-to-neural-net
[6] Kuznetsov D., Syryamkin V. 2015. *AIP Conference Proceedings* **1688** 040004 doi: 10.1063/1.4936037
[7] Artificial Neural Network Shodhganga URL: www.shodhganga.inflibnet.ac.in/jspui/bitstream/10603/148236/6/06_chapter%201.pdf
[8] Kornilov S., Rempe N., Shidlovskiy S. 2016 *Technical Physics* **61**(6) 841–848 doi: 10.1134/S1063784216060116

IOP Conf. Series: Materials Science and Engineering **516** (2019) 012042 doi:10.1088/1757-899X/516/1/012042

Extreme effects in field localization of acoustic wave: super-resonances in dielectric mesoscale sphere immersed in water

I V Minin[1,2], **O V Minin**[1]

[1] Tomsk Polytechnical University, Tomsk, Russia
[2]Siberian Institute of Metrology, Novosibirsk, Russia
E-mail: prof.minin@gmail.com

Abstract. By the numerical simulation we predict acoustic super-resonance modes with several tens of thousands of times field-intensity enhancement (order of magnitude: 10^4 - 10^5) supported by dielectric mesoscale spheres immersed in water. The super-resonances are related to internal scattering at specific values of the Mie and the particle material parameters and they are responsible for generation of giant fields within the particles near its surface. Taking into account the analogy between electromagnetic and acoustic waves this phenomenon is valid in electromagnetic (optical) waveband.

1. Introduction

The phenomenon of photonic nanojet is well known now from the optical [1] and THz [2] bands, allow to form localized light in subwavelength area by scattering of light on low loss dielectric particle.

In the acoustic, the scattering of sound from a spherical fluid obstacle of size comparable to a wavelength was considered in [3, 4]. It has been shown that the amplitude of the scattered wave in the backward direction from a fluid sphere a few wavelengths in diameter exceeds twice that from a rigid sphere of the same size for the case of the relative sound velocity 0.8 and density equal to that of the surrounding medium [4]. Focusing effect in the sound scattering by a spherical balloon with high diameter filled with carbon dioxide was investigated in [5, 6].

Recently, it has been theoretically demonstrated for the first time [7, 8] that an existence of acoustic analogue of photonic jet phenomenon [9], providing for subwavelength localization of acoustic field in shadow area of arbitrary 3D penetrable mesoscale particle, is possible. It is important to note that the principle difference between optical and acoustical materials properties is a shear speed of sound, i.e. acoustic materials are anisotropic due to the two speed of sound [10].

In our analysis of acoustojets phenomenon [7], in simulations we use the rigorous partial-wave expansion method [11], which depends on the beam-shape and scattering coefficients, to obtain the scattered pressure around the solid elastic spherical particle, where both compressional and shear waves were taken into account. Simulations show [10] that for a 5λ radius sphere with relative refractive index of about 1.6, the acoustic jet remains under the diffraction limit by approximately few wavelengths in depth, with an intensity gain close to 20 dB relative to the incident intensity. Some of the examples of acoustojet by a homogeneous sphere made of lead, polyethylene and silver were illustrated and characterized according to subwavelength beam waist, intensity gain, and propagation depth.

The simulation in the first approximation in the harmonic case was based on the Helmholtz equation [12]. It was demonstrated [12] the influence of the main parameters of dielectric penetrable spherical cavity, immersed in water, to transformation of whispering gallery mode into acoustojet (acoustic jets) by interaction of acoustic plane wave scatterer.

For investigation of resonant scattering of ultrasound in dielectric spherical cavity immersed in water as a material we have selected Rexolite© that closely match the impedance of water [13, 14].

The main parameters of Rexolite© are: sound velocity 2337 m/s, density 1.04 g/cm^3. According to [12] the initial parameters of the particle near WGM resonance (sometimes also referred to as a morphology-dependent resonance) were selected as: radius of particle $R=4\lambda$ (at frequency of 1 MHz in water), relative density contrast is 1.0402 and the speed of sound contrast is 1.570.

2. Results of investigation

Lord Rayleigh first reported on circumferentially circulating acoustical resonances that he named “whispering gallery modes [15].” In a whispering gallery mode, acoustic waves do not penetrate particle close to the center but bounce around the circumference of the sphere.

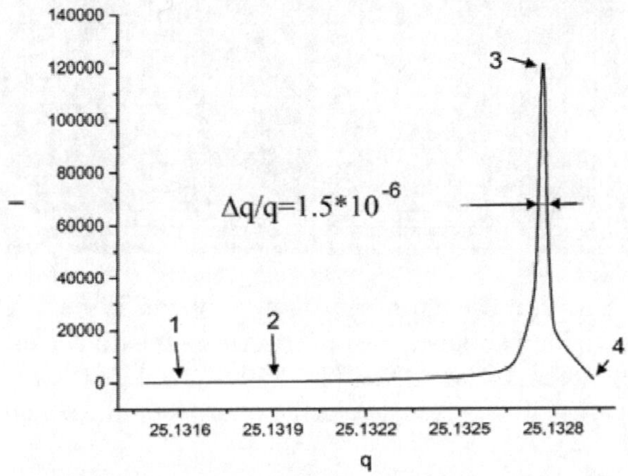

Figure 1. Resonant scattering on Rexolite© sphere immersed in water.

Analysis of simulations shown that for a permeable sphere, impedance leads to the dissipation of acoustic power in addition to a phase change in the scattered wave. Figures 2*a* and 2*d* shown that the increased focusing properties (acoustojet formation) near the shadow surface in the interference pattern and the hot spot lobes is due to that for the uniformly dielectric sphere a positive impedance phase angle significantly increases the contribution both the octopole and the quadrupole modes, and decreases the contribution of the dipole mode [16].

Figures 2*b* and 2*d* correspond to the conditions for the beginning of the formation of an acoustic jet (Figure 2*a* shown the generation of a nonresonant acoustojet) immediately before and after superresonance, respectively. The normalization is relative to the illumination.

IOP Conf. Series: Materials Science and Engineering **516** (2019) 012042 doi:10.1088/1757-899X/516/1/012042

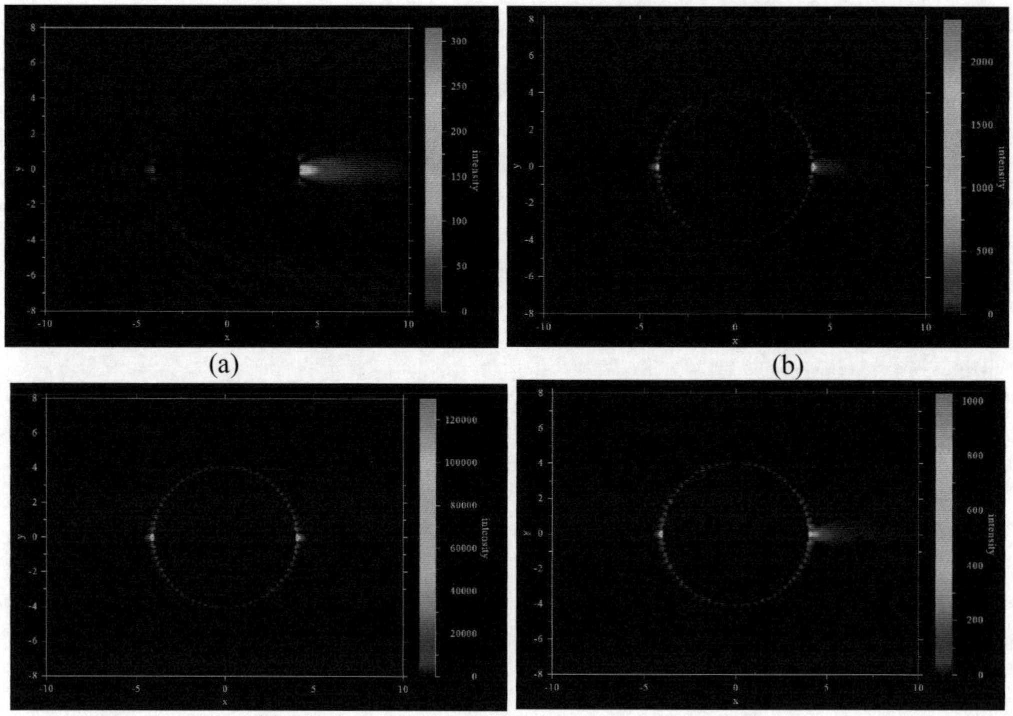

(c) (d)

Figure 2. Near superresonant scattering on Rexolite© sphere immersed in water. The particle illuminated from the left to right. The labels of (a–d) corresponds to points of (1–4) from Figure 1, respectively. Maximal field intensity are: 325 (a), 2500 (b), 125000 (c), 1000 (d). Figure 2(c) visualizes the near-field intensity of the resonance mode of the Rexolite sphere in water.

A giant field enhancement mode caused by microsphere's internal partial waves we define as a 'super-resonance' effect. Characteristic features of superresonance can be seen in Figure 2*c*, where the two "hot spots" represent the distribution of the field intensity inside a spherical particle with $q = 25.132766$ and with $\ell = 33$ multipole at the super resonance condition (q is size (Mie) parameter defined as q=2πR/λ). The enhancement factor can be extremely large at these "hot spots", reaching the order of 10^4 –10^6. The pronounced whispering gallery mode and example giant field intensity are clear visible in Figure 2*c*. Inside the particle there is high intensity with maximal field intensity enhancement about 125 thousand. It could be noted that at the superresonance condition the hotspot spots has a super resolution about of 0.21…0.23λ, which exceeds solid immersion resolution limit (λ/2n) of acoustojet. It is noticeable that the field in the resonance mode band passes fleetingly from the spherical particle to the background water environment.

The quality factor of resonator by analogy with simple oscillator can be defined as

$$Q=ka/\Delta(ka)=q/\Delta q.$$

From the data of Figure1 it is followed that Q=6.7×10^5.

It could be noted that superresonance mode exist only in 3D case (microspheres) in contrast to WGM which can be excited in both cylinder (2D case) and sphere.

3. Conclusion

In this work, it was shown that the giant localization of the field and the superresolution mode can be excited in a mesoscale sphere submerged under water under certain conditions of resonance. This effect is observed in the parameter area, next to those characteristic of an acoustojet.

Studies of such superresonances are of interest for the following main reasons:

- it gave the ability to create highly localized acoustic fields, both, inside and outside the particle near it surface.
- it is a possibility for optimization of several applications e.g. in acoustojet [7, 8], surface optomechanics [17], acoustic superlens microscopy [18, 19] and surface-enhanced acoustic Raman spectroscopy [20, 21], etc.
- optimization of the whispering gallery mode at super resonance is of interest for optofluidic resonators [22] where such microfluidic cavities were made in the form of a water submerged solid micro resonator [23, 24].
- this effect is not limited to consider above relative density contrast and the speed of sound contrast. In fact it valid for all dielectric particles but superresonant conditions depends on a function of relative density and the speed of sound contrast and the Mie parameter q.

We show giant field enhancement ($10^4 \ldots 10^6$) can be achieved with mesoscale dielectric sphere immersed in water under super resonance condition. The super-resolution ($0.21 \ldots 0.23\lambda$) is observed at hot spots. We call this new effect as super resonance effect.

We believe the discovery of superresonance is of important to mesoscale particle super-resolution acoustic imaging [18, 25, 26], biosensors and hold new promise for applications in different scientific directions.

It could be noted that the effective refractive index of the polymer sphere increases when it is immersed in water due to the water molecules adsorbed on its surface, which increases the penetration depth of the evanescent wave and the effective radius of sphere [27].

In common, taking into account the analogy between electromagnetic and acoustic waves [28, 29], this phenomenon is valid in electromagnetic (optical) waveband [30]. However, in the scattering by elastic penetrable solid particles, the presence of two different sound velocities, the longitudinal and transverse ones, strongly affects the modal characteristics within the scatterer due to mode conversion. Therefore, the effect of the ratio of these two speeds on the super resonance effect is the subject of further research.

References

[1] Heifetz A. *et al* 2009 Photonic Nanojets *J Comput Theor Nanosci* **6**(9) 1979–1992
[2] Minin O.V. *et al* 2017 *APL Photonics* **2** 056106
[3] Toulis W. J. 1963 *J. Acoust. Soc. Am.* **35** 3 286-292
[4] Anderson V. C. 1950 *The Journal of the Acoustical Society of America* **22** 426
[5] Thomas C *et al* 2009 *Am. J. Phys.* **77** 197
[6] Borrero M. A. P. 2011 *Am. J. Phys.***79** 401
[7] Minin I.V. *et al* 2016 *Acoustojet: acoustic analogue of photonic jet phenomenon* URL: https://arxiv.org/ftp/arxiv/papers/1604/1604.08146.pdf
[8] Minin I.V. *et al* 2017 *Optical and Quantum Electronics* **49** 54
[9] Lukiyanchuk B.S. *et al* 2017 *Optical material express* 7**(6)** 1820–1847
[10] Lopes J.H. *et al* 2016 *A theoretical analysis of jets* URL: http://www.ica2016.org.ar/ica2016proceedings/ica2016/ICA2016-0943.pdf
[11] Williams E.G. *et al* 1999 *Fourier Acoustics: Sound Radiation and Near field Acoustical Holography* (Academic Press)
[12] Minin I. V. *et al* 2017 *Journal of Physics: Conf. Series* **881** 012025
[13] Selfridge A. R. 1985 *IEEE Transactions on Sonics and Ultrasonics* **SU-32** (3) 381–394
[14] Wang H. *et al* 1999 *Proc. the SPIE Conference on Ultrasonic Transducer Engineering* 3664 35-42
[15] Rayleigh L.1910 *Philos. Mag* **20** 1001–1004.
[16] Treeby B. E. *et al* 2007 *J. Acoust. Soc. Am.* **122**(1) 46.
[17] Matsko A. *et al* 2009 *Physical Review Letters* 103 257403

[18] Minin I.V. *et al* 2017 *A Method to Improve the Resolution of the Acoustic Microscopy* URL: https://arxiv.org/abs/1712.01638
[19] Minin I.V. *et al* 2018. *MATEC Web of Conferences* **155** 201815501029
[20] Wheaton S. *et al* 2015 *Nat. Photon.* **9** 68-72
[21] DeWolf T. and Gordon R. *2016 Phys. Rev. Lett.* **117** 138101
[22] Bar-David D. *et al* 2018 *Opt. Express* **26(15)** 19115
[23] Armani A. M. *et al* 2005 *Appl. Phys. Lett.* 87(15) 151118
[24] Teraoka I. *et al* 2017 *Optics Communications* **393** 244–251
[25] Rubio C. *et al* 2018 *EPL* **123** 64002
[26] Rubio C. *et al* 2019 *Physics Letters A* **383** 396–399
[27] Teraoka I. and Arnold S. 2006 *J. Opt. Soc. Am.* B **23** 1434
[28] Carcione J. M., 2015 *Wave Fields in Real Media. 3rd editions* (Elsevier Science: Amsterdam) doi:10.1016/C2013-0-18893-9
[29] Carcione J. M., Cavallini F.1995 *Wave Motion* **21** 149–162
[30] Hoang T. X. *et al* 2015 *Optics Express* **23(9)** 12337–12353

IOP Conf. Series: Materials Science and Engineering **516** (2019) 012043 doi:10.1088/1757-899X/516/1/012043

Antenna system of ultrawideband borehole radar based on loaded electric dipoles

E V Balzovsky, Yu I Buyanov, V I Koshelev

Institute of High Current Electronics SB RAS, Tomsk, Russia
E-mail: bev@lhfe.hcei.tsc.ru

Abstract. The design of an ultrawideband borehole radar antenna system based on loaded electric dipoles is presented. The antennas are designed to operate in a dielectric medium in the frequency range of 0.25–1 GHz. The results of calculations and measurements of antenna characteristics are presented. The results of sounding the objects when placing antennas in a medium with a relative permittivity of 4 are presented as well.

1. Introduction

When drilling exploratory boreholes, an important task is to determine the position and dielectric properties of oil-saturated reservoirs or water-filled cavities. It is also necessary to determine the boundaries of fractures and high hardness rocks to prevent contact of drilling tools with them. The use of ultrawideband (UWB) electromagnetic pulses allows measuring the parameters of the medium and determining the presence and position of inhomogeneities near the well. This is achieved due to the high resolution of UWB pulses and possibility of time selection of signals reflected from areas with a high gradient of conductivity and/or permittivity.

There are known designs of UWB borehole radars in which the transmitter and receiver are in the same dielectric body - probe. When sounding the medium for reflection, the probe moves over a predetermined distance [1]. The receiver and the transmitter can also be located in different wells, allowing cross-hole studying the environment [2].

The use of semiconductor pulse generators of nano- and sub-nanosecond duration with an amplitude of several kilovolts is promising for increasing the range of UWB borehole radar sounding. At such voltages, the antennas of the receiver and transmitter should be spaced apart and the input port of the receiver should be protected from a high-power radiation pulse. The dipole antennas have the minimum of the pattern in the axial direction, the choice of coaxial dipoles as antennas of such radar reduces the effect of high-voltage radiation pulse on the input port of the receiver.

In this paper the antenna system of the UWB borehole radar is presented. It has been designed to operate in a medium with a dielectric constant $\varepsilon = 4$ when sounding by short pulses with a frequency spectrum of 0.25–1 GHz. A pulse generator with a half-height duration of 0.8 ns, an amplitude of 2.5 kV, and a repetition rate of up to 10 kHz was used as a pulse source. The used generator is specified to be loaded on 50 Ohms at low frequencies and direct current. This feature of the generator determined the design of the antenna system of borehole radar. The antenna should be matched to the generator from the lowest frequencies. The paper presents a description of the design and the results of modeling and measuring the characteristics of the antenna.

2. Antenna system design

To test the functionality of the receiving and transmitting system, a model of the UWB borehole radar was created. The model includes a pulse generator, two antennas in a common dielectric housing, and a digital stroboscopic oscilloscope. The radar is based on two identical dipole antennas 1 and 2 (Figure 1) located in a dielectric housing 3 with an external diameter of 120 mm, internal diameter of 65 mm, and a length of 3.5 m. The housing is placed in a box 4 filled with a model medium. We used the wet sand with a relative permittivity characteristic of sand-oil and sand-drilling fluid mixtures [3]. The distance *D* between the centers of dipoles varied to find the optimal value. Both antennas can move along the dielectric body axis with a fixed *D* to scan and record the reflected pulses from the objects on the top surface of the box.

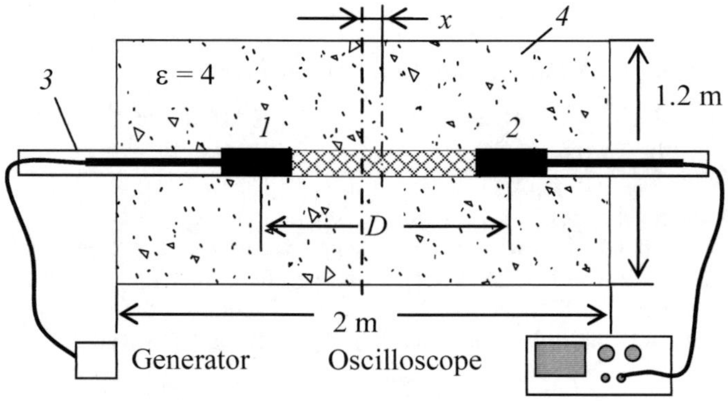

Figure 1. Drawing of experimental setup.

The borehole radar antenna is an electric dipole with an extended bandwidth. The arms of the dipole 1 and 2 (Figure 2) have a cylindrical shape with a cone at central feed point. The length of the dipole is 135 mm. The internal wire of the RG213 coaxial cable is attached to the arm 1 of the dipole, and the cable braid is connected to the arm 2. To reduce the frequency dependence of the input impedance of the dipole, its transverse dimensions are increased due to the metal ribs 3. The entire space between the plates 4 and 5 is filled with a dielectric with $\varepsilon = 3.5$.

To expand the frequency band of antenna matching to the low frequencies, the plates 4 and 5 of the dipole are connected to each other by four metal rods 6 with series-connected resistors 7 and 8 having a nominal value of 100 Ohms. The rods and ribs form segments of strip lines. The choice of the diameter of the rods and the distance to the ribs and cylindrical surface of the dipole arms has improved antenna matching in the frequency range of 0.5–1 GHz.

To reduce the influence of the currents induced on the outer surface of the tube 9 in which the coaxial cable is mounted, the short-circuited segments of coaxial lines (so called "locking cups") 10 of the length 125 mm and wave impedance of 95 Ohms are used. They are located at a distance of 30 mm from each other and from the plate 4 of the dipole. The external diameter of the antenna is 64.5 mm, and the length is 620 mm including three locking cups. The antennas are placed in a fiberglass tube 11 with a dielectric permittivity $\varepsilon = 4.6$ and a wall thickness of 27 mm.

3. Transceiver parameters

Characteristics of the borehole radar transceiver were measured with a stand which simulates the borehole. The antennas were placed horizontally in a fiberglass tube along the axis of a wooden box filled with moistened sand. Internal dimensions of the box were the following: length 2 m, height and width 1.2 m. The dielectric permittivity $\varepsilon = 3.9$ of the model medium was determined by measuring the delay time of propagation of radiation pulses with a duration of 0.7, 1.5 and 3 ns emitted by UWB combined antenna [5] through the medium and in the free space at the same distance. The frequency dependence of the electromagnetic wave attenuation in the medium was calculated from the ratio of the spectra of the received pulses that passed through the medium and through free space at a distance of 1.2 m between the receiving and transmitting antennas. The results are shown in Figure 3.

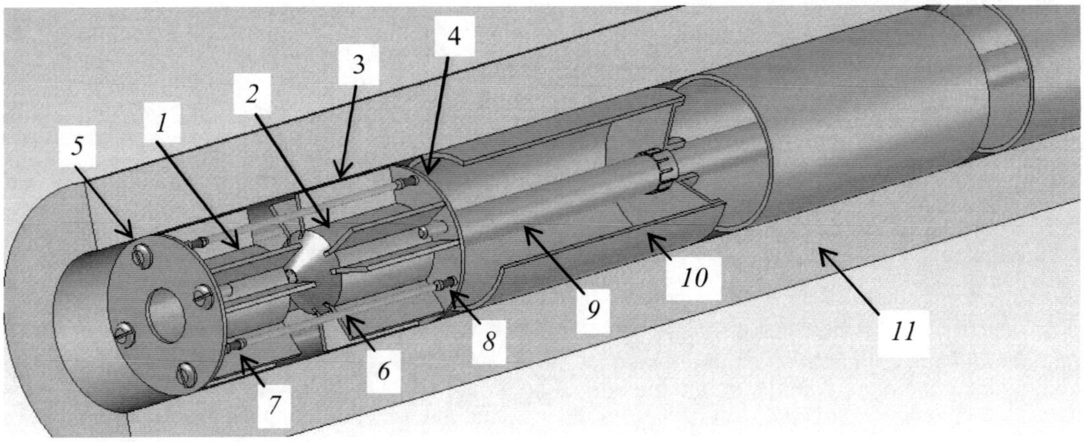

Figure 2. Design of the antenna.

The voltage stand-wave ratio (VSWR) of a single antenna placed in a medium with a 5 m long cable, measured using a vector network analyzer is shown in Figure 4 (curve 1). The VSWR calculated from the ratio of the spectrum of the pulse reflected from the antenna and the spectrum of pulse of the generator is shown in Figure 4 (curve 2). The VSWR of the antenna obtained by the numerical simulation is shown in Figure 4 as well (curve 3).

The pulse waveform at the output of the generator is shown in Figure 5 (curve 1). The waveforms of the generated field strength on the lateral surface of the box are shown in Figure 5. Curve 2 corresponds to measured waveform by using the received UWB dipole [6, 7]. The receiving dipole has the effective height equal to 0.3 cm in the frequency range of 0.2–1.8 GHz. The waveform distortions of the recorded pulses do not exceed 24% compared to the standard TEM antenna. Curve 3 corresponds to calculated waveform.

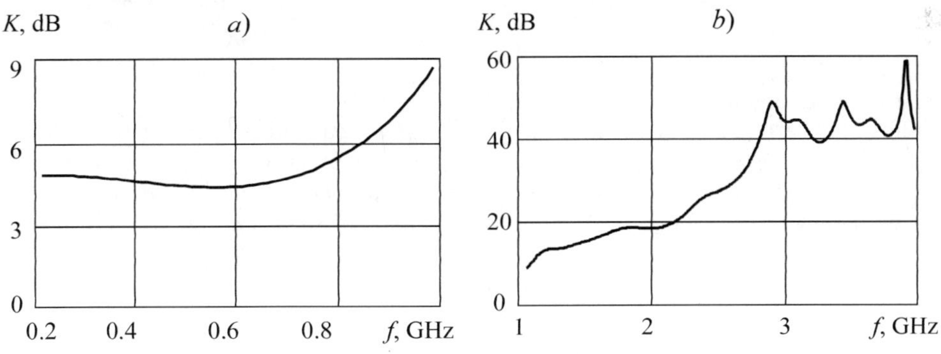

Figure 3. Wave attenuation in the medium at a distance of 1.2 m.

The calculated radiation patterns of the antenna in the medium are shown in Figure 6. The maximum of pattern is perpendicular to the antenna axis in the frequency range of 0.25–0.9 GHz. At higher frequencies, the pattern is distorted. Results of measurement (curve 1) and calculation (curve 2) of the interaction between the transmitting and receiving antennas are shown in Figure 7. We have chosen $M = |U_p^{rec}| / |U_p^{gen}|$ as an estimation of interaction, where U_p^{rec} is the peak value of the voltage at the output of the receiving antenna and U_p^{gen} is the voltage at the input of the transmitting antenna. Figure 7 shows the results of measurement (curve 1) and calculation (curve 2) of the interaction

IOP Conf. Series: Materials Science and Engineering **516** (2019) 012043 doi:10.1088/1757-899X/516/1/012043

between the transmitting and receiving antennas. At $D > 60$ cm and $|U_p^{gen}| = 2.5$ kV the peak output voltage of the receiving antenna does not exceed 3 V. The peak electric field strength at a distance of 60 cm from the dipole axis is 450 V/m.

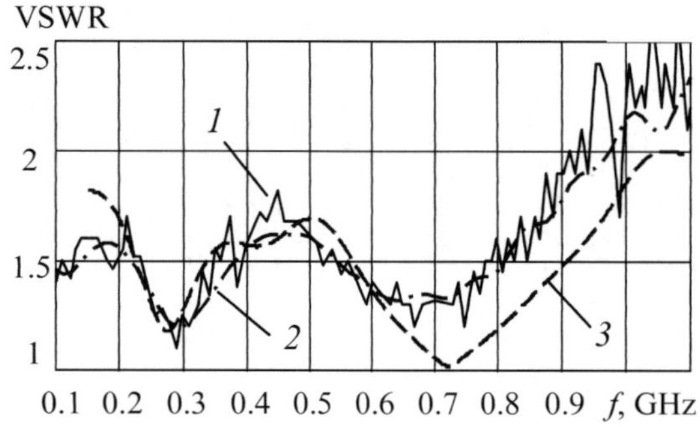

Figure 4. VSWR of the antenna. 1 – measured by network analyzer, 2 – measured in time domain, 3 – simulation.

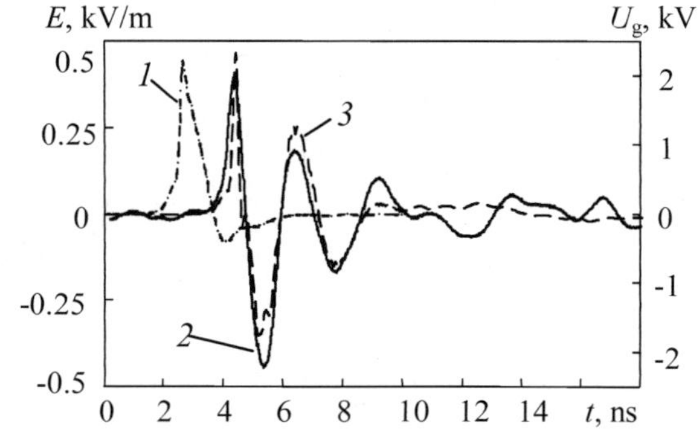

Figure 5. Generator pulse (1); measured (2) and calculated (3) antenna radiation field.

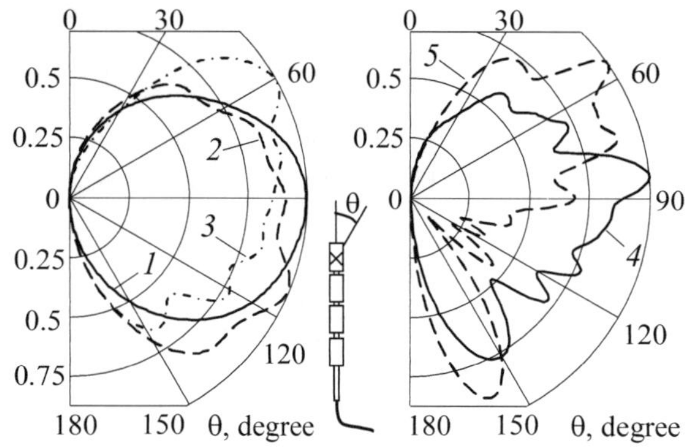

Figure 6. Calculated antenna patterns. 1 – 0.25; 2 – 0.4; 3 – 0.6; 4 – 0.9; 5 – 1 GHz.

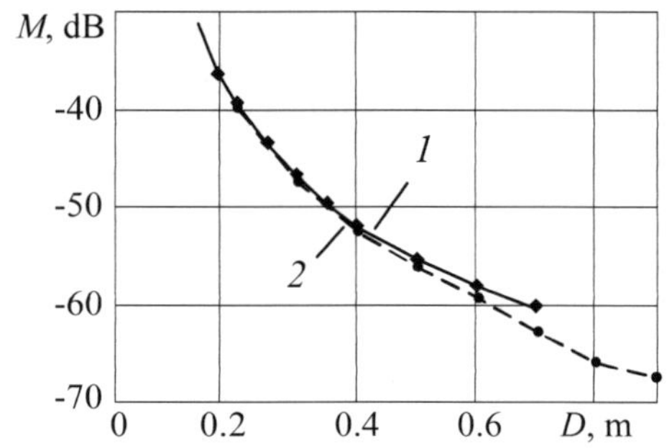

Figure 7. Interaction between transmitting and receiving antennas. 1 – measurements; 2 – simulation.

4. Object sounding results

The sounding was carried out by moving the receiving and transmitting antennas in 5-cm steps inside the fiberglass tube. The tube was placed horizontally in the container with the wet sand. The distance from the axis of antenna displacement to boundaries of the medium was equal to 60 cm. The distance between the antennas $D = 60$ cm was fixed by a polyester-styrene foam rod. The pulse at the output of the receiving antenna was recorded every 5 cm of antennas moving inside the tube. The results were displayed as a set of waveforms $U(t, x)$ in grayscale. Positive voltage corresponds to the light areas and negative one to the dark areas, respectively. The results of sounding of a 140-mm diameter and 400-mm length metal cylinder on the top surface of the box are shown in Figure 8. Area 1 in Figure 8*a* corresponds to the direct radiation from the transmitting antenna, area 2 – to the reflection from the ends of the container with medium which are perpendicular to the axis of scanning, and area 3 – to the reflection from the object, respectively. The result obtained by subtracting the *x*-averaged voltage $\langle U(t_0, x)\rangle$ from the original waveforms for each fixed time moment t_0 is shown in Figure 8*b*. The difference between waveforms in the presence of an object and waveforms without the object is shown in Figure 8*c*. The time of reflection appearance corresponds to the distance from the object to the axis of scanning. Localization along the *x* axis corresponds to the length of the object.

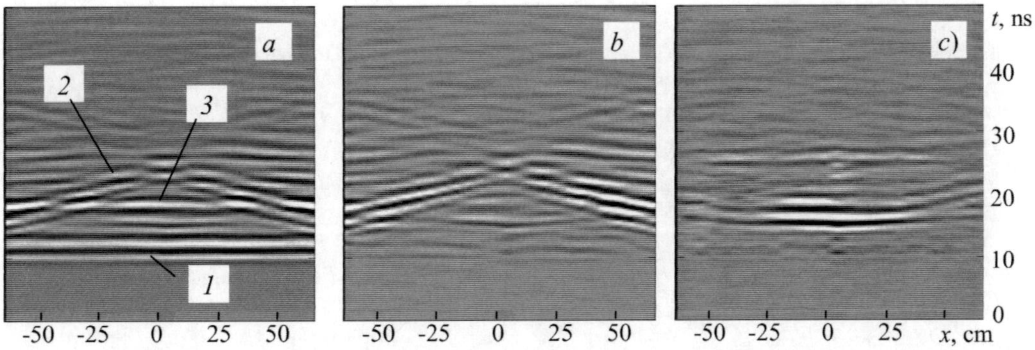

Figure 8. Results of scanning of a metal cylinder on the top surface of box with medium.

Results of scanning of a 105×85 cm metal sheet on the top surface of the container with dielectric medium are shown in Figures 9*a–c*. The sheet was oriented by its narrow side along the axis of scanning. The size and the time of appearance of the reflection area in the Figure 8 correspond to the predetermined place and size of the object. Results of sounding of two 100×35 cm metal plates at a distance of 30 cm from each other are shown in Figures 9*d–f*. Objects are come out at the specified distance to the axis of scanning. The results indicate the possibility to obtain information about the presence of inhomogeneities in the medium with minimal processing of the initial data.

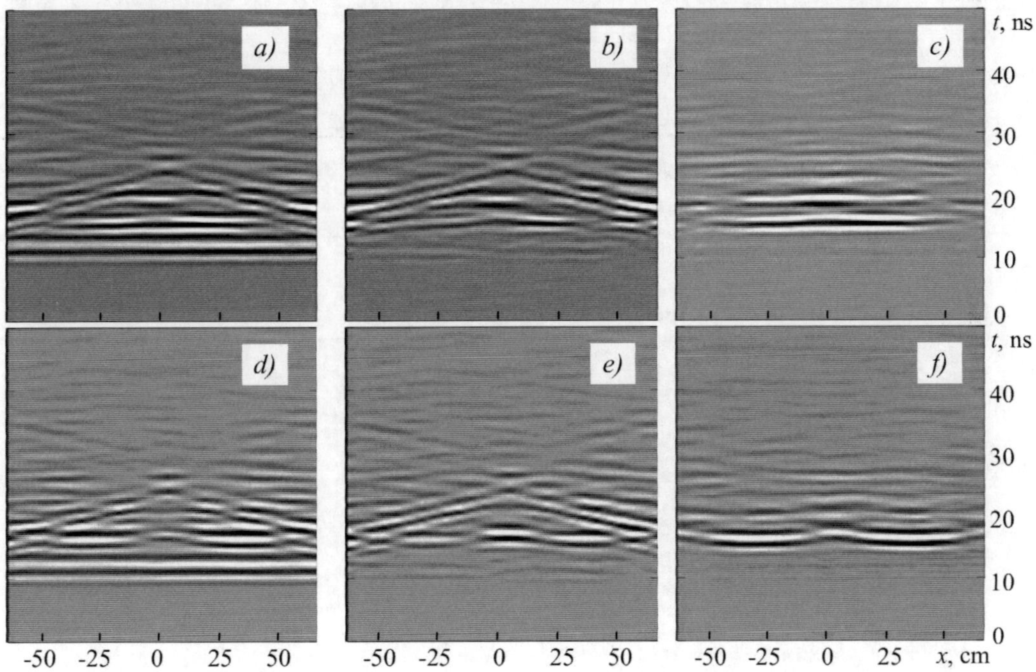

Figure 9. Results of scanning of objects on the top surface of box with medium: the metal sheet (a–c) and two metal plates (d–f).

5. Conclusion

The antenna system of ultrawideband borehole radar based on a dipole with resistive loads has been developed and investigated. The antenna system is intended to sounding the medium nearby the borehole with nanosecond pulses. If the antenna is excited by a monopolar voltage pulse with duration of 0.8 ns at half-height and amplitude of 2.5 kV, the peak electric field strength at a distance of 60 cm from the dipole axis has the value of 450 V/m.

IOP Conf. Series: Materials Science and Engineering **516** (2019) 012043 doi:10.1088/1757-899X/516/1/012043

References

[1] Boger M, Glasmachers A 2013 *Proc. 7th Int. Workshop on Advanced Groung-Penetr. Radar* 6601551 doi: 10.1109/IWAGPR.2013.6601551

[2] Zhou H, Sato M 2004 *IEEE Trans. Geosci. Remote Sensing* **42** 335 doi: 10.1109/TGRS.2003.817215

[3] Epov M., Mironov V., Bobrov P., Repin A. 2009 *Rus. Geol. and Geophys.* **50** 470 doi: 10.1016/j.rgg.2009.04.004

[4] Andreev Yu., Buyanov Yu., Koshelev V. 2005 *J. Commun. Technol. Electron* **50** 535

[5] Balzovsky E., Buyanov Y., Koshelev V. 2004 *J. Commun. Technol. Electron* **49** 426

IOP Conf. Series: Materials Science and Engineering **516** (2019) 012044 doi:10.1088/1757-899X/516/1/012044

Possibilities for applied joint speech processing and computer vision systems

E S Romeu, V I Syryamkin

National Research Tomsk State University, Tomsk, Russia
E-mail: edgarsolisromeu@gmail.com

Abstract. This report addresses the current state of the Computer Vision and the Speech Processing fields. It explains some of the most used algorithms in these fields and provides an overview of the current use of these technologies in the market. The objective of this report is to find new possible industrial and commercial applications that use both Computer Vision and Speech Processing.

1. Introduction

Now a day people live with gadgets and technology that would be unthinkable some 20 years ago like tablets, Smart watches, an ubiquitous internet with ever growing capabilities and many other things, but the truth is that this is only the beginning, and there are many more developments that are being worked on to this day. Speech processing and Computer Vision are two of the fields of computer science that are having huge advances, with many of them already part of modern life, and with many more still to come.

2. Speech Processing

2.1 General Aims of the Speech Processing Field

Computer Science has many fields and one of them is Speech Processing. The aim of this discipline is to create computer systems capable of recognizing spoken words. The main idea is to detect the words spoken by a person and process that data to identify what the person is saying [1].

2.2 Brief History of the Field

This field that has been in existence since the 18th century with Wolfgang Von Kempelen's Acoustic Mechanical Machine. A quick review of some major breakthroughs in Speech Processing would include Edison's dictation machine in 1879, a computer system names Audrey capable of recognizing spoken digits with 90% accuracy developed in 1952 at Bell Labs and IBM's Tangora that could be adjusted for particular voices and recognized a wide array of English words and phrases. The speech recognition software of the 1980's required the speaker to speak slowly with space between words and without background noises. In 1997, Dragon's Naturally Speaking Software was released and it allowed for its user to speak complete phrases, instead of only words [2].

During the first phase of speech processing software development, these systems worked by translating the sound waves into numbers and storing them, then the speech recognition would be triggered when an identical sound was introduced as input. This method is generally known as Template Matching.

IOP Conf. Series: Materials Science and Engineering **516** (2019) 012044 doi:10.1088/1757-899X/516/1/012044

Eventually Machine Learning was introduced and it revolutionized the field. The main player that used ML to its advantage was google in 2008 with its Google Search App for Iphone. Eventually that company developed an algorithm called "Hummingbird" that allowed better use of language when it is used [2].

2.3 Connectionist Temporal Classification (CTC) Algorithm, How Does It Work

A common used algorithm for Speech Processing is the Connectionist Temporal Classification (CTC). Here is a fast and simple overview of how it works:

1. Transform the analog sound waves into digital information. To do this you record a word and you sample a tiny piece of a recording by marking the height of a sound wave at a specific point in time. The idea is that we end up with an array of numbers that represent a sound wave's amplitude every $1/16000^{th}$ of a second intervals [3].

2. The information is pre-processed, one good option is to group the sampled audio into 20 millisecond long pieces, when this information is plotted the resulting graph looks like a sound wave. Sound waves are formed by the combination of many waves. It is possible to deconstruct the complex sound wave into simple sound waves by adding the amount of energy that is contained in each, this is done through Fourier Transform [3].

3. The neural network is fed the data and it tries to find out the letter that is being spoken. It is a good idea to use a recurrent neural network, since they use memory to influence their predictions, this is useful at recognizing patterns in words and guessing which letters will be next. The result is a mapping of the most likely spoken letters. Eliminating possible spaces and ignoring the repeated characters can refine this output [3].

4. Using large databases that record the likelihood of a word appearing in a language, the neural network can be trained to predict the word that is being constructed. To do this, companies like Google and Apple record their users to build great collections of thousands of recordings of words to feed them to neural networks and get even better results. That is why they install their apps freely, to gain access to users [3].

Researchers working for Microsoft used Deep Neural Networks to create a Speech Recognition system that was tested against the Switchboard Speech Recognition task which is an evaluation to find out if these systems are up to the standards of the industry. The result was a Word Error Date of 6.3%, at the time of the test in 2016, this was the best result ever obtained on NIST 2000 Switchboard which made the system the most accurate system ever developed [4].

2.4 The State of Speech Processing in the Industry

To learn how speech recognition can be applied in the Industry let's take the example of Apple's Siri. It uses microphones to detect speech and then transcribes the text using Automatic Speech Recognition. The text is then transformed into parsed text and gets evaluated locally in the device, if it can't be done properly then Siri uses Cloud web services to query the request. Once Siri has executed the command and has the results, it provides them to the user either by text or by a verbal response [5].

These companies have more ambitious plans for these products. The idea is to use speech recognition software to push for devices in houses in what is called the Internet of Things. By doing this, a customer uses a search engine by only asking a question without having to open a computer and he can also do online sales using this technology.

Customers have responded to this business model and adopted the products mentioned. For example, Tech Crunch has published a report with data captured from May 2016 to May 2017. This report found that the demographics of a "super user", a person that spends twice the amount of time with personal assistant technologies on a monthly basis than average, is a 52-year-old woman spending 1.5 hours per month with assistant apps. This is an interesting fact that shows the possibilities for market expansion of these speech based assistants, especially because the users demographics are not dominated by young tech savvy people [6].

It is also good to point out that it is projected that by the year 2020 half of all search engine queries will be done by speech, also on 2017, 7% of all US households had smart speakers (assistants like Alexa, Siri, etc), but by 2020 the percentage is projected to raise to 75%. This has convinced major companies to invest in this technology since more people are accepting the concept of communicating with businesses through smart speakers. We have to keep in mind that these are only the possibilities of Speech Recognition through the lens of retail [6].

According to market intelligence firm Tractica, Speech Recognition software market will grow to 5.4 billion dollars by 2025 from a 136 million USD in 2016. SR is also expected to drive an expansion of sales in professional services and hardware. Tractica estimates that by 2025 the market opportunity for SR will be of 22.3 billion. That is why it is important to protect Intellectual Property related to this sphere as well as assessing the costs for related products and services [7].

Researchers working for Microsoft used Deep Neural Networks to create a Speech Recognition system that was tested against the Switchboard Speech Recognition task which is an evaluation to find out if these systems are up to the standards of the industry. The result was a Word Error Date of 6.3%, at the time of the test in 2016, this was the best result ever obtained on NIST 2000 Switchboard which made the system the most accurate system ever developed [4].

Table 1 contains some interesting startups that are using this technology and their place in the market.

Table 1. List of some Speech Processing Start Ups [8]

Company	Description	Advantages	Disadvantages	Business Plans and cost
Tetra	Software that uses speech recognition to take notes on phone calls.	Synchronization with Google Calendar. Lets you organize the conversation using keywords, tag highlights, label speakers and creates summaries.	The process is not instant; it takes 10 minutes to generate the notes. Laws regarding recording business and private phone calls, the app notifies the parties involved that they are being recorded, as a way of dealing with it.	Free release with subscription plans and business support being introduced in the future. They have obtained 1.5 million USD in seed investment.
Alexa	An intelligent assistant that uses speech recognition to do online searches.	Makes a good use of speech recognition to create an easy to use interface.	Concerns from privacy advocates that have issues with the way amazon saves the voice recordings of the users for future neural training.	15 billion USD in projected revenue in the next three years (2019 - 2021).
NeuroLex [9]	A startup founded in 2016 that is developing an app that analyses the voice of a patient for signs of depression, schizophreni	This startup points the way to using processing language in the medical profession.	The technology is still in development. More factors and tests are needed to determine if a	They have invested $210,000 USD and are looking for 1.5 million USD for

IOP Conf. Series: Materials Science and Engineering **516** (2019) 012044 doi:10.1088/1757-899X/516/1/012044

	a, Alzheimer and Parkinson.		person has one of the diseases listed.	investment at seed level.
Ummo	A speech coach application that helps its user to improve their oratory skills.	The app analyses the filler words used and also the pace and volume of the speech, as well as pauses, clarity and accents.	The app focuses on analyzing and presenting the data of the speech, but doesn't present solutions for the users. Only works on Apple.	It costs 1.99 USD in the Apple Store. It has a monthly download rate of 9049 per month.

3. Computer Vision

3.1 What Is the Aim of Computer Vision?
Computer Vision tries to create computer systems capable of reconstructing properties of images fed into the system mainly shapes, illuminations and color distributions [10].

3.2 Computer Vision Algorithms
There are a multitude of Computer Vision related algorithms and their effectiveness depends on the type of image that is being subjected to analysis. Histograms of Oriented Gradients (HOG) are a method to transform an image into data that can be more easily interpreted by a computer, by getting rid of information that is not important.

1. Turn the image being detected black and white.
2. Look at every pixel and focus on how dark it is compared to surrounding pixels and mark the gradient (draw an arrow on the direction the pixel is getting darker, gradients show the flow of from dark to light).
3. Break the image into small 16×16 pixel squares.
4. In each pixel count how many gradients point to a major direction.
5. Replace the squares in the images with an arrow showing the strongest arrow directions.
6. The result is a HOG image pattern.

HOG images are compared to other patterns that were acquired while training the neural network with other faces [11]. Figure 1 shows the result of transforming a color image into gradients [12].

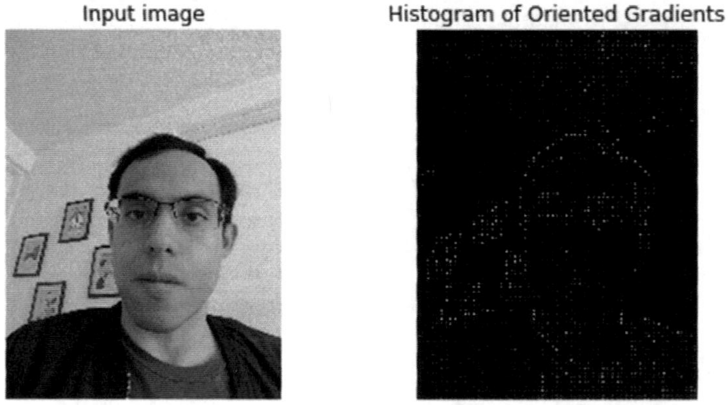

Figure 1. Histograms of Oriented Gradients.

HOG can be used with other algorithms like Face Landmark Estimation to generate programs that recognize faces.

Face Landmark Estimation finds 68 specific landmarks that can be found in any face (eyes, chin, eyebrows) and maps them. The results of this algorithm are rotated, scaled and sheared, so as to present the landmarks as a 2d face viewed from the front. This is useful since many times people are turning their heads when photographed and for a computer its difficult to recognize the same face viewed from different angles [11]. An example of this is presented on Figure 2.

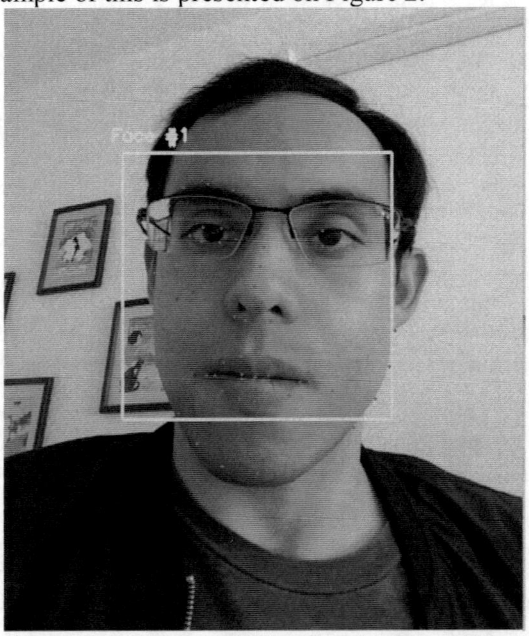

Figure 2. This image shows the specific landmarks detected by the machine [13].

Deep Convolutional Neural Networks are used to train computers to find useful 128 measurements to recognize each face, this is known as embedding. The process is done by loading two different pictures of a same person and then a picture of a different face, the algorithm looks at the measures it is currently creating for the images and tweaks the neural network so the measurements for the pictures of the same person are closer, than the ones of the other face [11].

3.3 Computer Vision Related Start Up

Table 2 shows a Computer Vision Start Ups.

Table 2. Computer Vision Start-ups [8]

Company	Description	Advantages	Disadvantages	Costs
Omnify	Analyzes outdoor audiences and creates behavioral insights about the audience.	Applies computer vision for marketing and retail purposes.	It is a new company, hasn't even been around for a year so it yet has to prove itself on the market.	Founded on march 2018, it has gained 2 million USD on founding.
Blippar	European company founded in 2011 that specializes in Augmented Reality	They focus on applying AR to consumer experience with virtual try-ons, loyalty schemes, gamification, etc.	Technology hasn't been adopted in a massive scale.	Has had 4 founding rounds the latest on august 2018. It has obtained 131.7

				million USD in investment.
Apprentice	AR applied for laboratories and assembly lines. Founded in 2014 in New York City.	Uses smart glasses to create an AR environment. Workers can follow instructions that are displayed to them on the glasses.	The idea can be adapted to more industries.	Has obtained 10.5 million USD in two rounds of investment.
Magicleap	Founded in USA in 2011, offers wearables (special glasses) to create an AR platform.	Developers create software programs that users access through the Magic Leap One specs. The software offers focuses on games, but it also has interior design software, an AR based internet browser, and a social network.	It has yet to gain massive exposure and adoption.	Has obtained 2.3 Billion USD in founding in 5 investment rounds.

4. Some Patents Examples

Table 3 shows some examples of international patents that make use one of these technologies. Note that patents number WO 2014/107076 and number WO 2015/176950 use both technologies in their invention.

Table 3. Example of Patents with information taken from the WIPO Patent Database [14]

Patent Holder	Description	Date of Publication	Number of Patent
Vorwerk & Co. Interholding GMBH (Germany)	A speech recognition device for kitchen appliances, so that it receives commands. It also includes a device for optical detection.	05.05.2015	WO 2015/176950
Ford Global Technologies, LLC. (United States)	Portable speech recognition device for vehicles.	04.05.2017	WO 2017/074328
Samsung Electronics Co Ltd (South Korea)	Speech recognition system that recognizes user's voice and commands and looks in its data base if the commands match others saved in memory and if they do then follows the instructions related to the pre-established command.	10.07.2014	WO 2014/107076
Samsung Electronics Co Ltd (South Korea)	A device that detects the gaze of its users and once detected launches an interaction interface where the user can speak commands to the system.	24.04.2014	WO 2014/061916
Xiaomi Inc. (China)	Device that recognize user's face and that prompts an internet fueled search for the user's information.	20.11.2015	WO 2017/000486

5. Results

What type of products can be developed now? There are fields that seem to be ready for a product that uses both CV and SP. One is the entertainment field where products like videogames can be developed. An example could be a game where the user gives commands to the system and the game could detect the user's face and access his data without the need of a password. That data could be stored in cloud services making easier user mobility between systems.

Other useful way these technologies could be implemented is in the field of marketing. Agencies could implement interactive interfaces that promote brands. Earlier Blippar was mentioned as a company already working in the VR realm, so its possible that they could implement speech recognition to their products.

Finally another obvious choice would be the field of Security. Devices that use voice and image patterns are being developed for the purpose of user access to systems as diverse as cars, computers, bank accounts and smartphones without the need of passwords or keys. As the technology develops these security solutions will be used more extensively and will become the most basic way of enforcing security for a device, building, archive, etc.

6. Summary

Speech Processing (SP) and Computer vision (CV) are two disciplines that have seen great advances in the last decade, specially thanks to the constant development of Machine Learning methods like the Neural Networks. Now a day these technologies are being adopted massively by companies and the public and are becoming ubiquitous in daily life.

This situation will only grow in the future as adoption grows and newer enterprises arise seeking to capitalize in the state of the technologies. New methods of Computer Recognition for both Audio and Image will be developed as Artificial Intelligence capabilities advance and more complex products become feasible.

Acknowledgments

The paper was supported by "The Tomsk State University competitiveness improvement programme" under grant (No 8.2.24.2018) and the Russian Foundation for Basic Research (grant No 16-29-04388).

The authors are grateful to Professor Stanislav V. Shidlovskiy and Tatiana B. Rumyantseva from Tomsk State University for their assistance in preparing the paper.

References

[1] Shikhman M., Shidlovskiy S. 2018 *IOP Conference Series: Materials Science and Engineering* **363** (1) doi: 10.1088/1757-899X/363/1/012019

[2] Abhang P.A., Gawali B.W. 2016 *British Journal of Applied Science & Technology* **10**(5) 1–13 doi: 10.9734/BJAST/2015/19000

[3] Lawson S 2018 The Past, Present and Future of Speech Recognition Technology *ClickZ* URL: https://www.clickz.com/report-past-present-future-speech-recognition-technology/209060/ af

[4] Machine learning is Fun Part 6: How to do Speech Recognition with Deep Learning Medium URL: https://medium.com/@ageitgey/machine-learning-is-fun-part-6-how-to-do-speech-recognition-with-deep-learning-28293c162f7a

[5] Luxton D 2015 *Artificial Intelligence in Behavioral and Mental Health Care* (Academic Press)

[6] Szeliski R 2011 *Computer Vision: Algorithms and Applications, Texts in Computer Science* (Springer-Verlag London Limited) doi: 10.1007/978-1-84882-935-0_14

[7] Dalal N., Triggs B. 2005 *IEEE Computer Society Conference on Computer Vision and Pattern Recognition (CVPR'05)* doi: 10.1109/CVPR.2005.177

[8] Shidlovskiy S. 2016 Boolean Differentiation Equations Applicable in Reconfigurable Computational Medium *MATEC Web of Conferences* **79** doi:10.1051/matecconf/20167901014

[9] Kazemi V., Sullivan J. 2014 *IEEE Conference on Computer Vision and Pattern Recognition* doi:10.1109/CVPR.2014.241

IOP Conf. Series: Materials Science and Engineering **516** (2019) 012045 doi:10.1088/1757-899X/516/1/012045

Using the Monitoring Network and Geoportal Technologies to Analyze the State of the Atmosphere in Krasnoyarsk (Russia)

A Tokarev[1], N Shaparev [1, 2]

[1]Institute of Computational Modeling SB RAS, Krasnoyarsk, Russia
[2]Siberian Federal University, Krasnoyarsk, Russia
E-mail: tav@icm.krasn.ru

Abstract. This study examines the state of the atmosphere in Krasnoyarsk in terms of sustainable development, based on the analysis of data presented in state reports on the state of the environment from 2007 to 2017, and data obtained from automated observation posts located on the city territory.

1. Introduction

Being a natural mixture of surface layer gases, the atmospheric air is an essential component of the environment. During the active industrial activity, large quantities of pollutants are emitted into the atmosphere, worsening the quality of atmospheric air both near the source of pollution, causing the local environmental problems, and influencing the environmental processes. According to WHO, about 7 million people worldwide died of the air pollution in 2012 [1]. This result is about 2 times higher than the previous showed estimates [2]; it confirms that the air pollution is currently the world's largest environmental health risk.

This study examines the state of the atmosphere in Krasnoyarsk in terms of sustainable development, based on the analysis of data presented in State Reports "On State and Protection of the Environment in the Krasnoyarsk Territory" for the period from 2007 to 2017, and data obtained from automated observation posts, located in Krasnoyarsk.

2. Object of Study

Krasnoyarsk was founded by the Russian Cossacks in 1628. To protect them from winds, they chose a place in the valley of the Yenisei River, partly framed by mountain ranges. Subsequently, the Moscow state route passed through Krasnoyarsk, connecting the central part of Russia with its Far Eastern regions. In 1832, Krasnoyarsk became the administrative center of the Yenisei Province. The favorable geographical location and presence of rich natural resources predetermined (since 1934) the intensive development of chemical, metallurgical, pulp and paper, machine-building, fuel and energy industries in the city, which eventually led to a large amount of atmospheric emissions. The situation worsened in the 1990s for the reason of sharp increase in the number of imported used cars. The security of the city from the winds additionally worsened the ecological situation and led to occasional appearance of the city smog. Currently, the city is located on both sides of the Yenisei extending for 25 km along the river, its size in the transverse direction is about 10 km. According to the Krasnoyarskstat for 2018, the city's population is about 1 million.

IOP Conf. Series: Materials Science and Engineering **516** (2019) 012045 doi:10.1088/1757-899X/516/1/012045

3. Methods and Tools of Research

3.1. Indicators of Sustainable Development for the Atmosphere
There are several systems of indicators of sustainable development in the world practice, among them one of the most comprehensive in terms of coverage is the UN system of indicators of sustainable development (132 indicators) [3].

The indicators of sustainable development to protect the atmospheric air (recommendations of the UN Commission) are determined by the following indicators [3]:

- *Input impact*:
  - Emissions of gases causing the greenhouse effect (CO_2, CH_4, N_2O); measured in gigagrams (Gg).
  - Emission of sulfur oxides (SOx); measured in tons / year per person, GDP, DE
  - Emissions of nitrogen oxides (NOx); measured in tons / year per person, GDP
  - Consumption of ozone depleting substances; measured in tons / year
- *States*:
  - Concentration of pollutants in the city's atmosphere (ozone, carbon monoxide, suspended solids, sulfur dioxide, nitrogen dioxide, nitrogen oxide); measured in mg / m^3
- *Management*:
  - Expenses on the air pollution abatement; USD

The use of sustainable development indicators to analyze the water use in Krasnoyarsk was shown in [4, 5].

3.2. Air Quality Index (AQI)
For an integrated assessment of the degree of the air pollution in many countries, standardized indicators are used, i.e. atmosphere quality indexes. Among them, the most widespread is the air quality index AQI, developed by the US Environmental Protection Agency [6]. This is 1st of all because it is informative and the easiest to understand, it can provide information on the atmospheric air pollution to the general public in a simple and clear way. AQI can take values from 0 to 500, the higher its value, the higher is the level of the air pollution and the greater is the health hazard. The scale of values is divided into a number of categories (Table 1): Good, Moderate, Unhealthy for Sensitive Groups, Unhealthy, Very Unhealthy, Hazardous. Each category has a visual color designation.

AQI is calculated based on concentrations of several pollutants: suspended particles with a diameter of less than 10 microns (PM_{10}) and less than 2.5 microns ($PM_{2.5}$), carbon dioxide CO, sulfur dioxide SO_2, nitrogen dioxide NO_2 and ozone O_3. Calculation of this index is carried out sequentially for each substance according to the formula of piecewise linear interpolation [6]:

$$AQI = \frac{I_{high} - I_{low}}{C_{high} - C_{low}}(C - C_{low}) + I_{low} \qquad AQI^* = \max[AQI_k] \tag{1}$$

In equation (1), C is the average concentration of the pollutant, [C_{low}, C_{high}] are the limits of the concentration interval from Table 1, in which C falls, [I_{low}, I_{high}] is corresponding AQI value interval. The integral index AQI* is calculated as the maximum value for all pollutants AQIk, where k is the type of substance.

IOP Conf. Series: Materials Science and Engineering **516** (2019) 012045 doi:10.1088/1757-899X/516/1/012045

Table 1. Conversion Factors for Individual Pollutants.

O_3 (ppb)		$PM_{2.5}$ (µg/m³)	PM_{10} (µg/m³)	CO (ppm)	SO_2 (ppb)		NO_2 (ppb)	AQI	
8 hr	1 hr	24 hr	24 hr	8 hr	1 hr	24 hr	1 hr		
$C_{low} - C_{high}$ (avg)								$I_{low} - I_{high}$	Category
0 – 54	–	0.0 – 12.0	0 – 54	0.0 – 4.4	0 – 35	–	0 – 53	**0 – 50**	Good
55 – 70	–	12.1 – 35.4	55 – 154	4.5 – 9.4	36 – 75	–	54 – 100	**51 – 100**	Moderate
71 – 85	125 – 164	35.5 – 55.4	155 – 254	9.5 – 12.4	76 – 185	–	101 – 360	**101 – 150**	Unhealthy for Sensitive Groups
86 – 105	165 – 204	55.5 – 150.4	255 – 354	12.5 – 15.4	186 – 304	–	361 – 649	**151 – 200**	Unhealthy
106 – 200	205 – 404	150.5 – 250.4	355 – 424	15.5 – 30.4	–	305 – 604	650 – 1249	**201 – 300**	Very Unhealthy
–	405 – 504	250.5 – 350.4	425 – 504	30.5 – 40.4	–	605 – 804	1250 – 1649	**301 – 400**	Hazardous
–	505 – 604	350.5 – 500.4	505 – 604	40.5 – 50.4	–	805 – 1004	1650 – 2049	**401 – 500**	

3.3. Air Quality Monitoring Network in Krasnoyarsk

According to the data of 2017 [7], Krasnoyarsk is included into the priority list of 20 large Russian cities with the highest level of air pollution.

Now there are 7 automated observation posts (AOP) in Krasnoyarsk and its suburbs, i.e. 1-Cheryomushki, 2-Severny, 3-Solnechny, 4-Berezovka, 5-Kubekovo, 6-Vetluzhanka, 7-Pokrovka. The posts make automatic measurement of mass concentrations of contaminants in the atmospheric air, meteorological parameters (wind direction and speed, temperature, humidity, atmospheric pressure), as well as collection, processing and transmission of the accumulated data to a remote server. The set of measurable indicators differs in different AOPs, but most of them transmit data on such pollutants as nitric oxide, nitrogen dioxide, sulfur dioxide, carbon monoxide, suspended particles $PM_{2.5}$. The collected data comes into the regional departmental information and analytical system on the state of the environment of the Krasnoyarsk Territory (RDIAS) [8].

The Institute of Computational Modeling (ICM SB RAS) also conducts research and development of software and hardware to control the level of the air pollution. Based on the geoportal software and technological platform [9] a scientific and research monitoring block was created [10, 11]. The system provides data collection from external sources, data storage and aggregation, automatic calculation of derived indicators, data upload and presentation via the web interface. Data sources can be either individual sensors, or external databases, or information systems through additional adapters. An API is also supported for integration with external information systems.

Currently, several data streams are connected, such as open sources of climate data, autonomous meteorological stations, and systems to control the state of the environment. Apart from that, a software module is developed aimed at regularly download of the data from observation posts from the KVIAS system through a specialized web service. This facilitates processing and analysis of data owing to the geoportal functions and software tools.

4. Results

4.1. Analysis of Pollutant Emissions According to UN Indicators

The initial data on the total emissions of pollutants into the atmospheric air of Krasnoyarsk from stationary and mobile sources were obtained on the basis of public information provided by the Territorial body of the Federal State Statistics Service for the Krasnoyarsk Territory and the Federal Service for Supervision of Natural Resources [7].

Industrial Krasnoyarsk enterprises, with the highest emissions of pollutants into the atmosphere [7] for 2017 are JSC RUSAL Krasnoyarsk – 56.8 thousand tons, Krasnoyarsk CHP-1 – 17.0 thousand tons, Krasnoyarsk CHP-2 – 14.0 thousand tons, Krasnoyarsk CHP-3 – 9.6 thousand tons.

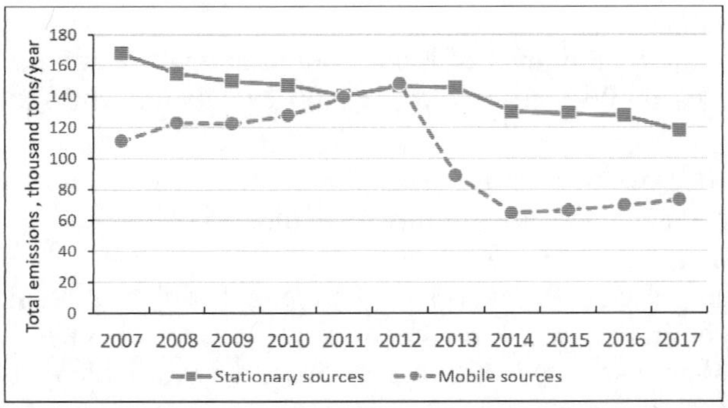

Figure 1. Gross Emissions of Pollutants into the Krasnoyarsk Atmosphere.

The gross volume of pollutants is a total characteristic of the amount of pollution entering the city's atmosphere. Figure 1 shows change in gross emissions from stationary and mobile sources to the air in Krasnoyarsk for 2007–2017. In 2017, the contribution from the mobile sources to the total emissions was about 38%.

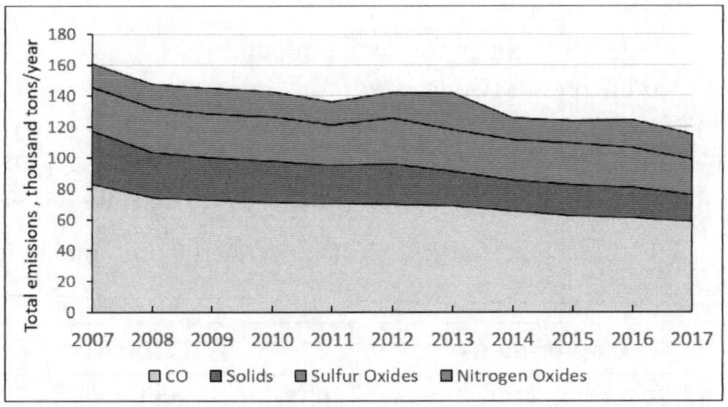

Figure 2. The Structure of Emissions of Pollutants From Stationary Sources Into the Krasnoyarsk Atmosphere.

Since 2013, changes have been added to the methodology for estimating the air emissions from the road transport [12]. In connection with the use of the new methodology, emissions from road transport are obtained almost 1.5 times lower than according to the previously used method.

Figure 2 shows the structure of emissions of various pollutants into the Krasnoyarsk atmosphere for 2007 – 2017. Figure 3 shows the gross emissions of various pollutants per person.

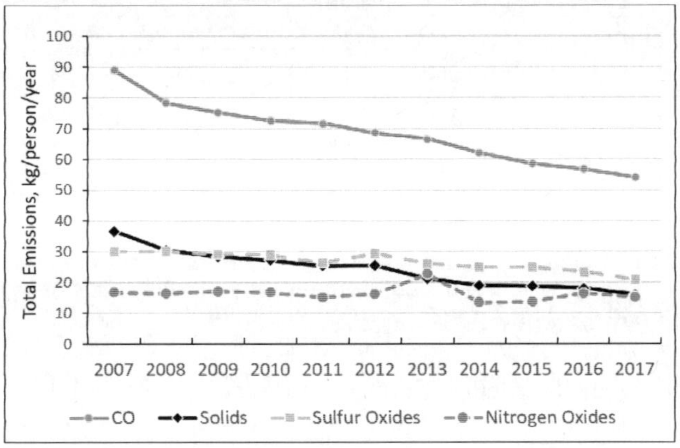

Figure 3. Dynamics of Emissions of Solids, Sulfur Oxides, and Nitrogen into the Krasnoyarsk Atmosphere per Person.

IOP Conf. Series: Materials Science and Engineering **516** (2019) 012045 doi:10.1088/1757-899X/516/1/012045

4.2. Air Quality Index Analysis

For AQI calculation, the data on the level of the air pollution from 7 automated observation posts of the KVIAS system were used. With regard to the AQI calculation, the source data set has several disadvantages:

- ozone and dust PM_{10} data are not available;
- measurements on $PM_{2.5}$ appeared only from the end of 2017;
- gaps in the data.

The full year 2018 was selected for analysis. The method of the AQI calculation uses the hourly data, so the source data coming every 20 minutes was transformed into per hour statistics data. Since the influence of individual components in the integral index was estimated, only hours containing data on most of the available substances were selected, i.e. $PM_{2.5}$, NO_2, CO, SO_2. The lack of data on suspended particles PM_{10} was compensated by data on $PM_{2.5}$, since they correlate and effects of the latter on health are more significant [13]. To assess the city's atmosphere state as a whole, the values were averaged at all observation posts.

The processing results are presented in Table 2, with Figures given per individual months and on average for 2018, where:

- *AQI** are average values for 2018.
- *Components* are average AQI values for each pollutant.
- *Influence* is the percentage of time when the corresponding pollutant was decisive for calculating the total air quality index (the sum for all components may exceed 100%, if there were moments in time when different substances had the same effect on the result).
- *Number of hours* is the number of hours for which there is data available on all pollutants.

Table 2. The Average AQI Values and the Effect of Individual Components for the Year 2018, by Months.

Month	AQI*	Components				Influence, %				Number of hours
		$PM_{2.5}$	NO_2	CO	SO_2	$PM_{2.5}$	NO_2	CO	SO_2	
January	94.6	94.5	20.0	5.4	8.4	99.5	0.4	0	0.1	2280
February	118.1	117.8	31.0	7.3	9.0	99.2	0	0	0.8	2381
March	80.7	78.2	29.6	4.2	7.7	91.0	8.8	0.1	0.4	2619
April	50.9	49.5	18.6	2.4	6.3	92.8	7.2	0	0.3	2722
May	38.8	37.8	11.8	2.3	8.6	94.2	2.8	0	3.4	2855
June	66.1	66.0	14.1	2.9	8.4	99.2	0.8	0	0	2443
July	84.5	84.2	13.9	3.7	2.9	98.7	1.4	0	0	2286
August	54.4	53.0	16.0	3.4	5.1	93.3	6.9	0	0.1	3405
September	53.7	50.5	23.3	4.4	4.0	85.5	14.8	0	0.3	2286
October	61.3	56.4	29.3	5.9	6.5	81.7	18.8	0	0.2	2534
November	57.9	53.9	23.3	6.0	7.8	76.3	20.7	0	4.0	3071
December	85.9	79.4	36.8	7.8	9.3	80.8	19.5	0	0.1	2131
2018 year	**68.9**	**66.8**	**21.9**	**4.5**	**7.0**	**90.9**	**8.6**	**0**	**0.9**	**31013**

The obtained data show that the main influence on the air quality index is caused by pollution by suspended particles – 91%. A small contribution comes from nitrogen dioxide pollution — about 8.6% and sulfur dioxide — less than 1% respectively.

4.3. Analysis of Correlation between Adverse Weather Conditions and AQI

Adverse meteorological conditions (AMC) are a short-term special combination of meteorological factors contributing to accumulation of pollutants in the surface layer of the atmospheric air. In accordance with Article 19 of the Federal Law "On Protection of the Atmospheric Air" No. 96-FZ, legal entities having sources of pollutant emissions into the atmospheric air, must take measures to reduce the emissions of pollutants into the atmosphere when receive AMC forecasts.

Depending on the expected level of atmospheric pollution, the FSBI Central Siberian DSMS, based on long-term observations of weather conditions and the level of atmospheric pollution using a specially developed forecasting scheme, provides warnings about the AMC periods.

Table 3 shows the list of announced periods with the AMC regime in the city of Krasnoyarsk for 2018, according to the data of the Federal State Budgetary Institution "Central Siberian DSMS".

Table 3. Announced AMC Regimes for 2018.

№	Begin	End	Duration, hours
1	03.01.2018 19:00	05.01.2018 19:00	48
2	20.01.2018 19:00	27.01.2018 15:00	164
3	02.02.2018 19:00	06.02.2018 19:00	96
4	12.02.2018 19:00	16.02.2018 19:00	96
5	28.02.2018 19:00	02.03.2018 19:00	48
6	06.04.2018 19:00	07.04.2018 16:00	21
7	26.06.2018 19:00	28.06.2018 19:00	48
8	19.10.2018 19:00	22.10.2018 7:00	60
9	06.12.2018 19:00	07.12.2018 19:00	24
10	25.12.2018 19:00	28.12.2018 19:00	72
11	28.12.2018 19:00	31.12.2018 19:00	72

Figure 4 shows the dynamics of changes in the average value of AQI and the periods of AMS in January – February 2018. Bold lines at the top of the graph indicate the AMS periods.

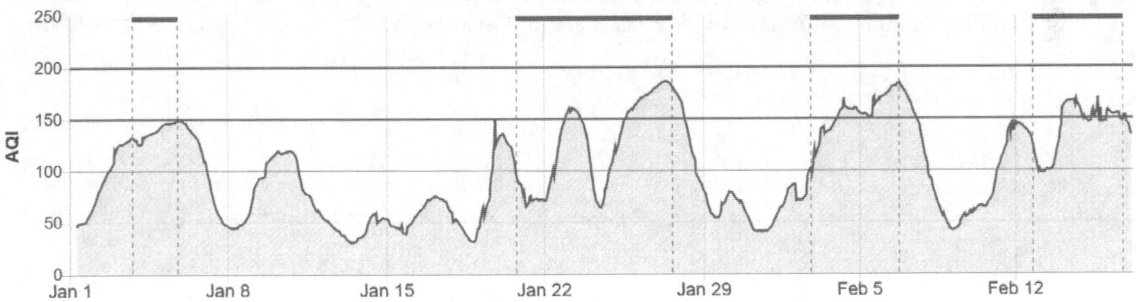

Figure 4. Comparison of Average AQI and AMS Modes in January – February, 2018.

5. Conclusion

The analysis of the state of the atmosphere in Krasnoyarsk in indicators of sustainable development was carried out on the basis of the data obtained from the State Reports "On State and Environmental Protection in the Krasnoyarsk Territory" for 2007 – 2017. The results showed that gross emissions in 2017 from stationary sources amounted to 62%, and from mobile sources – to 38% respectively. The structure of emissions of various substances in 2017 from stationary sources is as follows: carbon monoxide – 50%, sulfur dioxide – 19.3%, solids – 14.7%, nitrogen oxides – 14%. It should be noted that specific emissions per person for all pollutants decreased during 2007 – 2017.

IOP Conf. Series: Materials Science and Engineering **516** (2019) 012045 doi:10.1088/1757-899X/516/1/012045

During 2018, the value of the AQI air quality index was determined online, based on the data received from the KVIAS system and our calculations. The data on individual posts and on average in the city showed that the determining component of the AQI value is the concentration of suspended particles $PM_{2.5}$. In normal periods, the AQI was not exceeding 100 (good / average level on the scale), while during AMC it was about 150 (an unfavorable level on the scale).

References

[1] 2014 *Burden of disease from the joint effects of Household and Ambient Air Pollution for 2012.* (WHO: Geneva)

[2] 2009 *Global Health Risks* (World Health Organization: Geneva)

[3] 1996 *Indicators of Sustainable Development Framework and Methodologies* (United Nations: New York)

[4] Shaparev N., Astafiev N. 2008 *International Journal of Sustainable Development & World Ecology* **15(6)** 574–583 doi: 10.1080/13504500809469853

[5] Shaparev N. 2009 *Resources of Krasnoyarsk Region in indicators of sustainable development* (KSPU publishing house: Krasnoyarsk)

[6] *Technical Assistance Document for the Reporting of Daily Air Quality – the Air Quality Index (AQI)*. U.S. Environmental Protection Agency (May 2016)

[7] *State reports «On the state and protection of the environment in the Krasnoyarsk Territory for 2007–2017»*. Krasnoyarsk (2008 – 2018)

[8] *Subsystem for monitoring air pollution.* URL: http://krasecology.ru/Air

[9] Shaparev N., Yakubailik O. 2016 *MATEC Web Conf.* **79** 01081 doi: 10.1051/matecconf/20167901081.

[10] Shaparev N., Tokarev A., Yakubailik O., Soldatov A. 2018 *IOP Conference Series: Materials Science and Engineering* **363**. 012034 doi: 10.1088/1757-899X/363/1/012034

[11] Yakubailik O.E., Kadochnikov A.A., Tokarev A.V. 2018 *Optoelectronics, Instrumentation and Data Processing* **54**(3) 243–249

[12] Order of the Federal Service for the Supervision of Natural Resources of November 1, 2013 No. 6-p «*On approval of the procedure for organizing work on the assessment of emissions from certain types of mobile sources*».

[13] *WHO Air quality guidelines for particulate matter, ozone, nitrogen dioxide and sulfur dioxide.* Global update 2005: summary of risk assessment. Geneva: World Health Organization (2006)

IOP Conf. Series: Materials Science and Engineering **516** (2019) 012046 doi:10.1088/1757-899X/516/1/012046

Simulation of the behavior of an unmanned aerial vehicle in virtual 3D scenes

A A Taganov, D V Shashev

National Research Tomsk State University, Tomsk, Russia
E-mail: asablindme@gmail.com

Abstract. The object of study is a computer-based solid-state model of an unmanned aerial vehicle. In the process of developing a simulation model, a computer-based solid-state model of the environment was created. The test flight of 3D model was made on mission points. The developed simulation model allows to accurately assess the performance of an unmanned aerial vehicle under certain flight modes.

1. Introduction

Before you start developing a simulation model, you need to select tools. It should be carefully documented; it should have a large number of libraries and extensions allowing developers to solve not only immediate tasks but also problems arising during the development of other modules of an unmanned aerial vehicle (UAV).

The simulation of the motion dynamics of a computer-based solid-state model of an unmanned aerial vehicle was carried out in the operating system for ROS robots using the Gazebo software package.

2. Robot Operating System and Gazebo

ROS (Robot Operating System) is a robot-programming framework that provides functionality for distributed work. The operating system provides the following services: hardware abstraction, low-level control of devices, the introduction of widely used functions, the transfer of messages between processes, and package management. The operating system is based on graph architecture, where data processing takes place in nodes that can receive and transmit messages between themselves. The library is focused on Unix-like systems, such as Ubuntu Linux, Fedora and Mac OS. ROS contains many open source implementations of common functionality and algorithms for robotics. These open source implementations are organized into ‘packages’. ROS distributions include many packages, while other distributions can be developed by individuals and distributed through code-sharing sites such as github [1].

Gazebo is a software package that simulates the interaction of a robot or even a population of robots with the real world. After the detailed description of a robot, it is possible to test both the work of the algorithms and the physical implementation of the robot in a virtual environment, before building it in hardware. The physics engine used in Gazebo allows taking into account such details of the robot’s interaction with the environment as friction, engine power, imitation of sensors and cameras, etc. [2]. A well-designed simulator allows you to quickly test algorithms, create robots, perform regression testing, and train your AI system using realistic scenarios. Gazebo offers the possibility to simulate robot populations in complex indoor and outdoor environments accurately and

IOP Conf. Series: Materials Science and Engineering **516** (2019) 012046 doi:10.1088/1757-899X/516/1/012046

efficiently. The software package provides a robust physics engine, high-quality graphics, user-friendly software and graphical interfaces [3].

Using auxiliary tools, we created an open space map, which was later integrated into the simulation model using the Gazebo software package. This allows testing algorithms, performing regression testing using realistic scenarios without using real robots. In Gazebo, the required physical 3D model of a robot is described using the SDF file format, which is usually used in ROS [4].

First, we create a model of the robot in the xml-format; each functional part (base, lidar, camera, screws, etc.) is described separately, and then these parts are combined together into a hexacopter model. Such parts of a robot like cameras, motors, sensors, etc. need connection to plugins that are written in C ++ or Python and describe the functionality of these parts. Plugins may be connected not only directly to the model and sensors, but to the whole world as well [5]. Figures 1, 2, 3 present models in simulated environment.

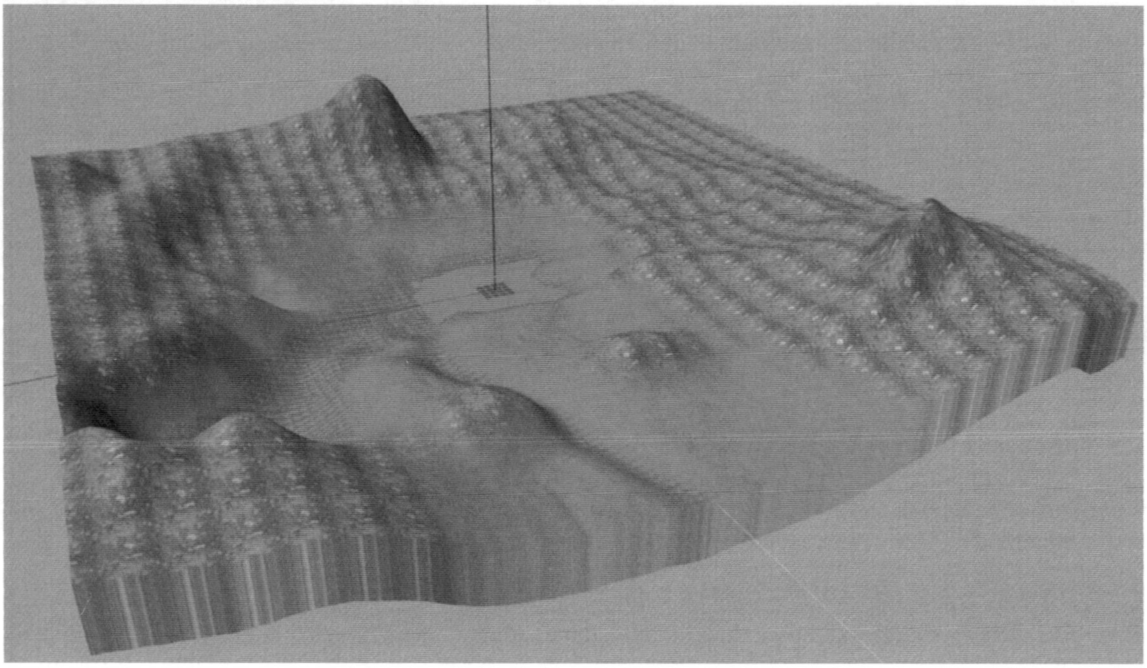

Figure 1. Ground model in Gazebo.

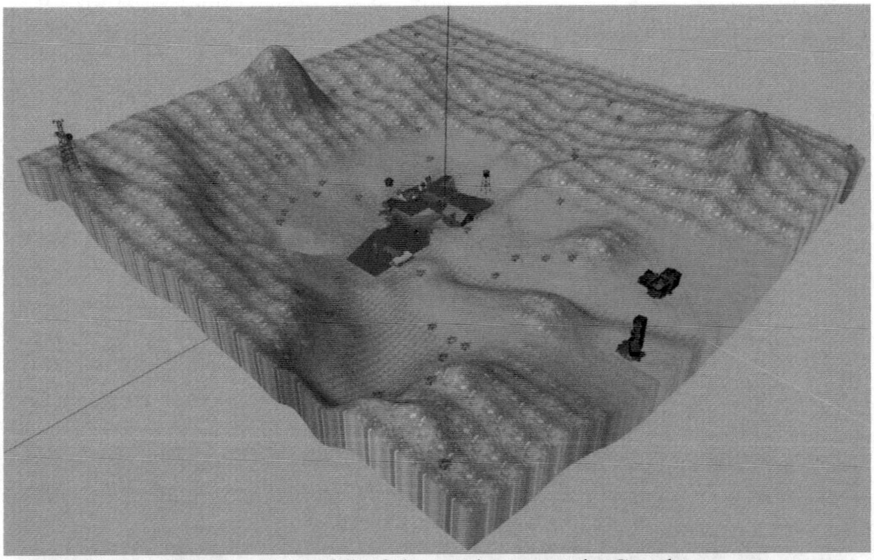

Figure 2. Model of the environment in Gazebo.

Figure 3. Hexacopter model in simulated environment.

3. MAVLink protocol and MAVROS

The unmanned aerial vehicle was controlled using an information protocol for interaction with drones or small MAVLink unmanned aerial vehicles. For information interaction with the device, the ROS package 'mavros' is used. This package implements the ability to send machine control commands through the MAVLink protocol using standard ROS data types [6]. MAVROS - MAVLink extensible communication node for ROS with a proxy for a ground control station. ROS can be used with the PX4 and the Gazebo simulator. It uses the MAVROS MAVLink node to communicate with the PX4 [7].

Lidar Velodyne VLP-16 and camera Microsoft Kinect were selected as tools of vision. Velodyne Simulator allows simulating the Velodyne VLP-16 in Gazebo. This is an open source software. The Velodyne Simulator is a workspace for models, which means that it can be built using the catkin build command in the root folder of the package [8, 9]. Working with lidar simulators requires some kinds of auxiliary work. After building the plugin and searching for its variables in your environment, we import the lidar to Gazebo. As an example of lidar operation, let us launch the UAV environment model without and with obstacles. Figures 4 and 5 present imitation of the lidar with and without obstacles on the UAV model.

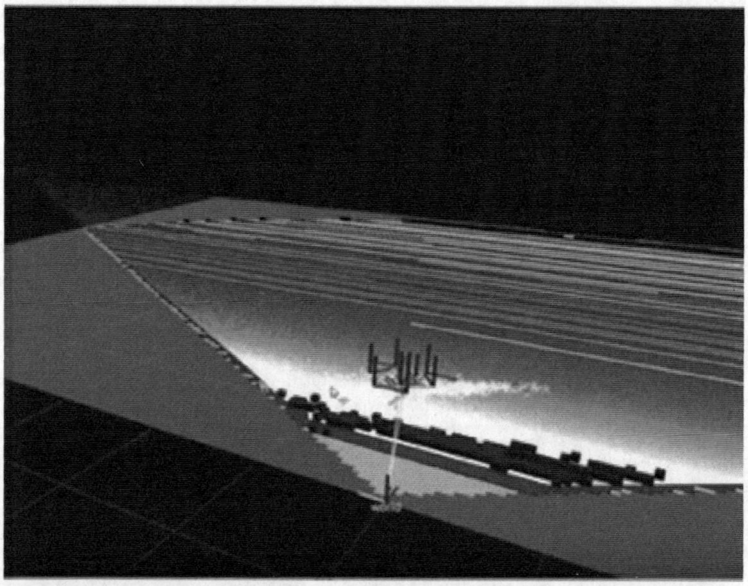

Figure 4. Imitation of the lidar without obstacles on the UAV model.

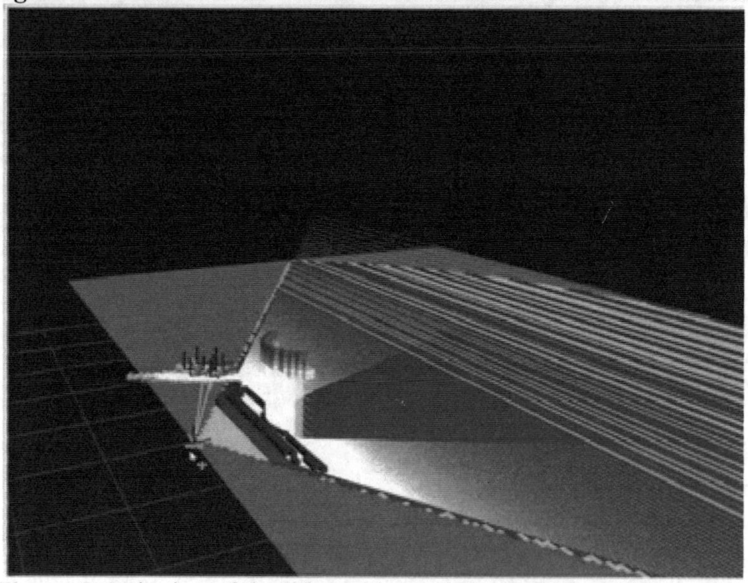

Figure 5. Imitation of the lidar with an obstacle on the UAV model.

After adding the Kinect camera model and sensor to the UAV model, it becomes possible to see through Rviz what the UAV sensors currently see. Figure 6 presents imitation of the Kinect camera on the hexacopter.

Figure 6. Imitation of the Kinect camera on the hexacopter.

4. Summary

Currently, this simulation model is in the process of optimization and planning for the possible introduction of newer technologies. In parallel, modules for testing the Sense and Avoid algorithm are under development. We plan to introduce two multi-rotor UAVs to the simulation model for testing algorithms for capturing moving objects.

Acknowledgments

The research was carried out in Tomsk State University with the financial support of the Ministry of Education and Science of Russia, a unique project identifier: RFMEFI57817X0241.

The authors are grateful to Tatiana B. Rumyantseva from Tomsk State University for English language editing.

References

[1] ROS. [Electronic resource]. URL: http://www.ros.org
[2] Shikhman M., Shidlovskiy S. 2018 *IOP Conference Series: Materials Science and Engineering* 363 (1) doi: 10.1088/1757-899X/363/1/012019
[3] Gazebo. [Electronic resource]. URL: http://gazebosim.org
[4] Specification – SDF. [Electronic resource]. URL: http://sdformat.org/spec
[5] Gazebo plugins in ROS. [Electronic resource]. URL: http://gazebosim.org/tutorials?tut=ros_gzplugins
[6] PX4 Autopilot User Guide. [Electronic resource]. URL: https://docs.px4.io/en/
[7] Gafurov O. *et al.* 2012 *Telecommunications and Radio Engineering* 71(17) 1565–1574 doi: 10.1615/TelecomRadEng.v71.i17.40
[8] Kuznetsov D., Syryamkin V. 2015 *AIP Conference Proceedings* 1688 (040004) doi: 10.1063/1.4936037
[9] Shidlovskiy S. 2016 *MATEC Web of Conferences* 79 doi:10.1051/matecconf/20167901014

IOP Conf. Series: Materials Science and Engineering **516** (2019) 012047 doi:10.1088/1757-899X/516/1/012047

Calculation of the radar cross-section of complex objects for the problem of modeling a synthetic aperture radar

A N Leukhin, A A Voronin, V I Bezrodnyi

Mari State University, Yoshkar-Ola, Russia
E-mail: leukhinan@list.ru

Abstract. A synthetic aperture radar is usually a complex software and hardware package that allows obtaining images in the radio range comparable in resolution with optical systems. The advantage of radio waves is that high-quality photography takes place despite cloudiness and dark time of the day. The development of algorithms for this system type is rather complicated process; mathematical modeling is used for reducing its cost. This paper shows the applicability of approximation methods to calculate the scattered field for imaging of synthetic aperture radar. The disadvantages of these methods are also discussed.

1. Introduction

At present, mathematical modeling is an important element of the development of complex technical systems; it can reduce the cost of conducting real experiments, speed up of hypothesis testing and find previously unknown relationships. The advantage of mathematical modeling is evident for each experiment in the case of synthetic aperture radar (SAR), which is usually installed on board an aircraft. It should be noted that even a simple synthesis of a radar image requires a large computational power and/or processing time because of the large amount of processed data. It is necessary to solve the problem of the formation of a radio hologram, each line of which represents a reflected signal from all objects that fall into the antenna pattern at modeling in addition to the image synthesis.

As a rule, calculating method of the scattered field for the simulation task of a synthetic aperture radar must meet two requirements of the simulation task. First, a reasonably accurate scattering model is needed to observe diffraction, diffuse and specular reflection effects. Secondly, a model is required, which is capable of calculating large areas of the underlying surface in a reasonable time.

Vehicles and anthropogenic structures, which are of particular interest to SAR, usually consist of metal structures. Only ideal conductors can limit the exact calculating model; its scattered field is determined mainly by the geometry. One of the most well-known methods for calculating the scattered field is the method of physical diffraction theory (PTD), developed by P. Ya. Ufimtsev [1].

On the other hand, the underlying surface usually consists of dielectric materials, calculations for which are more complicated than for an ideal conductor. The area of the underlying surface is many times larger than the area of interest objects. In this case, a statistical description of the reflecting properties of materials is used, taking into account the polarization and the angle of irradiation.

Despite the fact that reflective models have been developed for a long ago, their use has not yet become widespread except for the calculation of the effective of a surface of scattering of radar cross-section (RCS). This paper presents an approach for modeling a radar scene with the formation of a radio hologram and the synthesis of a radar image from the simulation data.

IOP Conf. Series: Materials Science and Engineering **516** (2019) 012047 doi:10.1088/1757-899X/516/1/012047

2. Research Method

The calculation of metallic objects is a combination of methods based on physical optics. It includes physical optics and the physical theory of diffraction. The principles of these methods are described in [1–3]. In addition, it should be noted works [4–6]; the calculation equations for arbitrary polarization are given there.

The approach, described in [4], assumes that the resulting scattered field is a superposition of electromagnetic fields, caused by different surface currents:

$$\vec{E} = \vec{E}_1 + \vec{E}_2 + \vec{E}_3 + \vec{E}_4, \tag{1}$$

where $\bar{E}_1$ is the vector of strength of electric field corresponding to the contribution from the smooth elements of the object; $\bar{E}_2$ is the vector of strength of electric field corresponding to the contribution from sharp edges; $\bar{E}_3$ is the vector of strength of electric field corresponding to the contribution from multiple reflections; $\bar{E}_4$ is the vector of strength of electric field corresponding to the contribution from underlying surface.

The facet surface model was used to simulate the reflection of a signal from the surface; it represents the surface as a set of elementary reflecting elements, namely plates of finite size, coinciding with the surface of large-scale irregularities. The Huygens-Kirchhoff's principle is the physical justification of such a model: each point of the surface, irradiated by an electromagnetic wave, is a source of a secondary spherical wave [7].

The use of the facet model is most appropriate for surfaces with multi-scale irregularities. The following conditions are necessary at dividing the surface into facets (reflectors) [8]:

- ✓ the sizes of the facet must be greater than the wavelength by several times;
- ✓ the signal, reflected from one elementary facet, does not depend on signals, coming from other reflectors;
- ✓ the radius of curvature of the average facet is larger than the facet sizes, i.e., the average facet level is almost flat;
- ✓ number of facets, forming the reflecting surface should be large, and their reflective properties are about the same.

The surface of the object should be represented by a set of reflecting areas (facets) of the same area, evenly distributed over the surface, to simulate the operation of the radar station with a given resolution in azimuth and inclined range. In this case, the size of the facets should be less than the parameters of the resolution of the radar station.

The reflected signal from the surface is the sum of the signals from all irradiated facets. Moreover, each partial signal has its own amplitude, determined by the orientation of the local diagram of backscatter (DBS), and its random phase.

The phase of the signal, reflected from the k-th elementary reflector (facet), is represented as

$$\phi_k = \phi_k^{(\text{reg})} + \phi_k^{(\text{rand})}, \tag{2}$$

where $\phi_k^{(reg)} = \omega\tau_k = (4\pi/\lambda)R_k$ is regular phase shift due to the time of wave propagation from k-th elementary reflector (facet) and back; h_k is ordinate of facet relatively to average surface level; $\phi_k^{(rand)}$ is random phase shift, caused by reflection from the facet and inhomogeneities along the propagation path (troposphere, ionosphere, etc.). It is usually considered that $\phi_k^{(rand)}$ has a uniform distribution law from 0 to 2π.

The signal on the input of the receiving antenna of radar station is the sum of the partial signals reflected from all facets in the irradiated area:

$$u(t) = \operatorname{Re} \sum_{k=1}^{N} A_k \dot{U}_M (t - \tau_k) \exp[i(\omega_0 t - \varphi_k)], \tag{3}$$

where A_k is amplitude of signal; $\tau_k = 2R_k/c$ is delay of the signal, reflected from the k-th facet; $\dot{U}_m(t)$ is law of modulation of probe signal; ω_0 is cyclical frequency of carrier oscillation.

The amplitude of signal, reflected from the facet, depends on the distance from the facet and the mutual orientation of diagram of antenna directivity and diagram of back scattering; it is defined:

$$A_k = \sqrt{\frac{2P_{\text{trans}}G_{\text{trans}}G_{\text{rec}}\lambda^2\eta_{\text{a}}L_{\text{loss}}}{(4\pi)^3 R_k^4} g_{\text{trans}}(\theta_k,\alpha_k) g_{\text{rec}}(\theta_k,\alpha_k)\sigma_k^0 S_k}, \tag{4}$$

where P_{trans} is transmitter power; G_{trans}, G_{rec} are amplification factors of power of transmitting and receiving antennas respectively; $g_{trans}(\theta_k, \alpha_k)$, $g_{rec}(\theta_k, \alpha_k)$ are normal diagrams of directivities on power of transmitting and receiving antennas respectively $g(0,0)=1$; η_a is the efficiency coefficient of the antenna-feeder system for transmission and reception; L_{loss} is loss coefficient in the microwave paths and on the spread; R_k is distance from phase center of antenna to k-th reflector; σ_k^0 is specific RCS of k-th reflector; S_k is area of k-th reflector.

The power of a signal, entering in receiving antenna of radar station for given type of underlying surface, depends on the polarization of the radiation and the mutual direction of polarization during radiation and reception. If the polarization of radiation and reception coincide, the levels of the reflected signal for horizontal and vertical polarization are close for most surfaces (arable land, terrain covered with vegetation). Exceptions are smooth surfaces (concrete, asphalt, gravel, calm water surface). The reflected signal at horizontal polarization is less than at vertical (up to 16 dB at small slip angles) for them [9].

The reflectivity of underlying surface is characterized by the scattering coefficient, which is the specific effective area of scattering (SEAS): the ratio of the RCS of the permit element on the terrain to the value of its geometric area:

$$\sigma^0 = \frac{\sigma}{S}, \tag{5}$$

where σ is the value of the RCS of the element of resolution (facet), S is area of surface of the resolution element.

The RCS ratios in polarization channels (for example, σ^0_{HV}/σ^0_H or σ^0_{HV}/σ^0_V), the ratio of RCS of polarization component to the averaged RCS $\sigma^0_{AVER}=(\sigma^0_H + \sigma^0_V+2\sigma^0_{HV})/4$ or difference of phases between vertical and horizontal polarizations or between agreed and cross polarizations are informative features at working with different polarizations.

The strict solution of the problem of reflection from uneven surfaces faces with serious mathematical difficulties in interesting cases for practice. Therefore, the reflection characteristics are usually determined experimentally at certain conditions. Thereafter an empirical model of the surface is obtained so that the calculation of the reflected field is relatively simple; the calculated reflection characteristics were close to those obtained experimentally.

The entire underlying surface is marked by one of the prepared materials, such as asphalt, concrete, dry dirt, wood, etc. Description of the materials is built on the works [9, 10], in which the experimental evaluation of the scattering characteristics is carried out. Below is a radar image, which shows the scene with the underlying surface (Figure 1).

3. Results and Discussion

Objects with well-known estimates were selected as test objects with a complex geometry: heavy V-52N bomber and BGM-109 Tomahawk cruise missile. Result of calculation of research model is below (Figure 2).

A rather old model of the V-52N bomber is best suited for estimating the RCS calculation. This is due to the fact that at the time of the release of the aircraft the methods of lowering radio-visibility were rather weakly developed. The all-metal body is well suited for calculation in accordance with the physical theory of diffraction. A typical estimate of RCS of V-52N is about 100 m^2 [11], which is equivalent to 20 dB and has an average value of 15.33 dB. In this case, the deviation is less than 5 dB.

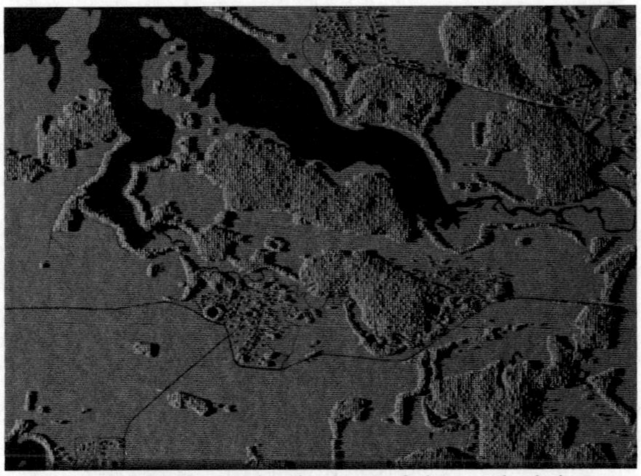

Figure 1. Radar image of the underlying surface.

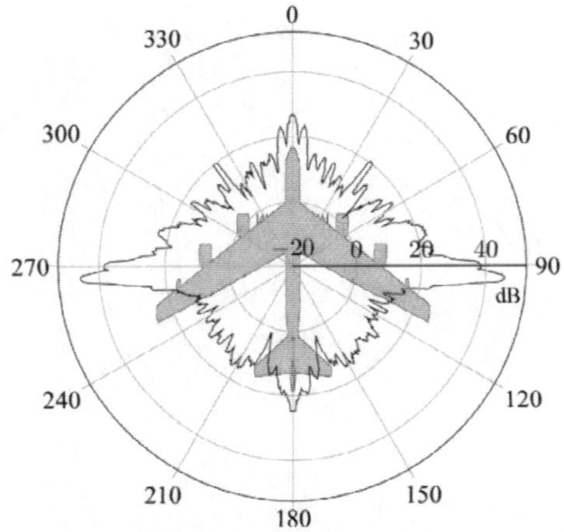

Figure 2. An example of calculating the RCS of V-52N bomber) at frequency of 1250 MHz.

Figures 3*a-e* show the diagrams of back scatter for different wavelengths. It can be said for each of them, that irradiation from the lateral direction increases the probability of irradiation of a cruise missile. At the same time, the rear hemisphere is also characterized by a smaller RCS. It looks rather unnatural because of the real cruise missiles are characterized by extremely low RCS in the direction of motion. In this case, the model may give an erroneous result due to the lack of the ability to calculate dielectric materials. Figure 3*e* shows the general dependence of the RCS [12, 13].

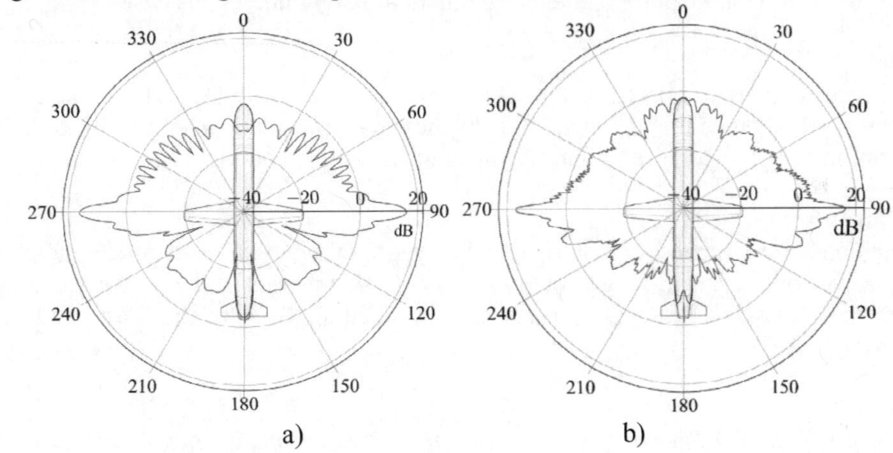

a) b)

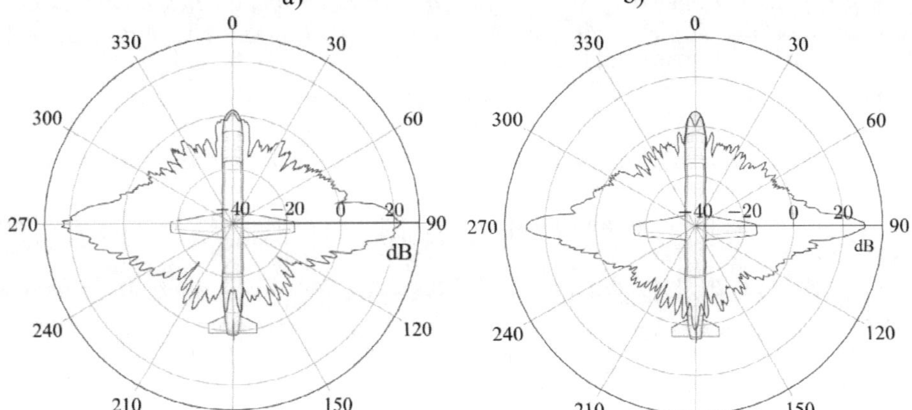

c) d)

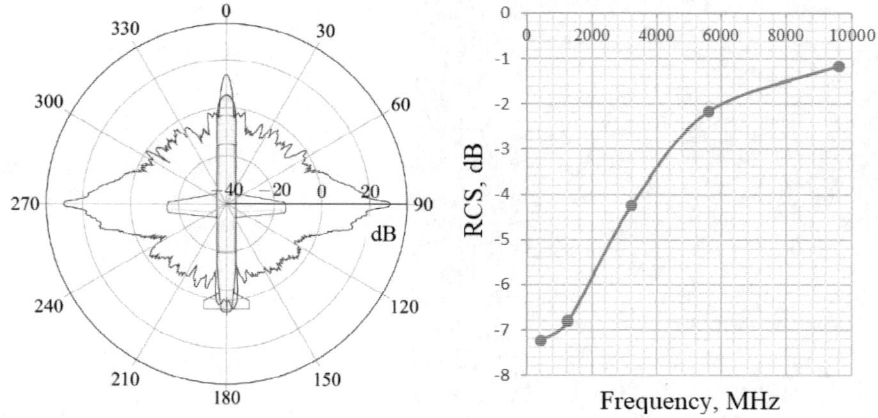

e) f)

Figure 3. The RCS of the BGM-109 Tomahawk cruise missile at a different frequencies, where a) 430 MHz; b) 1 250 MHz; b) 3 200 MHz; d) 5 600 MHz; e) 9 600 MHz; and f) dependence of the average RCS on frequency.

In turn, the real cruise missile has more developed means of lowering radio-visibility. The body is covered by radio absorbing materials. In addition, the modification of BGM-109 block 4 uses carbon-fiber aerodynamic surfaces. Characteristic estimates of the RCS of BGM-109 Tomahawk cruise missile is 0.05–0.001 m^2 [todo], depending on the angle of shooting. The simulation results are in the range from -7 to -1 dB. The error is approximately two orders of magnitude in this case.

4. Conclusion

Using the example of two air objects, it was shown that the developed model shows the expected result for the problem of basic modeling of metal objects; however, this model shows itself to be insufficient for the task of modeling objects with a multilayer radio absorbing coating.

Acknowledgments

The investigations were carried out at financial support of grant of Russian Foundation of Basic Research in framework of scientific project No 15-07-99514 and grant of the Ministry of Education and Science of the Russian Federation (project No 2.2226.2017/Project Part and project No 2.9140.2017/Basic Part).

References

[1] Ufimtsev P. Y. 2009 *Theory of Edge Diffraction in Electromagnetics Elektronische Ressource Origination and validation of the physical theory of diffraction* (NC. SciTech Publishing Inc.: Raleigh) doi: 10.1049/SBEW054E

[2] Gordon W. B. 1975 *IEEE Transactions on Antenna and Propagation* **23** 590–592 doi: 10.1109/TAP.1975.1141105

[3] Ufimtsev P. Y. 1997 *Method of edge waves in the physical theory of diffraction* (National technical information service: US)

[4] Borzov A. B., Suchkov V. B. and Sokolov A. V. 2004 *Journal of Radio Electronics* **2** 1–13 (in Russian)

[5] Borzov A. B. *et al* 2014 2014 *Journal of Communications Technology and Electronics* **59(12)** 1356–1368

[6] Akhiyarov V. V. *et al* 2015 *Journal of Communications Technology and Electronics* **60(12)** 1297–1304

[7] Boev S. F. *et al* 2018 *IOP Conf. Ser.: Mater. Sci. Eng.* **457** 012010 doi: 10.1088/1757-899X/457/1/012010
[8] Shkolnyj L. A. *et al* 2008 *Radar Systems of Aerial Reconnaissance, Decoding of Radar Images*. (Zhukovsky Air Force Engineering Academy: Moscow).
[9] Cloude S. R. 1996 *IEEE IEEE Trans Geosci Remote Sens* **34(2)** 498–518
[10] Pottier E. 1997 *Proc SPIE* **3120** 132–143 doi: 10.1117/12.278958
[11] Il'in V. E. *et al* 1996 *Bombers* (Victoriya, AST: Moscow)
[12] Yurchenko A. *et al 2016 MATEC Web of Conferences* **79** 01085
[13] *Air & Space Power Journal* 2003 **XVII(1)** (DIANE Publishing: US)

IOP Conf. Series: Materials Science and Engineering **516** (2019) 012048 doi:10.1088/1757-899X/516/1/012048

The influence of skin-electrode contact on the quality of ECG recording for personal telemedicine systems

I A Lezhnina, A A Boyakhchyan, M A Ivanov, M I Samolutchenko

National Research Tomsk Polytechnic University
E-mail: bojahchyan@yandex.ru

Abstract. In this article are presented the results of the study of the Tele-ECG system. Presented system automatically records the ECG signal and transmits it to the server, where occurs process of the decryption (automatic detection of R-waves and ST-segments on the ECG). A distinctive feature of the system is the use of it includes devices with capacitive sensors. The results quality research of the signal received through different dielectrics. The necessity of revision of sensors that will allow to avoid the interference caused by man and external factors. As a result, this refinement will allow to receive a signal for automatic diagnostics with the least loss of information due to the presence of interference in the signal.

1. Introduction

To date, telemedicine gadgets are tightly embedded in our daily lives. Fitness trackers have not only athletes, but also ordinary people who take care of their health. It appeared on the market a wide variety of portable cardiographs [1–6]. But because of the small diagnostic value of such devices have not found popularity among cardiologists and patients of medical institutions.

Tele-electrocardiography (tele-ECG) is one of the most important areas of preventive cardiology. In the Russian Federation, telemedicine is one of the state priorities in the development of medicine, as well as in other growing countries with a large territorial divide (India, Brazil, Indonesia, China). The widespread introduction of such systems is significantly complicated by both the lack of suitable technical means and the insufficient scientific and evidence base for the medical use of these agents. To date, such systems are not implemented everywhere.

The prototype of the Tele-ECG system developed at the Tomsk Polytechnic University has the following structure (Figure 1): the server ensures reception of records from ECG-Express devices and stores them in a database. The database also stores the results of an automatic ECG analysis, interpretations of doctors and other additional information.

The server module for the analysis of ECG records also provides processing of ECG records in order to extract informative parameters from the ECG curve, in particular, the calculation of time and amplitude parameters.

Figure 1. Structure of the Tele-ECG system.

The ECG-Express device used in this tele- ECG system is fully automatic and operates on the “one-button” principle. The user doesn’t need to think about the location of the sensors. Registration of an electrocardiogram, after pressing the start button is conducted within 30 seconds, then the data is sent to the server where the automatic ECG decoding occurs. As part of the prototype, were implemented the main functions: automatic detection of R-waves and ST-segments on the ECG.

2. Review of capacitive electrodes

Portable electrocardiographs use two types of electrodes: sticky disposable silver chloride electrodes or ordinary metal pads. Regular use of a cardiograph with sticky disposable electrodes causes difficulties, since for its use a person needs to take off clothes, glue electrodes to the right place, get rid of hair. For these reasons, such cardiographs not find mass use in the home. The advantage of such cardiographs is that the ECG signal has more information, compared with the signal obtained from metal electrodes. Portable cardiographs with conventional metal electrodes, due to the lack of a conductive gel, cannot provide the necessary contact and ensure the required accuracy.

An alternative solution is to use capacitive electrodes of company Plessey. These electrodes allow you to get an ECG that is comparable in quality to an ECG obtained from disposable sticky electrodes, despite the presence of interfering factors on human skin (hair, sweat, etc.), which complicate the process of ECG recording with other electrodes. Registered ECG obtained using Plessey [7] sensors through the ECG-Express cardiograph [8] were presented to doctors Rehabilitation Departments of the Research Institute of Cardiology of the Tomsk National Research Medical Center of the Russian Academy of Sciences in Tomsk [9], which confirmed the accuracy of the data.

The principle of operation of capacitive sensors (electrodes) is based on the ability of electric capacitors to conduct alternating current. Between the surface of the skin and the metal contacts of the sensor there is a dielectric layer, thus forming a flat capacitor. The ECG signal, the main power of which is concentrated in the frequency band 0.5 - 40 Hz, passes through a capacitor and causes a current in the input circuit of the signal amplifier. Figure 2 shows a simplified, universal model for a capacitive sensor, indicated its location on the device. A full description of the scheme is presented in the article "Wireless Non-contact EEG / ECG Electrodes for Body Sensor Networks" by Yu. M. Chi and Gert Cauwenberghs [10].

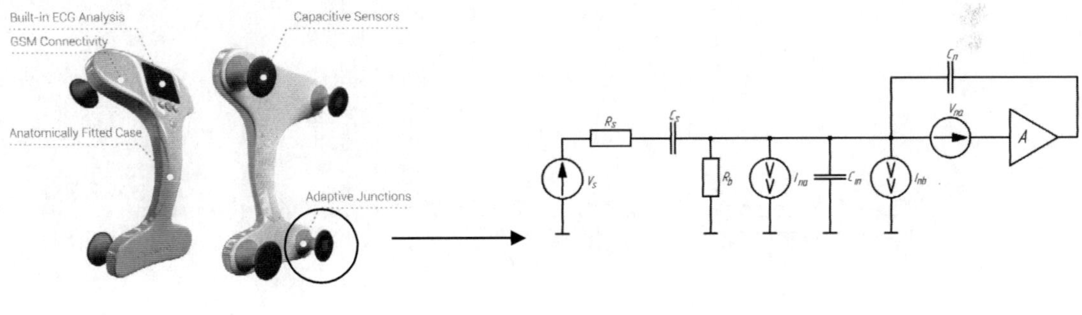

Figure 2. Simplified universal model for capacitive sensor.

Another distinctive feature of Plessey electrodes is the ability to record ECG through clothing, which will allow the use of the device in virtually any environment, even when not at home. The only drawback is the excessive susceptibility to myographic interference.

Since 2016, Plessey sensors have fallen into the category of electronics, the import of which is prohibited on the territory of the Russian Federation. In the course of the work carried out as part of the study, it was decided to develop their own analogs of Plessey sensors with detuning from the skin-electrode contact. This refinement will allow the sensors to avoid interference created by man and external factors. In order to find out the effect of various factors and conditions on the sensor readings were carried out a series of experiments.

3. Experimental methods

All electrocardiograms presented below were obtained using the ECG-Express portable electrocardiograph. Cardiograms were passed through special digital filters in order to get rid of the interference that arises from network interference. Filters were made in the program of mathematical modeling - MATLAB. Presented cardiograms are also visualized in MATLAB. Figure 3 shows an electrocardiogram from a "naked" chest that has no hair.

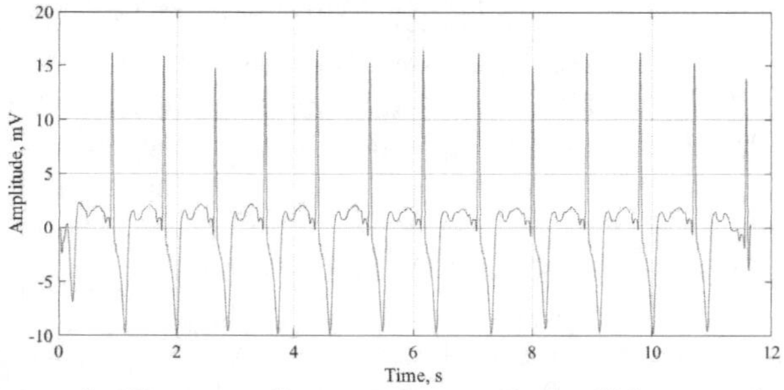

Figure 3. Electrocardiogram from a "naked" human chest.

For the study, two healthy students who have no disruption of the heart were selected. Proof of student health was a medical examination by a cardiologist. The main difference between the students was: the presence and absence of hair on places where electrodes are applied (on the chest). Figure 3 shows the electrocardiogram of a person with the presence of hair.

Figure 4 shows ECG from the chest of a man with the presence of hair, the cardiogram becomes difficult to read, but the QRS complex is clearly visible, which means that with the correct selection of digital filter values, it will be possible to improve the quality of the ECG.

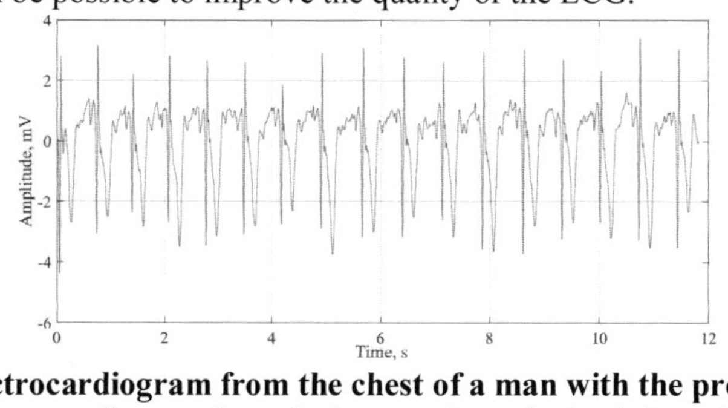

Figure 4. Electrocardiogram from the chest of a man with the presence of hair.

Figure 5 shows the electrocardiogram from the human chest obtained through the shirt.

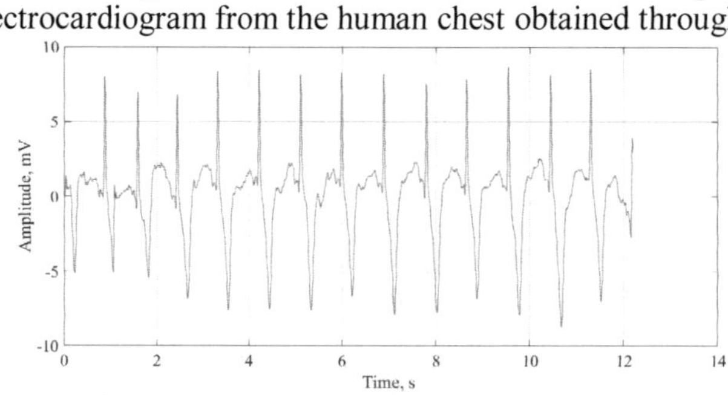

Figure 5. Electrocardiogram from a man's chest through a shirt.

The electrocardiogram shown in Figure 5 was obtained from a person's chest through a shirt (fabric material is cotton). This cardiogram demonstrates that the indications are of good quality, with the same disadvantages as when registering from the chest without hair.

The last experiment was the registration of the ECG through any tissue in addition to the material. Figure 6 shows an attempt to register an ECG through a sheet of paper.

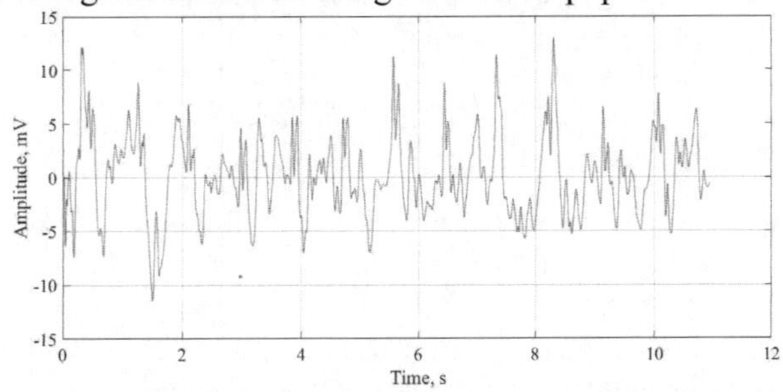

Figure 6. Electrocardiogram from a person's chest through a sheet of paper.

The signal received through a sheet of paper at the moment can't be filtered to find a useful signal. Probably, a significant distortion of the ECG signal when using paper as an additional dielectric is associated with a significant decrease in the capacity of the skin-electrode contact. Necessary to conduct detailed studies of the effect of contact capacity on the recorded signals, including using mathematical modeling methods. Thus, by means of field experiments, we managed to clarify the behavior of the Plessey sensors through various materials. The use of capacitive sensors allows you to build on the quality of electrical contact with the skin, as well as to measure ECG through clothing. In the course of the research, the reasons were found that caused irreducible distortions in the ECG signal - microvibration and inconstancy of contact with the body.

4. Description of the model

For the purpose of detailed studies of the effect of the capacitance of a contactless electrode on the recorded signal have been developed models of skin-electrode contact and the circuit for recording ECG signals. The simulation was held in the application package "OrCAD". The registration scheme is constructed using operational amplifiers and represents the instrumental amplifier of a difference signal from two capacitive electrodes. Figure 7 shows a circuit simulating skin-electrode contact.

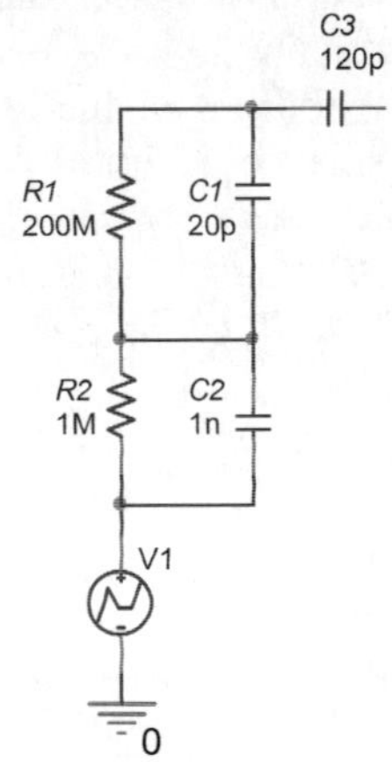

Figure 7. Equivalent circuit of the skin-electrode contact.

The elements R1, C1 represent the coupling layer. It's a layer between the electrode and the stratum corneum. The elements R2, C2 simulate the stratum corneum. The ECG signal comes from the V1 source. The capacitive electrode is modeled by a capacitor C3. Fig. 8 shows the simulation results.

In this case, the presence of a dielectric (tissue) in the simulation was carried out by lowering the electrode capacitance - C3 [8].

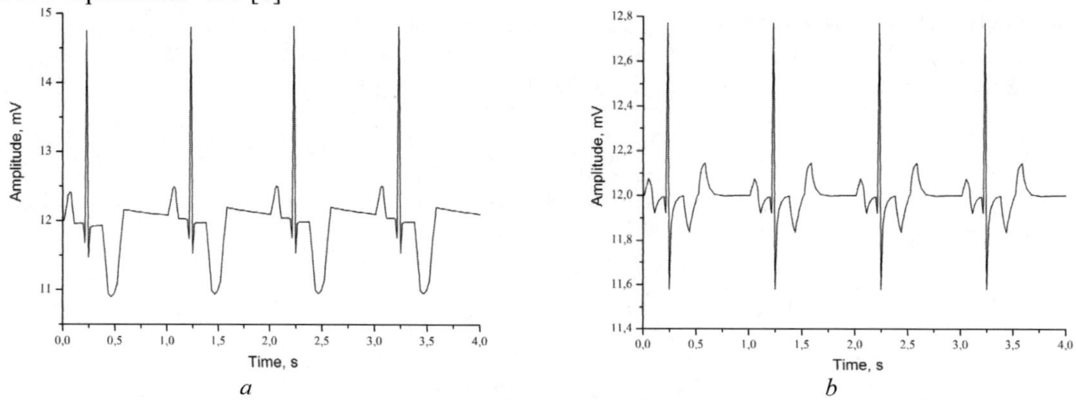

Figure 8. Simulation results: a - normal conditions (C3 - 600 pF), b - with a reduced capacity of the skin-electrode contact (C3 - 10 pF).

In the process of modeling the addition of an additional dielectric (Figure 8*b*) i.e. a decrease of capacitance of the electrode leads to a change in the ECG signal. Moreover, these changes that were similar to those in experimental studies (Figures 3 and 5).

5. Results

As a result of the work, a scientific hypothesis was formulated on how to improve the quality of measurement and reduce interference in the signal. This requires further research and development of a sensor with an electronic detuning system from the effect of sensor capacitance based on feedback in the measurement circuit.

In the further development of capacitive electrocardiogram measurement sensors will be the most promising technology for the development of portable and personal electrocardiography.

The reported study was funded by RFBR according to the research project № 18-38-00535.

References

[1] Heart monitor CardioQVARK URL: http://www.cardioqvark.ru/
[2] AliveCor. URL: https://store.alivecor.com/
[3] Kopylov F.Yu. *et al* Diagnostic capabilities of the new cardio monitor CardioQVARK URL: http://www.cardioqvark.ru/science/science_30.html/
[4] Rafalsky A. S. *et al* 2012 *Instruments and Experimental Techniques* **55** 283–287
[5] Aristov A. A. *et al* 2016 *Journal of Physics: Conference Series* **671** 012002
[6] *Odnokopylov I. G. et al 2016 IOP Conf. Ser.: Mater. Sci. Eng. 189(1) 012010 doi: 10.1088/1757-899X/189/1/012010*
[7] Bekmachev A. 2013 *Components and technologies* 1 130–133
[8] Boyakhchyan A.A. *et al* 2018 IOP Conference Series: Materials Science and Engineering **363** doi: 10.1088/1757-899X/363/1/012030.
[9] Scientific Research Institute of Cardiology URL: https://www.cardio-tomsk.ru/
[10] Chi Y. M. *et al* 2010 *IEEE Reviews in Biomedical Engineering* **3** 106–119

IOP Conf. Series: Materials Science and Engineering **516** (2019) 012049 doi:10.1088/1757-899X/516/1/012049

IR Orientation system for visual impaired people with tactile matrix

V L Khmelev, O A Killer, K V Zavyalova

National Research Tomsk State University, Tomsk, Russia
E-mail: Oleg.Killer.1997@mail.ru

Abstract. The article investigates the software and hardware system of tactile vision. Which will allow visually impaired people to navigate in around environment. The system has to base on a near-infrared active scanning optical module and information tactile matrix. The optical module receives depth profiles, and then a computer performs statistical processing and transmits the result to a tactile matrix, which make a user feel distance and direction to the obstacles. The system will operate a frequency of 30fps, in real time, with a scan distance of 4-5 meters.

1. Introduction

Visually impaired people has problems with movement and orientation in megalopolises. Cause cities expand not only in width, but they are grow up too. The basic problem for visually impaired is an orientation in unknowable space, now, they solve this task by using "white" sticks. A “white” stick is a plastic probe, that user sounds the way by using a manual scan. Another way is using ultrasound stick or infrared stick. These types of device are range meters with different types of radiation. Both of them is sounding a surface in front of a user and if their beam detects an obstacle than device will vibrate. The intensity of vibration will indicate a distance between device and obstacle. However, in cases of IR or US range meters, a visually impaired still have to use a manual scanning. Which will not allow moves fast enough. Obviously, people will using an instantaneous scanning system for 3-dimensional scenes. Moreover, our team start first step in area of design and development devices for visual impaired people, after work with electronic area last 5 year [1–3]. Such approach would not use the manual scanning. The development of the entertainment industry has led to the creation of active IR short-range scanning technology; such systems use in PlayStation and Xbox consoles for recognize and capture player movement. For experiments, we use the Kinect V1 sensor by Microsoft. The optical system provides fast scanning with a frequency of 30 frames per second and scanning distance up to 4 meters. The obtained data in the form of depth profiles, one channel image, where every pixel is a distance between a camera and some point of space. Figure 1*a* shows the scene, Figure 1*b* shows the distance profile of the same scene. The device consists of an IR projector, which radiates spatially modulated light and an IR camera, as a receiver of the light scattered by objects of the scene. The profile is transferred according to the USB 3.0 standard to the microcomputer running Windows 10. Where a computer statistically processes the depth image, after the data will transfer to the tactile matrix. The tactile matrix will be place on the user's body. The matrix informs the user about surrounding obstacles with a frequency of 30 Hz.

a b

Figure 1. a)origin scene; b) depth profile.

2. Processing of depth image

The depth profile transfers to a computer via USB 3.0. The intensity of a pixel in depth profile is the distance from the camera lens to the object, the closer the object than the lower pixels intensity. The further, the intensity is higher. The Kinect scans objects around a distance of 4 meters, the value of resulting pixels is in the range from 0 to 4096.For experiments, we were using a personal notebook (Intel Core i7-4700MQ, CPU 2400Ghz 12Gigabyte of RAM), but after will be using a one-board microcomputer. The first step of image processing is dividing a depth image into 24 sections (2 rows, 12 elements in each row), Figure 2*a*. We use multi-thread computing; each thread computes one zone of profile. The next step is computing of a histogram, the dependence between numbers of pixels inside the zone and pixels intensity, Figure 2*b*. We use an empirical value. The value is a bound of geometry size for obstacle. The next, we compare histogram pillars and value, from the closet distance to furthest. The first of pillars, which is bigger than bound, will recognize as an obstacle, red pillar shows a chose in Figure 2*b*. The distance between lenses and obstacle recalculates to voltage and packing for sending to the tactile matrix. Package will sent to tactile matrix

a

b

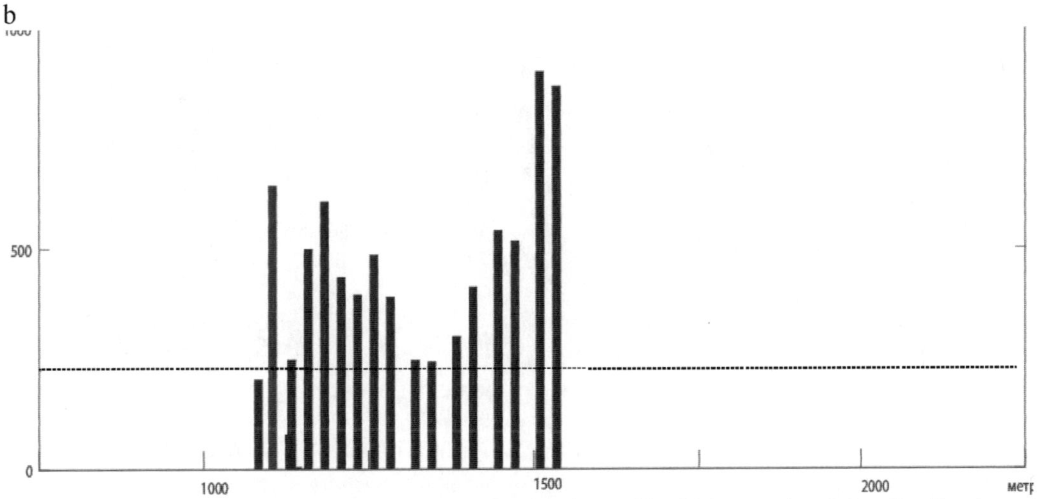

Figure 2. a) Dividing into sections of a depth profile; b) Example of the distribution of data within the section.

3. Tactile matrix

A tactile matrix is a device consisting of elements that can excite human skin receptors. In our matrix, we used vibro-motors. The vibration intensity of each motor depends on the voltage that applied to them. We use a network of microprocessors to control the voltage on the vibration motors. 3 cheap microprocessors control 24 vibration motors. In a serial network, three microprocessors are connecting in 1 – 2 system. Two Arduino microcontrollers (they connect directly to the vibration motors) paralel connect to last one, and one connected to a singleboard computer according to the USB standard. Information controller receive data from the microcomputer via USB and transform them into signals for the serial port according to the RS232 standard. We used Arduino NANO microcontrollers as voltage controllers. These microprocessors are easy to program and several pin ones can be programmed to work with PWM. PWM allows you to change the supplied voltage to the vibration motors without using a DAC. Figure 3*a* shows the scheme of the developed tactile matrix. We made it in the form of a mask. The user will put it on his forehead. Figure 3*b* shows the practical implementation of the scheme.

a

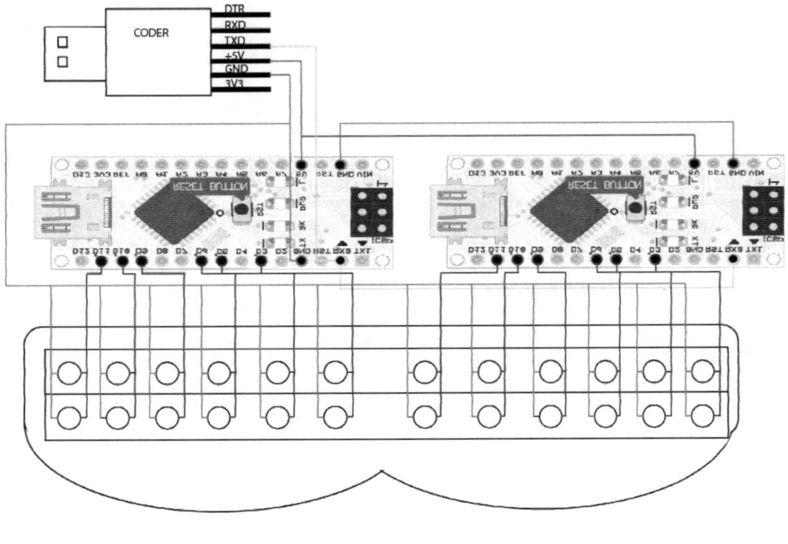

b

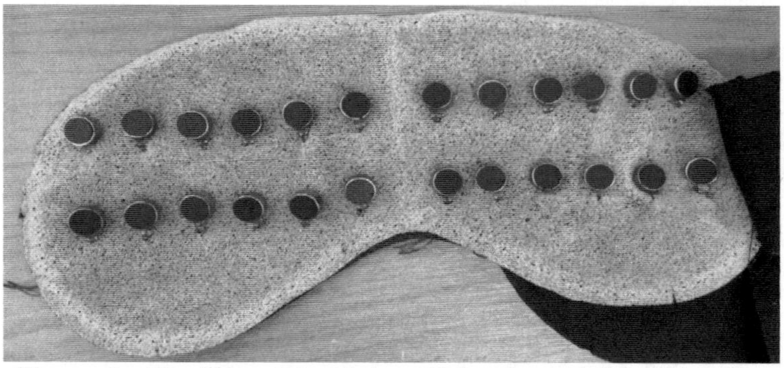

Figure 3. a)Tactile matrix scheme; b) Implementation of tactile matrix.

4. Summary

At this stage, we are testing a system to check the boundaries of the system. It is important to determine how quickly the human brain is able to get information through the skin. It is also necessary to determine how sensitivity of the skin changes with a change of vibration; therefore, we should build an experimental curve of change intension of voltage on vibro-motors from an object range. The next step is to reduce the size of the optical module. Such step will allow us move from testing the algorithms for sensitivity to full-scale experiments, where volunteers will be navigate through maze with closed-eyes. Next, we have to translate the prototype by change wire to wireless connections, so that each individual component of the software can work without a wire, and device start to be more comfortable. We think its really imrotant for last stage. The most important stage is testing the device together with visually impaired people and making constructive changes based on their comments.

Acknowledgments
This work was supported by the grant of the Russian Science Foundation №18-79-00294

References

[1] Shipilov S. E. *et al* 2018 *Russian Journal of Nondestructive Testing* **54**(6) 443–447 doi: 10.1134/S1061830918060086
[2] Satarov R. *et al* 2016 MATEC Web of Conferences **79** 01079
[3] Shipilov S.E. *et al* 2017 *Optical and Quantum Electronics* **49**(10) 1–12 doi: 10.1007/s11082-017-1172-7

Gesture based terrain mapping and recognition-overview

M Mondal, S I Syryamkin

National Research Tomsk State University, Tomsk, Russia
E-mail: mainakme2140@gmail.com

Abstract. Up until a few years ago, the only way to get an aerial overview of a designated area was to fly over it with a manned aircraft and physically inspecting the areas or taking pictures of it. Unmanned Aerial Vehicles, have been a blessing to all the industries which relied on aerial images or drawings for their work. Unmanned aerial vehicles also knows as drones, in more colloquial terms, is a blessing to all industries in the world. The cost of manufacturing and assembling drones have gone down significantly as majority of people adapted to using it. The film industries are using it to get excellent footage with accurate direction and lighting. Civilians can use it as toys. Students can use it to learn avionics and flight mechanisms. Youtubers use it for professional high resolution footage. Rescue teams use it for surveying the area. Armies use it to get a tactical advantage in a firefight and recently, Amazon is using drones for superfast delivery service. The possibilities are endless if properly invested on. Machine Learning and Artificial Intelligence are next best thing in the field of Computer Science and that's why I plan to merge these two fields and present something which fundamentally adds more features and a hive mind facility to even work without a commanding signal.

1. Introduction

The current real time terrain mapping, usually carried out by Self Drive Cars usually use the overlay of the map to find directions. The on board computer of the car, usually uses a bunch of sensor like: Proximity Sensor, Stereoscopic Cameras to identify objects and cars in its trajectory. This method very viable for cars because it has less to no space constraints to house the compute power. I propose a method to identify terrain defects for tactical combat and rescue operations by dividing the work into terrain recognition and terrain mapping.

The Operator is the person who controls the drones and provides the drones with directional data for it to follow. There is also an autopilot which takes over the controls when no gesture commands are received for a while.

2. Levels of Autonomy according to the United States Department of Defense

Level 1 - Human operated system in which a human operator makes all the decisions regarding drone operation.

Level 2 - This system can perform many functions independent of human control. It can perform tasks when delegated to do so, without further human input. Examples are engine controls, automatic controls, and other automation that must be activated or deactivated but a human.

Level 3 - The third level of autonomy is a human supervised system. This system can perform various tasks when it is given certain permissions and directions by a human. Both the system itself and the supervisor can initiate actions based on sensed data. However, the system can only initiate these actions within the scope of the current task.

IOP Conf. Series: Materials Science and Engineering **516** (2019) 012050 doi:10.1088/1757-899X/516/1/012050

Level 4 - The final level of autonomy is a fully autonomous system. This system receives commands input by a human and translates these commands in specific tasks without further human interaction.

3. Literature Survey

There is a lot of interesting research in the field of Drone technologies, and the major players are including 4k video recording, object tracking, directional sense, and a lot more. They offer a lot of features and have a high price tag associating with it.

The DJI Spark is one of the commercial drones available for the common public and it has gesture based control system. According to Paul Pan, the Senior Product Manager at DJI, "Controlling a camera drone with hand movements alone is a major step towards making aerial technology an intuitive part of everyone's daily life, from work and adventure to moments with friends and family." There are numerous research papers associating with the gesture recognition in a drone system. The majority of them using Motion Controller or sensor like the Leap motion controller or the Kinect controller.

The pitch and yaw(movement) is controlled by the hands and the gaps in communication are filled with the autonomous drive controller. An important distinction within the concept of autonomy is the difference between automatic and autonomous systems. An automatic system is a fully preprogrammed system that can perform a preprogrammed assignment on its own. Automation also includes aspects like automatic flight stabilization. Autonomous systems, on the other hand, can deal with unexpected situations by using a preprogrammed ruleset to help them make choices.

As described by the levels of autonomy according to the United States Department of Defense, it is ideal to have a system which is at level 3 at all times and can initiate level 4 when no human input is received, i.e. the drone system will always be under gesture control and when no gesture input is received, the internal flight controller of the drone can take over its control.

3.1. Current Work on Flight Controllers

1. PX4 – PX4 is part of the Drone code Project, a shared and collaborative open source project to deliver a complete end-to-end platform for unmanned aerial vehicles (UAVs). A complete auto pilot system for large drones [1].

2. Flight Controller – This is a microchip which is programmed to use sensor data from the sensors in the drone and use it to keep the drone stable [2].

3. Autopilot Hangar - a multi headed/ Hybrid- controlled drone management system which has an autopilot system [3].

4. Motion Tracking in drones – Numerous Open Source organizations provide code to implement for motion tracking.

The second important task in this system is the mapping system. Many companies including google are doing this using various algorithms. The algorithms are primarily known as SLAM or simultaneous localization and mapping. These algorithms are not always known for their accuracy and use sensor data to make an approximate map of a local area.

3.2 Current Work on Terrain Mapping

1. Terrain Mapping in Real-Time: Sensors and Algorithms - this paper presents an overview of the work being conducted at the Australian Centre for Field Robotics (ACFR) into sensors and systems used for real-time terrain mapping [4].

2. Vision-based LIDAR (Light Detection and Ranging) Integration: - multiple sensor based mapping algorithms on a land traversing rovers for mapping the terrain in a very accurate manner [5].

3. A Set-Theoretic Algorithm for Real-Time Terrain Mapping of Mobile Robots in Outdoor Environments - rough terrain environments for mobile robots. A few experiments have been carried out in this method [6, 7].

4. Elastic Terrain Mapping – A terrain mapping system used by some projects in NASA, National Aeronautics and Space Administration for terrain mapping using land bots [8, 9].

All the mapping systems implementations, if implemented in real-time cost huge computing resources and cannot map accurately. The most accurate is the LIDAR based Mapping as it is used in self drive cars and other heavy vehicles with ample computing power.

4. Methodology

There are several attempts for real-time mapping using drones but most of the projects are pretty much at its infancy as they all try to map the terrain in real time. They all require huge computing power since terrain mapping requires huge resources.

My plan is to divide the work into 4 parts:

1. Gesture Recognition – handled by an independent flight controller.
2. Drone Pilot – the schedule to be run when no gesture commands are received.
3. Terrain recognition - the schedule to take high resolution images of the area and recognize the pits and outline of the area for tactical use.
4. Terrain Mapping - the Schedule which runs on the a separate high – end computer which uses its compute power and the high resolution images from the previous schedule and maps the area with accuracy.

For the actual system, there should be 2 working computer units.

1. Drone Controller System – a computer specifically to pilot the drone. This computer also listens to the gestures control commands given by the operator and decides on the which schedule to follow, Autopilot or Gesture.
2. Mapping System – the Computer which takes in all the sensor data and segregates the data into the what's required immediately and what will be sent to the EC service for further processing. This unit also communicates to the operator and will relay the terrain recognition data to consumption device.

5. Results

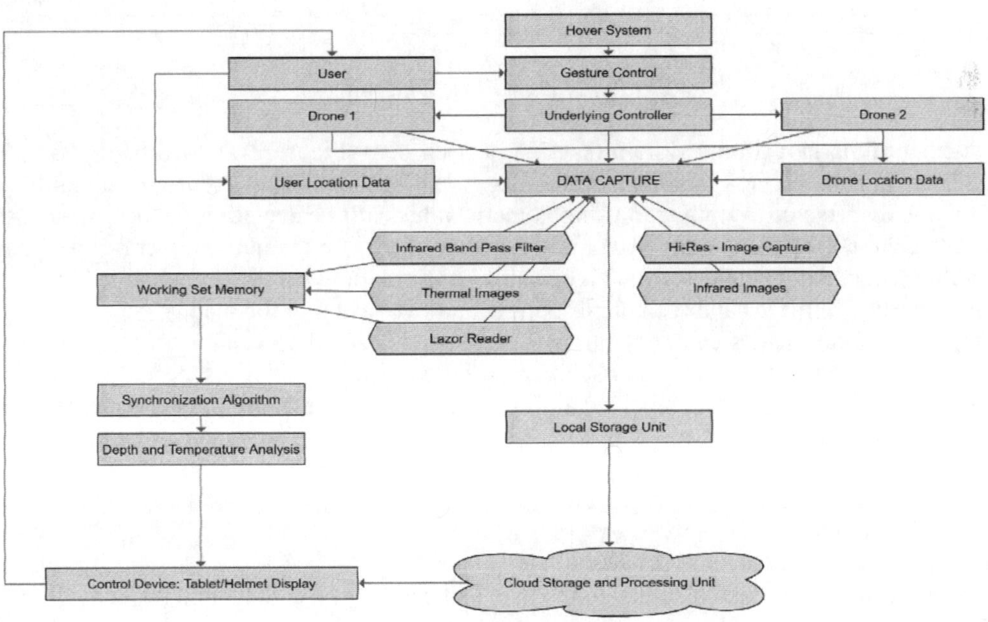

Figure 1. Structure Diagram.

As we see here (Figure 1), there are 2 major zones of work.

1. Which goes to the working set memory (after storing the data to the non-volatile memory) which is required for Real-time data processing to assist the operator of the system navigate through the terrain.

2. The second zone revolves around the data capture and processing the captured data using an elastic cloud service (Scalable Computing). The Scalable Compute Service will process all the data and produce a 3d map of the area which can be used for future endeavors.

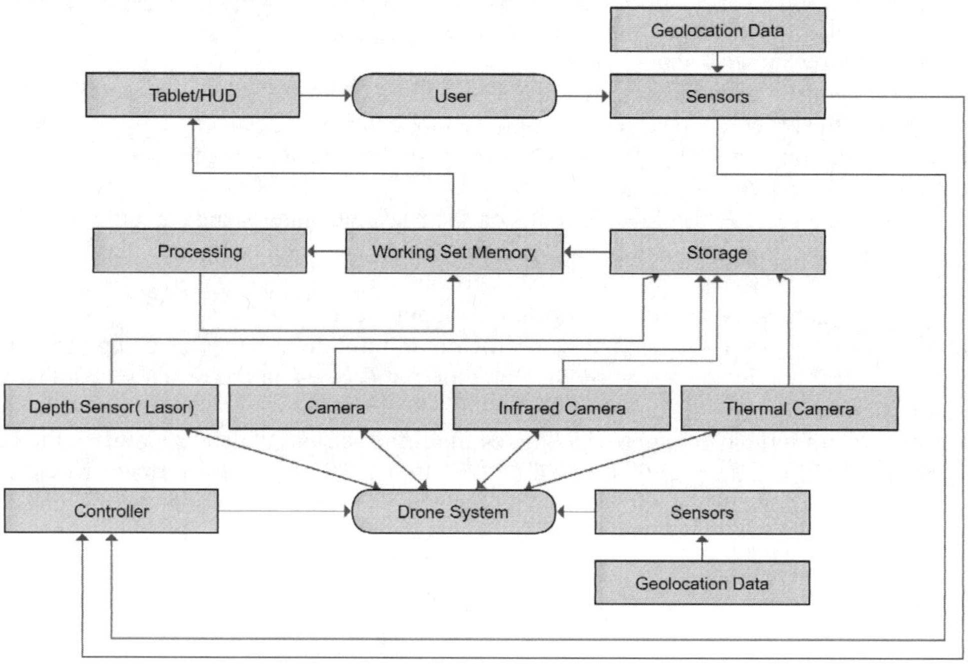

Figure 2. Data Flow Diagram.

The distinguishing part of this system that can be seen here (Figure 2). It will be the fact that this device does not have to process high resolution images in real-time. The computer handling the terrain detection will only produce a depth map using numeric values after calculating the depth and a thermal imaging data can be used for overlay. This data can be sent over to the human operator for physically exploring the terrain. An object detection algorithm like tensor flow maybe implemented in the middle which takes in camera images and recognizes objects and overlays it in the map.

The high resolution images captured can be uploaded to a cloud server for processing. The cloud server can use a commercial or an open source mapping software which used geolocation data and combines it with all the sensor data received and spits out an accurate map of the given area.

Categorization of data in the given concept:

1. Thermal Images.
2. Stereoscopic camera images (high-resolution image capture).
3. Depth sensor data.
4. Infrared Camera Data (for night vision).
5. Generic Camera footage (for relay to the user and use for object detection by Tensor flow).
6. Geolocation data of the Drone as well as the operator.

With 6 the types of data taken into account and then being divided into terrain recognition as well as terrain mapping handled by two different controllers, we can recognize terrain in real time as well as produce a map of that area if needed.

6. Summary

The sole idea of this project was to differentiate between mapping an area and terrain detection to utilize computing resources in an efficient way. The research currently being done in this sphere is phenomenal and when we combine all the research we get different views on how to aliment an idea and the sphere of Unmanned Ariel Vehicle is just at its infancy. We have many different challenges ahead like increasing flight time, increasing signal range and many other things but with the insight I provided, it is certain that all the problems can be solved given time. This article shows the current work being done in the sphere of terrain mapping and drone automation, and combines the two for the inception of a new concept. Further, these tasks can be sub-divided into more tasks for a simpler and faster approach to tackle the problem of terrain mapping and recognition.

Acknowledgments

The paper is prepared within the framework of the competitiveness improvement programme of National Research Tomsk State University and grant of the Russian Foundation for Basic Research No 16-29-04388.

The authors are grateful to Professor Stanislav V. Shidlovskiy and Tatiana B. Rumyantseva from Tomsk State University for their assistance in preparing the paper.

References

[1] PX4 latest stable release v1.8.2 URL: https://px4.io/
[2] Beginners guide to drone autopilots (flight controllers) and how they work URL: https://www.dronetrest.com/t/beginners-guide-to-drone-autopilots-flight-controllers-and-how-they-work/1380
[3] Autopilot URL: https://autoflight.hangar.com/
[4] Scheding S., Leal J., Bishop M., Sukkarieh S. 2001 *1st Australian Geospatial Information and Agriculture Conference.*
[5] Yongwei Zhang *et al* 2017 *29th Chinese Control And Decision Conference (CCDC)* doi:10.1109/CCDC.2017.7978402
[6] Shikhman M., Shidlovskiy S. 2018 *IOP Conference Series: Materials Science and Engineering* **363** (1) doi: 10.1088/1757-899X/363/1/012019
[7] Shidlovskii S. 2006 *Journal of Computer and Systems Sciences International* **45** (2) 282–286 doi: 10.1134/S1064230706020122
[8] Bailey T., Durrant-Whyte H. 2006 *IEEE Robotics & Automation Magazine* **13(3)** 108–117
[9] Sarkar A. *et al* 2016 *Conference on Industrial Informatics and Computer Systems (CIICS)* doi: 10.1109/ICCSII.2016.7462401

IOP Conf. Series: Materials Science and Engineering **516** (2019) 012051 doi:10.1088/1757-899X/516/1/012051

Ultrasonic Signal Generator with digital feedback

T Muksunov, I Kuzmenko, K Zavyalova

Tomsk State University, Tomsk, Russia
E-mail: zkv@mail.tsu.ru

Abstract. It is proposed generator of power ultrasonic signal with digital feedback and digital control of generating signal. Such approach will allow for the implementation of a rather complex control algorithm that is not feasible by circuit engineering means. A functional diagram of the layout of the control unit and generating a powerful ultrasonic signal has been developed. The digital microcontroller generates periodic pulses that trigger the generator at an appropriate frequency and controls the voltage at the output of the regulated power supply. The voltage level of the power source is set according to required power of the generated signal. The algorithm and software for the control of the signal generator with digital feedback has been developed.

1. Introduction

Generators of powerful ultrasonic signals are widely used in cutting, coagulation and processing of various materials. One of the broad applications of powerful ultrasonic vibrations is surgery [1–3]. To generate a powerful ultrasonic signal with an amplitude of hundreds of volts with a supply voltage less than 30 volts and a power up to 200 watts, is preferable to use the resonance effect. At resonance, there is an accumulation of energy in the oscillatory system and an increase in the amplitude of oscillations. As a result, the generator must adjust its frequency to the resonant frequency of the controlled system. The resonant frequency will depend on the mechanical properties of the ultrasonic instrument, the electrical properties of the piezoelectric element and the matching electrical components of the generator circuit. It is extremely difficult to take into account all these factors with sufficient accuracy in advance. Analytical solutions and numerical simulations can approximately estimate the spectrum of resonant frequencies, however, such estimates will not exactly match the characteristics of the real device. It is necessary to provide a feedback system and generator frequency control. It is proposed to carry out digital feedback and digital control of the generator, which will allow for the implementation of a rather complex control algorithm that cannot be implemented by circuit engineering tools.

The development was accompanied by numerical simulation of various variations of the circuit in the NI Multisim software package, which made it possible to select the parameters of the electrical components to maximize the amplitude of the output signal.

2. Development of the layout of the control unit and generating a powerful ultrasonic signal

There are already quite developed analog technologies for generating high-power ultrasonic signals [4–6]. Generators with an auto-tuning of the resonant frequency based on the inclusion of a piezoelectric transducer in the feedback circuit already exist. In this case, the voltage switches, usually on field-effect transistors, are controlled by an electrical signal from the oscillating system being excited. The system has the maximum response at its resonant frequency, and therefore the control

IOP Conf. Series: Materials Science and Engineering **516** (2019) 012051 doi:10.1088/1757-899X/516/1/012051

signal through the feedback acquires the resonance frequency. A similar scheme was assembled and studied. It was possible to observe the effects of cavitation in the water bath when connecting the piezoelectric element. However, this circuit does not have digital or analog external control, and we cannot control the transition to the optimal resonant frequency. As a rule, there are several resonances, and there is no guarantee that this system will adjust to the optimal frequency.

Next, a push-pull excitation circuit of the piezoelectric element on transistors IRF640, controlled by an IR2153 chip, specifying a sequence of command pulses, was assembled. The frequency was changed manually using a trimming resistor in the control circuit IR2153. Manual tuning of the frequency made it possible to investigate the operation of the generator at different frequencies. Further, by adjusting the resistor, an increase in the power consumption at the resonant frequencies was detected. It was found that it is advisable to replace the transistors with more powerful IRFZ44N, and, in addition, transistors need radiators. Also, experimental studies of the generator were carried out on the more powerful transistors IRFP4229. It is necessary to apply a shielding metal case into which the generator should be placed, and the signal lines from the controlling microcontroller should be shielded by using a coaxial cable to establish electromagnetic compatibility.

Also, it was found that due to the effect of unilateral saturation of the transformer, there is a loss of control over the operation of the piezoelectric element and the exit of the system from the operating mode. To eliminate the effects of this effect, it is necessary to complicate the control algorithm and digital feedback.

3. Development control unit and powerful ultrasonic signal generator

A functional diagram of the layout of the control unit and the generation of a powerful ultrasonic signal, shown in Figure 1 is proposed. The digital microcontroller generates periodic pulses that trigger the generator at an appropriate frequency and controls the voltage at the output of the regulated power supply. The voltage level of the power supply will determine the power of the generated signal. The signal from the generator is fed to the piezo elements of the ultrasonic instrument, which are electrically the combination of connected capacitors, coils and resistors. The values of the current and voltage at the instrument input are measured using analog-digital converters of the microcontroller via feedback lines. Based on the feedback signals, the microcontroller can estimate the amplitude of oscillations, and their spectral composition, and respond to changes in the resonant frequency by changing the output pulses and the supply voltage.

This design of the functional circuit was chosen for the reason of providing digital feedback in the presence of full digital control of the frequency and phase of the generated signals, which will allow the implementation of the most complex control algorithms.

A scheme was developed for a digitally controlled generator of high-power ultrasonic signals (Figure 2). The circuit provides inputs for microcontroller (control 1, control 2), and outputs for providing digital feedback. Control signals are applied to field-effect transistors, which generate a meander signal. In such a switching circuit, transistor overheating is minimized. Through the step-up transformer, the signal is fed to the high-voltage circuit to which the ultrasonic instrument is connected. For matching with the capacitive load inductance is included in the circuit.

As a block of the microcontroller in this layout it is proposed to use the STM32F407 microcontroller, which provides the ability to transmit and receive real-time signals. The microcontroller control algorithm with digital feedback would optimize operating frequency.

The field-effect transistor IRFZ44N was chosen as the main elements controlling powerful electrical signals, since its parameters meet the requirements for power, heat dissipation, performance, and are widely available for mass production of products from it.

IOP Conf. Series: Materials Science and Engineering **516** (2019) 012051 doi:10.1088/1757-899X/516/1/012051

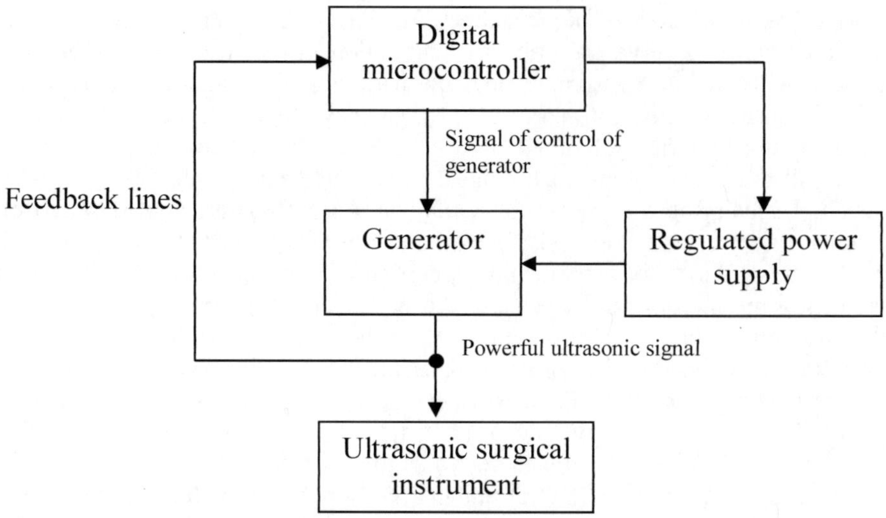

Figure 1. Functional diagram of the layout of the control unit and generating a powerful ultrasonic signal.

For electrical isolation of the surgical instrument and the generator, a transformer with a transmission coefficient of 10 is used, which makes it possible to increase the voltage on the instrument relative to the power supply 10 times - this is enough to excite piezoelectric elements PZT8 and their operation with the minimum mass of the transformer.

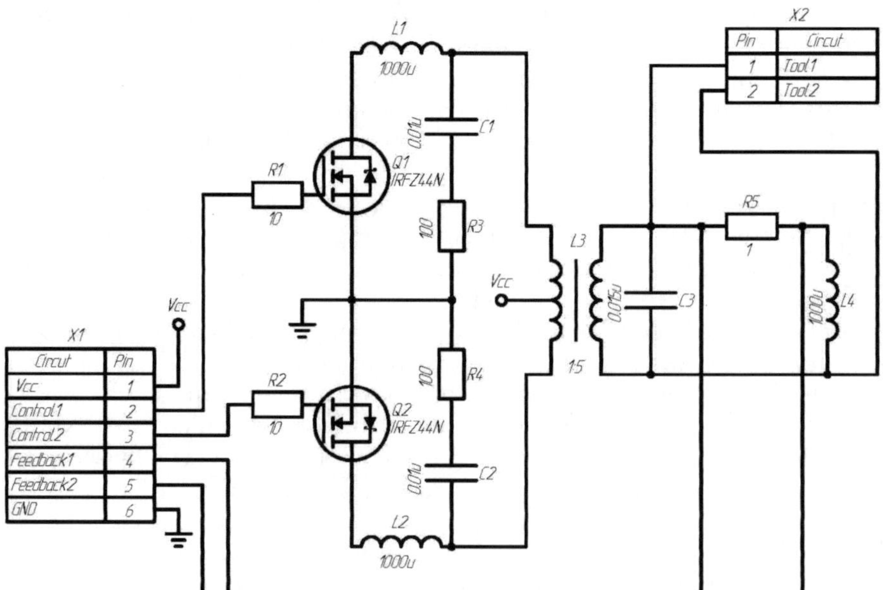

Figure 2. Electrical circuit of a controlled generator of high-power ultrasonic signals.

This layout reproduces the electrical diagram of the original generator without taking into account the housing, controls and the system of radiator cooling. Applying NI Multisim we have estimated resonance amplitude increasing (Figure 3).

IOP Conf. Series: Materials Science and Engineering **516** (2019) 012051 doi:10.1088/1757-899X/516/1/012051

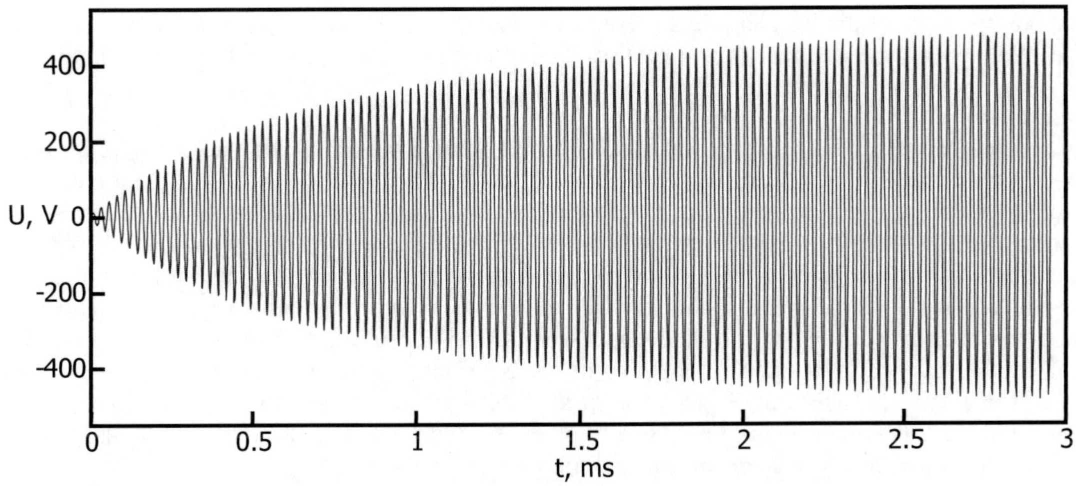

Figure 3. Simulated signal of generator on resonance frequency 40 kHz.

During the work on the model, the need was established to change the way the signal power was adjusted depending on the mode of operation. Figure 4 shows the modified and modified functional layout of the layout.

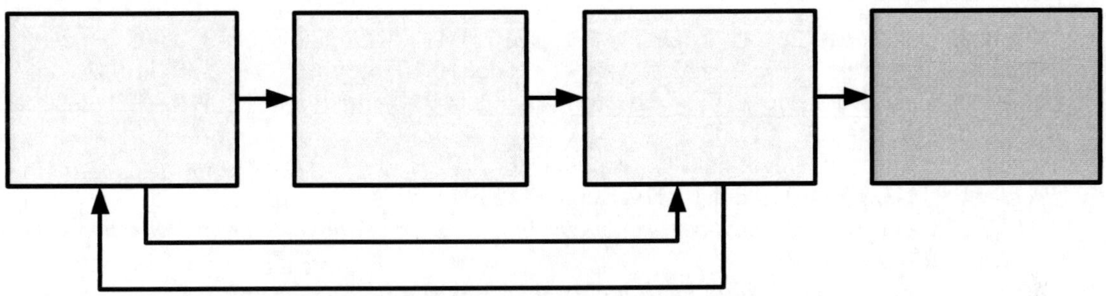

Figure 4. Modified functional layout scheme.

The microcontroller generates a rectangular signal of a given frequency. Since the outputs of the microcontroller do not provide sufficient current strength for the field-effect transistors to work optimally, it was decided to use a transistor control driver that provides the necessary gain of the control signal of the microcontroller.

For power control, it was originally proposed to use a pulse-controlled power supply. The advantage of this solution was the possibility of smooth adjustment of voltage within the specified limits. In the process, it was found that the exact setting of the output parameters is not necessary. At the same time, impulse control means the need to install additional integrating circuits, which leads to a significant deterioration in the weight and size characteristics, as well as a decrease in the overall efficiency of the installation.

In this regard, it was decided to abandon the pulse adjustment of the power source. In place of this, the power control is carried out by changing the transmission coefficient of the transformer in the matching circuit, by switching its secondary windings. This adjustment method does not produce nonlinear signal transformations, which eliminates the need to add additional integrating circuits to the system, and does not lead to a significant deterioration in the weight and size characteristics.

The feedback of the microcontroller and the power amplifier allows real-time analysis of the signal applied to the instrument, which allows for the dynamic adjustment of the control signal, if necessary, to ensure a given operating mode.

The power circuit of the installation is galvanically isolated from the controller both via the control channel and the feedback channel, which ensures the protection of the microcontroller from failure.

4. Conclusions

We proposed generator circuit of a powerful ultrasonic signal with digital feedback. It is proposed to adjust the power through switching the secondary winding of a transformer with a different number of turns. The control algorithm analyzes the current and voltage feedback signals to determine the optimal operating frequency. The generation of a powerful ultrasonic signal is carried out by switching large currents in the primary winding of a step-up transformer with powerful field-effect transistors. Thanks to the binary control of transistors, through the driver, which provides a large charge current of the gate, minimal overheating of the field-effect transistors is ensured. A sufficiently high voltage is formed on the secondary winding of the step-up transformer to excite ultrasonic vibrations in the piezoelectric elements based on PZT8. Measuring the current passing through the piezoelectric elements and the voltage on them allows you to receive a feedback signal indicating the presence of resonance. At resonance, the voltage and current increase many times.

Acknowledgments

The work was supported by the Ministry of Education and Science of Russia within the framework of the project "Creation of a high-tech complex of ultrasound surgery" (Unique identifier of the project RFMEF157517X0163).

References

[1] O'Daly B. J., Morris E., Gavin G. P., O'Byrne J. M., McGuinness G. B. 2008 *Journal of Materials Processing Technology* **200(1-3)** 38–58 doi:10.1016/j.jmatprotec.2007.11.041

[2] Petošić A., Ivančević B., Svilar D. 2009 *Ultrasonics* **49(6–7)** 522–531 doi:10.1016/j.ultras.2009.01.002

[3] Pantelic M., Ljikar J., Devecerski G., Karadzic J. 2015 *Energy systems in surgery. Medicinski Pregled* **68(11-12)** 394–399 doi:10.2298/mpns1512394p

[4] Titse U., Schenk K. 2008 *Semiconductor circuitry 12th ed. Volume I: Trans. with him* (DMK Press: Moscow)

[5] Dyakonov V.P. 2009 *Generation and signal generators* (DMK Press: Moscow)

[6] Dvornikov A.A., Koptev G.I., Panina T.A. 1990 *External excitation generators and high-frequency autogenerators* (Publishing House of Moscow Power Engineering Institute: Moscow)

IOP Conf. Series: Materials Science and Engineering **516** (2019) 012052 doi:10.1088/1757-899X/516/1/012052

Control of ordered group of particles based on ultrasonic levitation

D Sukhanov, E Sivkov, F Emelyanov

Tomsk State University, Tomsk, Russia
E-mail: sdy@mail.tsu.ru

Abstract. A method is proposed for controlling a group of particles under conditions of ultrasonic levitation in air. The possibility of placing particles in a stack of nodes of standing waves in a given order using a system of phased arrays of ultrasonic emitters is considered. Flat phased arrays are placed on the sides and on top of the levitation area. The side arrays create a plane in which the levitation of particles in an ordered grid is arranged. The top and bottom lattice ensures the placement of particles in an arbitrary order within the plane. The results of numerical simulation are presented.

1. Introduction

A control of particles in the conditions of ultrasonic levitation is necessary for a number of practical applications, such as levitation chemical reactors, soldering, fusion and coagulation. It is possible to calculate the particle trajectory under the action of a traveling ultrasonic field based on Newton's laws, and controlling the phase and amplitude of the wave, specify the desired acceleration vector for the particle. However, the further trajectory of the particle essentially depends on the friction forces, turbulent flows, temperature gradients. As a result, the behavior of the particle becomes poorly predictable. It is advisable to create a potential well for a particle at a given point in space, and then by moving the region of minimum potential to ensure controlled particle motion. The position of the particle in the potential well should be stabilized due to the friction force. As a function of the potential, it is suitable to apply the Gor'kov potential [1, 2].

In [3] authors proposed to use a combination of acoustic force and acoustic streaming. Side emitters are used to form a Cartesian grid of standing waves for arranging particles in a liquid. In this case, the emitters operated at the same frequency. They have shown an ability of proposed system to pattern for medium and large size particles.

Ultrasonic levitation technologies based on vertically emitting sources of plane waves, such as the near field [4] and in the far-field emitters [5, 6], are most widely used. The use of reflectors of the focusing form [5] allows localizing the region of levitation along horizontal axes. In a standing wave, it is possible to levitate particles up to 1/3 of a wavelength [5]. Researchers from the Polytechnic University of San Paulo, Brazil [7], by optimizing the shape and placement of the emitter and reflector, provided the levitation of steel balls in the ultrasonic field of standing waves. Later, they managed to provide levitation in a non-resonant mode on a standing wave formed by a single reflection from a focusing reflector [8]. Particles under the action of the sound field can not only move into a stable position, but also rotate at a given speed, as was shown in [9]. The control of particles in a standing wave by irradiating a flat reflector with the field of a phased array is shown in [10]. However, this approach does not allow arbitrary manipulation of particles along the vertical axis and does not allow the formation of raster images by particles. Theoretically forces acting on levitating spherical particle in an ultrasonic standing wave has been investigated in [11].

Here we propose to apply the ultrasonic fields at different frequencies to control the position of the particle in three-dimensional space. The difference in the frequencies of the radiators oriented orthogonally allows us to obtain more stable conditions for the levitation of particles. Changing the phase relationships of counter emitters provides control of a group of particles in three-dimensional space.

2. System overview

The layout of the emitters is shown in Figure 1. It is proposed to apply 6 plane phased array emitters: two oppositely directed emitters for the X, Y and Z axes. The X-axis emitters operate at frequency f_1, the Y-emitters at f_2, and the Z-emitters at f_3. Counter-emitters of waves at the same frequency create standing waves, and levitating particles are grouped into pressure nodes of standing waves [3].

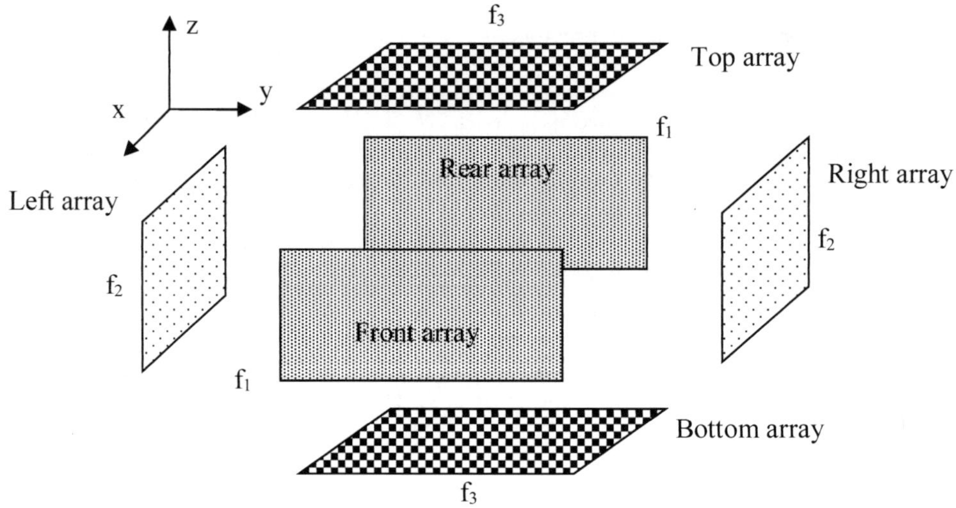

Figure 1. Placement of phased arrays.

It is proposed to focus left, right, front and rear phased arrays at by Z axis the center of system, but not localized horizontally (Figure 2). Such fields from horizontal grids of standing waves that form a rectangular area of levitation with potential holes in a periodic Cartesian grid. In total, with fields from phased arrays radiating in a vertical direction (top and bottom arrays), a field is formed that provides a given pattern of placement of particles (Figure 3).

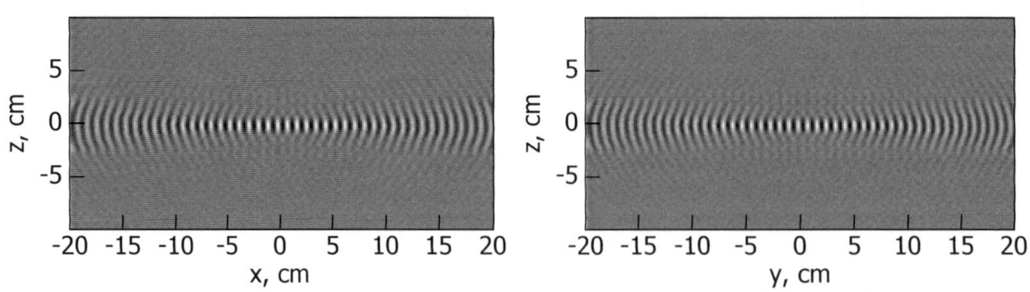

Figure 2. Field of horizontally radiating arrays (a – left and right arrays field at frequency 38 kHz, b – Front and rare arrays field at frequency 40 kHz).

The field of the upper and lower phased arrays form the image of the Greek letter "π" by focusing the field on the levitation plane at a frequency of 22 kHz. It is assumed that the particles will be grouped into the maximum field and repeat the shape of the specified image. To prevent particles from gathering at local maxima inside the image, the fields of the side grids form a rectangular grid that prevents particles from moving inside homogeneous areas of the image.

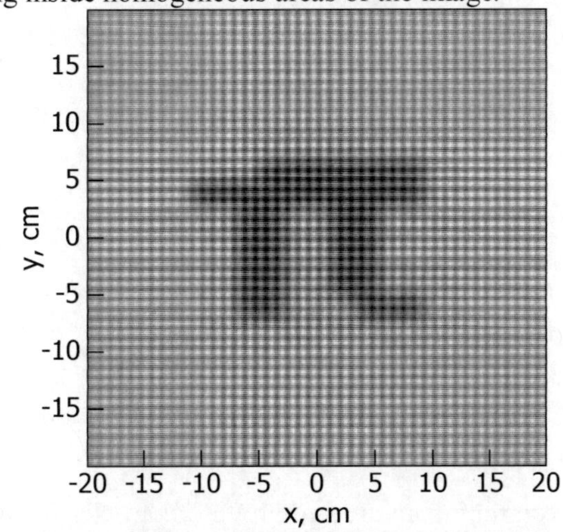

Figure 3. Sum of all arrays field forming an image of letter "π".

Oriented orthogonally grids emit at different and non-multiple frequencies to increase the stability of levitation of particles in an ultrasonic field. Numerical modeling shows that if frequencies are the same the behavior of particles is not stable.

3. Numerical simulation of particles levitation

We conducted a numerical simulation of levitation in an ultrasonic field in the air for Styrofoam particles with a density of 15 kg/m^3 with a diameter from 400 mkm to 600 mkm. The modeling was carried out on the basis of the Newton's motion equations, the Gor'kov formula [1] for taking into account the force of acoustic pressure. Also the friction force in air for spherical particles has been and taken into account. The initial distribution of particles was set randomly on a flat rectangular area of 20 cm by 20 cm. As a result of the simulation, it was found that after 100 ms the position of the particles stabilized in accordance with the specified field distribution (Figure 4).

According to the results of numerical simulation, one can see that the particles have stabilized in a given figure. By changing the phase ratio of the left and right gratings of the emitters, all structures are moved along the x axis. A change in the phases of the front and rare arrays allows you to move the structure along the y axis. Similarly, changing the phase difference top and bottom arrays allows you to move the structure along the vertical axis. Moving should occur, so that the particles have time to stabilize in the new position. Numerical simulation shows that the speed should not exceed, for the case under consideration, the value of 20 mm/s.

The side emitters ensured the formation of a grid of local field maxima in a Cartesian grid with a period of about 4 mm. Particles cannot pass from one local minimum to another unless sufficient radiation is applied from the upper and lower lattice. The upper and lower lattices form the required image of a layer of levitating particles. In the field of the gradient change of the field of vertical emitters, a sufficient force arises for the rearrangement of particles and the formation of the required stable image.

Not all particles have shifted to a given region of maximum field, since the potential gradient exists only near intensity changing regions. To displace all particles into the required image area, it is necessary to change the field of the upper and lower phased grids so that the field gradient piecemeal moves the particles into the area of the required image.

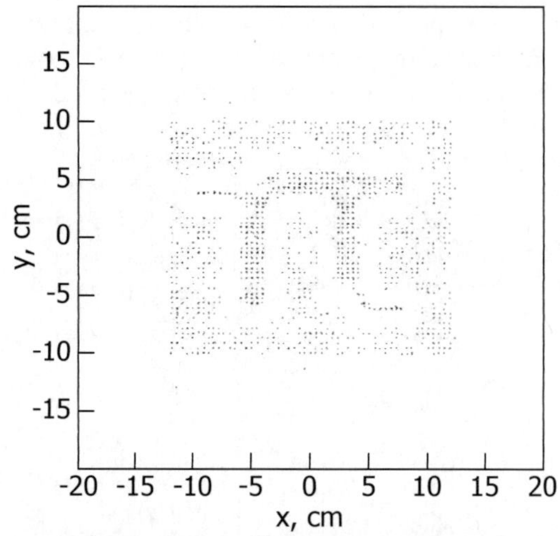

Figure 4. Stabilized Styrofoam particles distribution while levitating in air by influence of ultrasonic field (numerical simulation).

4. Conclusions

A method is proposed for controlling an ordered group of particles in ultrasonic levitation in air. It is proposed to provide levitation of particles in the Cartesian grid by creating a field of standing waves from orthogonal phased arrays emitting at different frequencies. The control of the fields of phased lattices allows you to create areas of stable levitation of particles of a given shape. The change in the phase relations of the counter lattices allows one to move ordered groups of particles along three orthogonal axes.. The results of numerical simulation on the formation of a flat bitmap image of levitating particles are presented.

Acknowledgments

The scientific research was carried out with the support of the grant of the Russian Science Foundation No. 17-79-20051.

References

[1] Gor'kov L.P. 1962 *Sov. Phys. Dokl.* **6(9)** 773–775
[2] Diego Baresch *et al* 2013 *The Journal of the Acoustical Society of America* **133** 25 doi:10.1121/1.4770256.
[3] Shilei Liu *et al* 2017 *Sensors* **17** 1664 doi: 10.3390/s17071664
[4] Jin Li *et al* 2011 *IEEE International Conference on Robotics and Automation* 12288561 doi: 10.1109/ICRA.2011.5979642
[5] Fuqiang Zhang *et al* 2018 *Open Access Library Journal* **05** 88024 doi: 10.4236/oalib.1104948
[6] Andrade M.A. *et al* 2011 *IEEE Transactions on Ultrasonics Ferroelectrics and Frequency Control* **58(8)** 1674 – 1683 doi: 10.1109/TUFFC.2011.1995
[7] Marco A. B. *et al* 2010 *IEEE Transactions on Ultrasonics, Ferroelectrics, and Frequency control* **57** 469–479 doi: 10.1109/TUFF c.2010.1427
[8] Marco A. B. Andrade, Flavio Buiochi, Julio C. Adamowski 2015 *Physics Procedia* **70** 68–71 doi: 10.1016/j.phpro.2015.08.044.
[9] Hong Z. Y. *et al* 2017 *Sci Rep.* **7** 7093 doi: 10.1038/s41598-017-07477-1
[10] Ayumu W., Koji H., Yutaka Abe. 2018 *Scientific Reports* **8** 10221 doi: doi.org/10.1038/s41598-018-28451-5
[11] Barrios G., Rechtman R. 2008 *Journal of Fluid Mechanics* **596** 191–200 doi: 10.1017/S0022112007009548

IOP Conf. Series: Materials Science and Engineering **516** (2019) 012053 doi:10.1088/1757-899X/516/1/012053

Tunable computing Slam navigation environments

Trang Hoang Thuy Nguyen[1,2], S V Shidlovskiy[1]

[1] National Research Tomsk State University, Tomsk, Russia
[2] Vietnam Maritime University, Hai Phong, Viet Nam
E-mail: trangnht@vimaru.edu.vn

Abstract. In many cases, to solve applied problems, the robot needs to know its real location, which is most often different from the data stored in the on-board system. For unmanned robotic devices, as well as for ground-based robots, it is most efficient to use local navigation algorithms, which consist in determining the coordinates of the device with respect to a certain starting point. The paper discusses localization algorithms provided that the map of the area is known in advance. Particular attention is paid to the Monte Carlo localization method because of several advantages. The paper presents an example of modeling the algorithm operation.

1. Introduction

Nowadays, simultaneous localization and mapping approach (SLAM) has become one of the most modern technical methods used for mobile robots to create maps in unknown or inaccessible locations and to update maps to a specific area while tracking your current location and distance. We built a highly structured system using mathematical models in conjunction with the Octomap environment to navigate a mobile robot.

2. Slam

In mapping and robotic navigation, simultaneous localization and mapping (SLAM) is one of the problems that scientists are interested in to create or update maps. The unknown environment also tracks the position of the internal agent. To overcome and improve positioning over short distances, synchronization simultaneous localization and mapping (SLAM) have made significant progress in the development of science and technology. Navigation can be divided into three main areas: the first is localization, the second is mapping, and the third is path planning. Localization performs the task of accurately determining the current position of the robot in the environment. Mapping is responsible for collecting sensor data and storing it in a form that is convenient for further processing. Path planning searches a route in an environment where the robot will move from start to goal using the data collected during a mapping process. These three tasks are dependent on each other. For example, localization will more accurately determine the location of the object on the map, and a good map will increase localization accuracy. In order to solve these dependencies better, simultaneous localization and mapping (SLAM) methods have been developed.

SLAM combines localization and mapping processes to make the combination both faster and more efficient. SLAM algorithms can currently solve specific environmental or hardware constraints. For example, lidar sensors and images combine Slam methods [1].

IOP Conf. Series: Materials Science and Engineering **516** (2019) 012053 doi:10.1088/1757-899X/516/1/012053

3. Octomap

Some applications for robots require a three-dimensional model of the environment. Although 3D maps are an integral part of many robotic systems, there are still a few flaws that are rarely deployed in a reliable and efficient system. The absence of such a deployment leads to the cloning of basic software components. Therefore, it can be considered as an important point in the study of robots.

Octomap is good for creating 3D probability maps. The matching process allows us to integrate measurements with a certain probability to explain sensor infiltration and other exceptions. In addition, it is a very efficient and relatively fast memory. Octomap is the process of mirroring with multiple resolutions; by changing the resolution and range we get a significant rate of change. In addition, the Octomap is a casting rail that allows you to distinguish between unidentified, free, and busy voxels that do not have other display processes (point clouds, elevation and multi-level maps, and surface).

The map to be navigated by the path planner will be in the form of an OctoMap, published in ROS and connected the Octomap server. The OctoMap library uses an octree based cubic grid system to represent three-dimensional space, with data structures and mapping algorithms provided in C ++.

An octree is a hierarchical data structure for spatial subdivision in 3D (Meagher 1982; Wilhelms and Van Gelder1992). Each node in an octree represents the space contained in a cubic volume, usually called a voxel. This volume is recursively subdivided into eight sub-volumes until a given minimum voxel size is reached (Figure 1) [2]. Cubes with octant components that are occupied or open can be saved as a single element. This results in a significant reduction in the size of the file on the map with large areas where all vertices are either occupied or open. The minimum voxel size determines the resolution of the octree. Since octree is a hierarchical data structure, a tree can be cut at any level to get a more complex part if the inner nodes are supported accordingly.

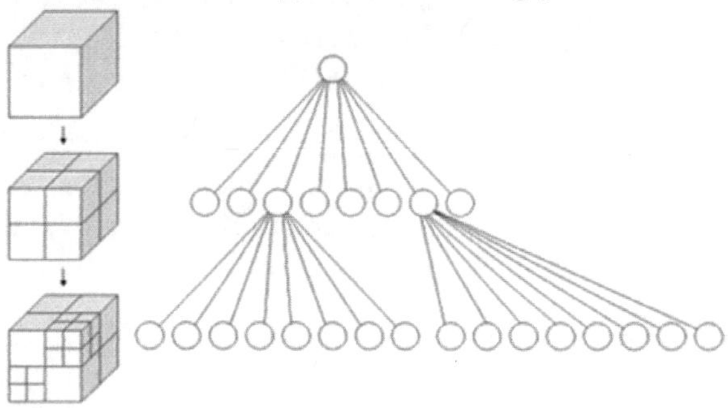

Figure 1. The structure of an octree and the cube shaped space it represents.

This allows Octomap to efficiently store models of 3D environments that can be quickly sent between ROS robots or between nodes on a single ROS device. This compression affects the Octomap messages that are stored and sent, but there is no real effect on the operation of the scheduling algorithm; Octomap consists of cubes of equal size that corresponds to the map resolution (Figure 2).

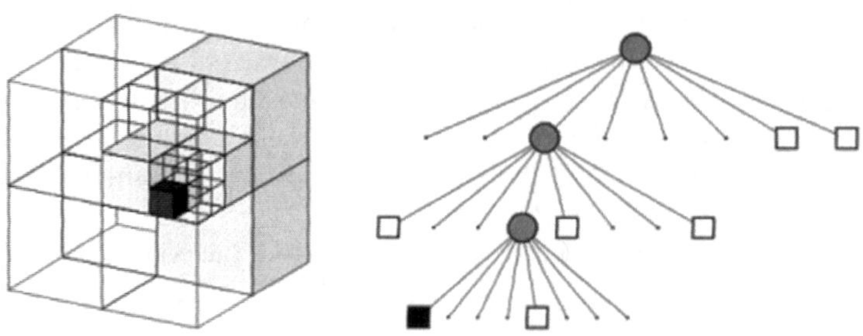

Figure 2. OctoMap storing one occupied vertex (black) within a larger free (white) space.

IOP Conf. Series: Materials Science and Engineering **516** (2019) 012053 doi:10.1088/1757-899X/516/1/012053

In the basic form, an octree can be used to model a logical attribute. In the context of mapping robots, volume is usually taken into account. If a certain volume is measured as occupied, the corresponding node in the octree will be initialized. Any uninitialized node could be free or unknown in this Boolean setting [2]. The OctoMap plugin for ROS can open an OctoMap from a file or create one of the point clouds. The point cloud is a list of coordinates in which a detector such as a laser scanner has detected a surface. When creating an OctoMap from a point cloud, it is possible to set any resolution that is considered appropriate for the task at hand, depending on factors such as the computing power available and the size of the robot. When loading maps from a file, a minimum cube size can be set to the size of a file, which enables faster processing of non-essential high resolution maps [3].

Here, the algorithm I am going to do is divided into point clouds, each of which contains four cubes, dividing the cubes until they stop dividing. The occupied areas will receive a value of 1 (black), the occupied area will receive a value of 0 (white) (Figure 3).

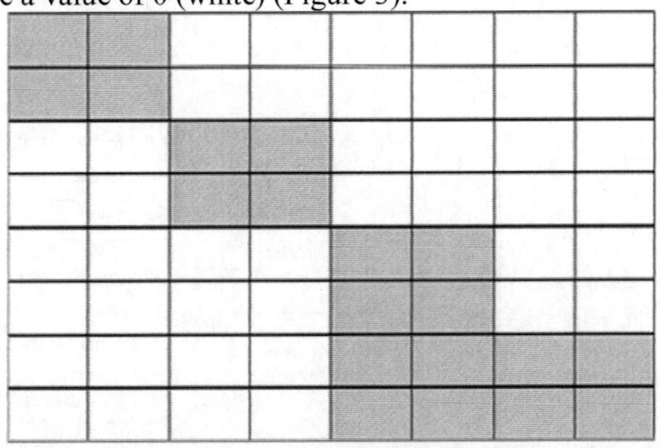

Figure 3. Octree structure: the white nodes of the octree leaves have a free state, eight gray leaf nodes are occupied by the status.

4. High-performance reconfigured systems

At present, with the continuous development of science and technology, the construction and design of intelligent control systems for various industrial automation objects to ensure coordination of their work are efficient with well-defined algorithms that are one of the important tasks of automation of production processes. In recent years, instead of using small and medium-sized microchips, people have switched to software logic devices, followed by large integrated circuits that are used as computer cores on industrial computers or in programmable logic controllers. Automating decision-making requires the use of modern mathematical methods and new technical means. Currently, when building control devices, boolean models are widely used, but some developers avoid direct use of logical algebraic functions. This is explained by the fact that considerable difficulties arose for the development of efficient algorithms for calculating complex functions of the algebra of logic of multiple arguments, and this also requires more intensive research and more thorough solving of them. The growing interest in the functions of the algebra of logic and its computational problems led to the creation of the theory of homogeneous structures. The paper proposes a logical model that can be adapted to a specific class of Boolean formulas. This model allows to solve the problem of the computing system of Boolean formulas from ordered and unordered iterative classes, as well as the class of Boolean formulas in the order of repetition and Boolean systems with and without arguments.

In mathematics and mathematical logic, Boolean algebra is a branch of algebra in which the values of variables are true and false truth values, usually denoted by 1 and 0, respectively. Instead, elementary algebra, where the variables are numbers and operation are simple addition and multiplication, the basic operations of Boolean algebra are the conjunction and are designated as $\wedge$, designated as a disjunction or $\vee$, and negation, and referred to as $\neg$.

The function whose values and arguments are binary is called Boolean.

Boolean functions (BFU) are defined as truth tables. Such a BFU entry is a formula that uses logical operators called the Boolean formula (BF).

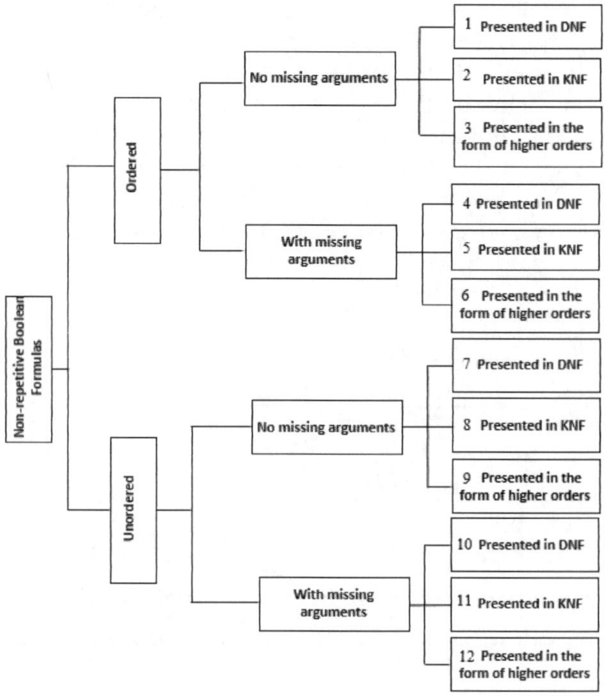

Figure 4. Classification of non-repetitive Boolean formulas.

In this article, we consider the formula in the basis of {AND, OR, NOT}. First, we consider the fully defined BFU of n variables defined on 2n input sets. The main metric of a BFU is the number of variables n. In principle, the boolean formula is divided into two groups, repeatless and repetitive. The classification of non-repetitive BF reduction is in Figure 4.

A formula will be called non-repetitive if each argument, taken in direct or inverse form, is included in it no more than once. In all other cases, the formula is repeated.

If in the notation of Boolean iteration formulas f (xi), where i = 1,2 ..., n, - n is the number of arguments, the index i is increased from left to right with the help of logical arguments, then we assume that the formula is sorted. We assume that the formula is ordered even with the same transformations, so we get a formula with incremental values and arguments that increase from left to right. If there are no arguments to any index in the record of a Boolean formula that does not repeat, we will assume that this formula contains the flaws of the corresponding arguments and calls it the formula with omitted arguments.

Thus, in accordance with the above classification (Figure 4), the set of Boolean formulas splits into pairwise disjoint classes — sets of the same type of formulas. Each formula of a particular class can be chosen as a representative of this class. Boolean formulas belonging to one class are realized by physically identical schemes, therefore, for each class, it suffices to implement only one scheme, the structure of which is described by the formula of a representative of a class.

We synthesize such an automaton, which will provide, at a certain setting, the calculation of all the BF groups provided in Figure 4, while the input arguments are not the same.

5. Tunable automatons

A tunable automaton can be represented as a set of automata with the same outputs, and the setting determines the automaton whose outputs are considered at this setting as the outputs of the entire

IOP Conf. Series: Materials Science and Engineering **516** (2019) 012053 doi:10.1088/1757-899X/516/1/012053

automaton being tuned. Therefore, tunable automaton cannot implement anything except automaton mappings.

If the tunable automaton at any settings generates the value of the signals at its outputs, which depends only on the values of the signals at its inputs, it is called a combinational tunable automaton, or a multifunctional logic module.

We form automaton mappings (Figure 5), which the tunable automaton should implement, in order to provide the calculation of the Boolean formula specified in the class. Let us write systems of Boolean formulas corresponding to each automaton mapping:

1) $z_1 = 0,\ z_2 = 0,\ z_3 = 0,\ z_4 = 0,\ z_5 = 0,$ (Figure 5*a*)

$f_r^1 = f_r^3 = 0,\ f_r^2 = i_1,\ f_r^4 = i_3;$

2) $z_1 = 0,\ z_2 = 0,\ z_3 = 0,\ z_4 = 0,\ z_5 = 1,$ (Figure 5*b*)

$f_r^1 = i_2,\ f_r^2 = f_r^3 = f_r^4 = 0;$

3) $z_1 = 0,\ z_2 = 0,\ z_3 = 0,\ z_4 = 1,\ z_5 = 0,$ (Figure 5*c*)

$f_r^1 = f_r^2 = f_r^4 = 0,\ f_r^3 = i_1 \cdot i_4;$

4) $z_1 = 0,\ z_2 = 0,\ z_3 = 0,\ z_4 = 1,\ z_5 = 1,$ (Figure 5*d*)

$f_r^1 = f_r^2 = f_r^3 = 0,\ f_r^4 = i_1 \vee i_2;$

5) $z_1 = 0,\ z_2 = 0,\ z_3 = 1,\ z_4 = 0,\ z_5 = 0,$ (Figure 5*e*)

$f_r^1 = f_r^2 = f_r^4 = 0,\ f_r^3 = i_1 \cdot i_2;$

6) $z_1 = 0,\ z_2 = 0,\ z_3 = 1,\ z_4 = 0,\ z_5 = 1,$ (Figure 5*f*)

$f_r^1 = f_r^2 = f_r^3 = 0,\ f_r^4 = i_1 \cdot i_2;$

7) $z_1 = 0,\ z_2 = 0,\ z_3 = 1,\ z_4 = 1,\ z_5 = 0,$ (Figure 5*g*)

$f_r^1 = f_r^2 = f_r^3 = 0,\ f_r^4 = i_1;$

8) $z_1 = 0,\ z_2 = 0,\ z_3 = 1,\ z_4 = 1,\ z_5 = 1,$ (Figure 5*h*)

$f_r^1 = f_r^3 = f_r^4 = 0,\ f_r^2 = i_1;$

9) $z_1 = 0,\ z_2 = 1,\ z_3 = 0,\ z_4 = 0,\ z_5 = 0,$ (Figure 5*i*)

$f_r^1 = f_r^3 = 0,\ f_r^2 = i_1,\ f_r^4 = i_2;$

10) $z_1 = 0,\ z_2 = 1,\ z_3 = 0,\ z_4 = 0,\ z_5 = 1,$ (Figure 5*j*)

$f_r^1 = f_r^2 = f_r^4 = 0,\ f_r^3 = i_1 \vee i_4;$

11) $z_1 = 0,\ z_2 = 1,\ z_3 = 0,\ z_4 = 1,\ z_5 = 0,$ (Figure 5*k*)

$f_r^1 = f_r^2 = f_r^4 = 0,\ f_r^3 = i_1 \vee i_2;$

12) $z_1 = 0,\ z_2 = 1,\ z_3 = 0,\ z_4 = 1,\ z_5 = 1,$ (Figure 5*l*)

$f_r^1 = f_r^3 = 0,\ f_r^2 = i_1 \cdot i_4,\ f_r^4 = i_2;$

13) $z_1 = 0,\ z_2 = 1,\ z_3 = 1,\ z_4 = 0,\ z_5 = 0,$ (Figure 5*m*)

$f_r^1 = f_r^2 = f_r^4 = 0,\ f_r^3 = i_1;$

14) $z_1 = 0,\ z_2 = 1,\ z_3 = 1,\ z_4 = 0,\ z_5 = 1,$ (Figure 5*n*)

$f_r^1 = f_r^3 = 0,\ f_r^2 = i_4,\ f_r^4 = i_2;$

15) $z_1 = 0,\ z_2 = 1,\ z_3 = 1,\ z_4 = 1,\ z_5 = 0,$ (Figure 5*o*)

$f_r^1 = f_r^2 = f_r^3 = 0, \ f_r^4 = i_2;$

16) $z_1 = 0, \ z_2 = 1, \ z_3 = 1, \ z_4 = 1, \ z_5 = 1,$ (Figure 5*p*)

$f_r^1 = f_r^3 = f_r^4 = 0, \ f_r^2 = i_4;$

17) $z_1 = 1, \ z_2 = 0, \ z_3 = 0, \ z_4 = 0, \ z_5 = 0,$ (Figure 5*q*)

$f_r^1 = f_r^3 = f_r^4 = 0, \ f_r^2 = i_1 \vee i_4;$

18) $z_1 = 1, \ z_2 = 0, \ z_3 = 0, \ z_4 = 0, \ z_5 = 1,$ (Figure 5*r*)

$f_r^1 = f_r^3 = f_r^4 = 0, \ f_r^2 = i_1 \cdot i_4;$

19) $z_1 = 1, \ z_2 = 0, \ z_3 = 0, \ z_4 = 1, \ z_5 = 0,$ (Figure 5*s*)

$f_r^1 = f_r^3 = f_r^3 = 0, \ f_r^4 = i_2 \cdot i_3;$

20) $z_1 = 1, \ z_2 = 0, \ z_3 = 0, \ z_4 = 1, \ z_5 = 1,$ (Figure 5*t*)

$f_r^1 = f_r^4 = 0, \ f_r^2 = i_4, \ f_r^3 = i_1;$

21) $z_1 = 1, \ z_2 = 0, \ z_3 = 1, \ z_4 = 0, \ z_5 = 0,$ (Figure 5*u*)

$f_r^1 = f_r^2 = f_r^4 = 0, \ f_r^3 = i_1 \cdot i_2 \cdot i_4;$

22) $z_1 = 1, \ z_2 = 0, \ z_3 = 1, \ z_4 = 1, \ z_5 = 0,$ (Figure 5*v*)

$f_r^1 = f_r^2 = 0, \ f_r^3 = i_1, \ f_r^4 = i_2;$

23) $z_1 = 1, \ z_2 = 0, \ z_3 = 1, \ z_4 = 1, \ z_5 = 0,$ (Figure 5*w*)

$f_r^1 = f_r^2 = f_r^3 = 0, \ f_r^4 = i_1 \vee i_4.$

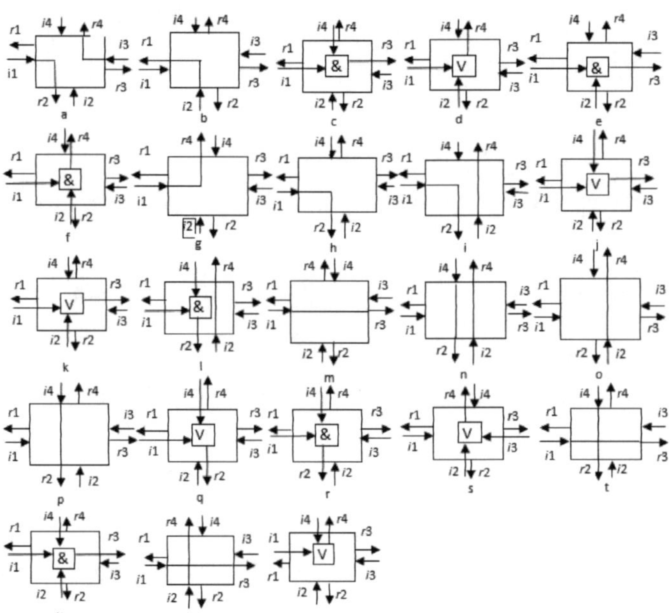

Figure 5. Automaton mapping.

Based on the structured automation approach [4, 5], we get the following boolean system that will describe the entire basic PVS

$$
\begin{cases}
f_r^1 = i_2 \cdot \left(\overline{z_1} \cdot \overline{z_2} \cdot \overline{z_3} \cdot \overline{z_4} \cdot z_5\right), \\
f_r^2 = i_1 \cdot \left(\overline{z_1} \cdot \overline{z_3} \cdot \overline{z_4} \cdot \overline{z_5}\right) \vee i_2 \cdot \left(\overline{z_1} \cdot \overline{z_2} \cdot z_3 \cdot z_4 \cdot z_5\right) \vee \left(i_1 \cdot i_4\right) \cdot \left(\overline{z_1} \cdot z_2 \cdot \overline{z_3} \cdot z_4 \cdot z_5\right) \vee \\
\quad \vee\ i_4 \cdot \left(\overline{z_1} \cdot z_2 \cdot z_3 \cdot z_5\right) \vee i_1 \cdot \left(\overline{z_2} \cdot \overline{z_3} \cdot \overline{z_4} \cdot \overline{z_5}\right) \vee i_4 \cdot \left(z_1 \cdot \overline{z_2} \cdot \overline{z_3} \cdot \overline{z_4} \cdot \overline{z_5}\right) \\
\quad \vee\ \left(i_1 \cdot i_4\right) \cdot \left(z_1 \cdot \overline{z_2} \cdot \overline{z_3} \cdot \overline{z_4} \cdot z_5\right) \vee i_4 \cdot \left(z_1 \cdot \overline{z_2} \cdot \overline{z_3} \cdot z_4 \cdot z_5\right), \\
f_r^3 = \left(i_1 \cdot i_4\right) \cdot \left(\overline{z_1} \cdot \overline{z_2} \cdot \overline{z_3} \cdot z_4 \cdot \overline{z_5}\right) \vee \left(i_1 \cdot i_2\right) \cdot \left(\overline{z_1} \cdot \overline{z_2} \cdot z_3 \cdot \overline{z_4} \cdot \overline{z_5}\right) \vee \left(i_1 \vee i_4\right) \cdot \\
\quad \cdot \left(\overline{z_1} \cdot z_2 \cdot \overline{z_3} \cdot \overline{z_4} \cdot z_5\right) \vee \left(i_1 \vee i_2\right) \cdot \left(\overline{z_1} \cdot z_2 \cdot \overline{z_3} \cdot z_4 \cdot \overline{z_5}\right) \vee i_1 \cdot \left(\overline{z_1} \cdot z_2 \cdot z_3 \cdot \overline{z_4} \cdot \overline{z_5}\right) \vee \\
\quad \vee\ i_1 \cdot \left(z_1 \cdot \overline{z_2} \cdot \overline{z_3} \cdot z_4 \cdot z_5\right) \vee \left(i_1 \cdot i_2 \cdot i_4\right) \cdot \left(z_1 \cdot \overline{z_2} \cdot z_3 \cdot \overline{z_4} \cdot \overline{z_5}\right) \vee i_1 \cdot \left(z_1 \cdot \overline{z_2} \cdot z_3 \cdot z_4 \cdot \overline{z_5}\right), \\
f_r^4 = i_3 \cdot \left(\overline{z_1} \cdot \overline{z_2} \cdot \overline{z_3} \cdot \overline{z_4} \cdot \overline{z_5}\right) \vee \left(i_1 \vee i_2\right) \cdot \left(\overline{z_1} \cdot \overline{z_2} \cdot \overline{z_3} \cdot z_4 \cdot z_5\right) \vee \left(i_1 \cdot i_2\right) \cdot \\
\quad \cdot \left(\overline{z_1} \cdot \overline{z_2} \cdot z_3 \cdot \overline{z_4} \cdot z_5\right) \vee i_1 \cdot \left(\overline{z_1} \cdot \overline{z_2} \cdot z_3 \cdot z_4 \cdot \overline{z_5}\right) \vee i_2 \cdot \left(\overline{z_1} \cdot z_2 \cdot \overline{z_3} \cdot \overline{z_4} \cdot \overline{z_5}\right) \vee \\
\quad \vee\ i_2 \cdot \left(\overline{z_1} \cdot z_2 \cdot \overline{z_3} \cdot z_4 \cdot z_5\right) \vee i_2 \cdot \left(\overline{z_1} \cdot z_2 \cdot z_3 \cdot \overline{z_4} \cdot z_5\right) \vee i_2 \cdot \left(\overline{z_1} \cdot z_2 \cdot z_3 \cdot z_4 \cdot \overline{z_5}\right) \vee \left(i_2 \cdot i_3\right) \cdot \\
\quad \cdot \left(z_1 \cdot \overline{z_2} \cdot \overline{z_3} \cdot z_4 \cdot \overline{z_5}\right) \vee i_1 \cdot \left(z_1 \cdot \overline{z_2} \cdot z_3 \cdot z_4 \cdot \overline{z_5}\right) \vee \left(i_1 \vee i_4\right) \cdot \left(z_1 \cdot \overline{z_2} \cdot z_3 \cdot z_4 \cdot z_5\right).
\end{cases}
$$

6. Modeling

Let us examine the PVS of the matrix type with the calculation nodes of the two-dimensional action, as shown in Figure 5.

We found, according to the classification (Figure 4) of a repeated, ordered, without missing the arguments, presented in the DNF formulas:

$$f_1 = x_1 \cdot x_2 \vee x_3 \cdot x_4 \cdot x_5. \tag{1}$$

A homogeneous computing environment that implements formula (1) is shown in Figure 6.

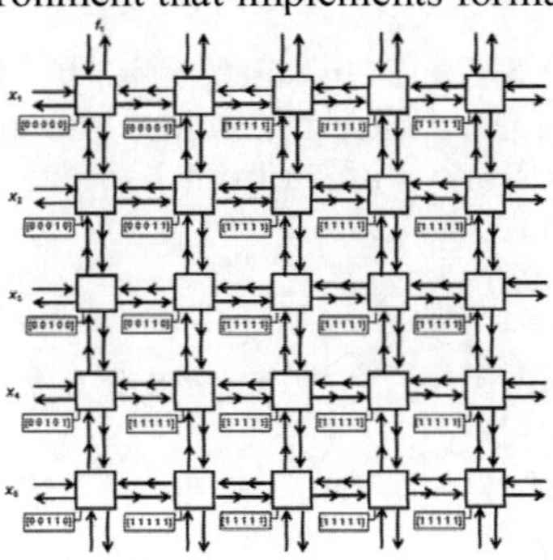

Figure 6. Expression Execution $f_1 = x_1 \cdot x_2 \vee x_3 \cdot x_4 \cdot x_5$..

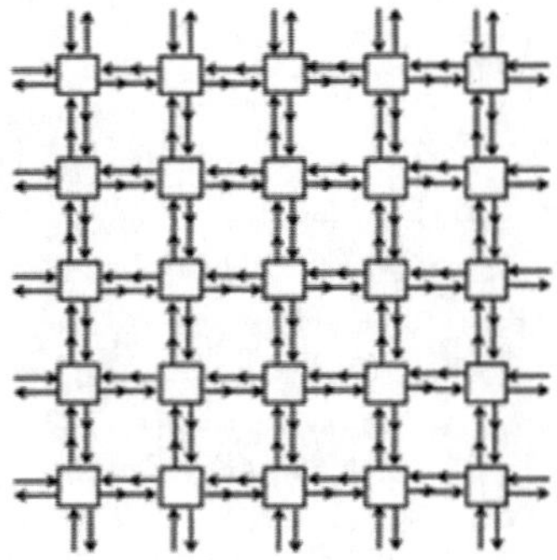

Figure 7. The following environment adjusts for the octree.

Let us build a tunable computing environment, to refine the octomap algorithm. In this case, you can use the matrix structure of the calculator (Figure 7). The elementary calculator for one of the subtasks can be described by the following formula:

$$\begin{cases} f_1 = x_1 \vee x_2 \vee x_3 \vee x_4 \vee x_5, \\ f_2 = x_1, \\ f_3 = x_1, \\ f_4 = x_1. \end{cases}$$

where as x_1 - x_4, data from a cloud of points is fed in, and x_5 is information about performed calculations in selected subgroups.

7. Conclusion

In this article, we used a point cloud to create an octree tree. The results show that when using uniformly structured models and parallel processing, the results and implementation time of the algorithm are faster at each stage of performing the tasks of image processing in three dimensions. This environment allows you to build a map based on octree very quick.

Acknowledgments

The research was carried out in Tomsk State University with the financial support of the Ministry of Education and Science of the Russian Federation, a unique project identifier: RFMEFI57817X0241. The authors are grateful to Tatiana B. Rumyantseva from Tomsk State University for English language editing.

References

[1] Lavrenov R., Zakiev A., Magid E. 2017 *International Conference on Mechanical, System and Control Engineering, ICMSC* 1-2

[2] Hornung A., Wurm K. M., Bennewitz M., Stachniss C., Burgard W. 2013 *Autonomous Robots* 189–206 doi: 10.1007/s10

[3] Bergström J. 2018 *Path Planning with Weighted Wall Regions using OctoMap* (Publishing House Luleå University of Technology:Lulea)

[4] Shidlovsky S. V. 2006 *Automatic control. Tunable structures* (TSU Publishing House: Tomsk)

[5] Kuznetsov D., Syryamkin V. 2015 *AIP Conference Proceedings* **1688** 040004 doi: 10.1063/1.4936037

IOP Conf. Series: Materials Science and Engineering **516** (2019) 012054 doi:10.1088/1757-899X/516/1/012054

Optimizing the A* search algorithm for mobile robotic devices

V V Maneev, M V Syryamkin

National Research Tomsk State University, Tomsk, Russia
E-mail: worklovery@gmail.com

Abstract. The purposes of this article are to analyze the pathfinding algorithms used to search and calculate the trajectory of movement of mobile autonomous robotic devices and to search for optimization methods in order to reduce the load on computing and memory units of the mobile devices. This paper analyzes the basic pathfinding algorithms including breadth-first search, depth-first search, Dijkstra's algorithm, greedy best-first search, and the A*algorithm, considers principles of their work, and proposes optimization methods. This work results in a new approach for using the A* algorithm for solving problems associated with pathfinding in real dynamic environments. The proposed approach speeds up the process of path calculation and the choice of the movement direction and reduces the requirements for computing power of autonomous mobile devices.

1. Introduction

The A* search algorithm is widely used in games and for planning routs of mobile robots [1]. This is a highly efficient heuristic algorithm for searching an acceptable and shortest path. It allows finding the least-cost path between the starting and ending points. The A*algorithm is considered one of the most acceptable algorithms because it combines the advantages of several search algorithms including breadth-first search and depth-first search algorithms and is already an optimized version of Dijkstra's algorithm. However, there is still a serious problem for small autonomous mobile devices because in some cases, the existing A* search algorithm may require significant computational power, which may adversely affect the speed of making decision about the direction and trajectory of motion [2]. In addition, in its pure form, the A* algorithm does not allow taking into account conditions that may arise in real environments such as dynamic objects or the final goal with an unpredictable motion path. In this paper, we consider methods and approaches that allow optimizing the A* algorithm, reducing the performance costs of computational units, and rebuilding the system to control dynamic changes and goals with an unpredictable motion trajectory.

2. Materials and methods

The pathfinding problem is well studied and the pathfinding algorithms are widely used in designing computer games and planning routes including auxiliary navigation systems. The simplest algorithm for solving the shortest path problem is the breadth-first search algorithm. It performs a sequential search of all possible paths evenly in all directions. Implementation of this algorithm considers neither the 'weight' of each segment nor the cost of calculating the shortest path.

The depth-first search algorithm is more advanced for our purposes. As the main direction of the recursive search, the depth-first search algorithm selects the direction of the target. The depth-first search algorithm searches all routes outgoing from the point under examination. If the route leads to a point that has not been considered earlier, the algorithm starts from this unexamined point, and then returns to the previous level and continues the search. This algorithm also does not consider the cost of passing a route [3].

Dijkstra's algorithm is a heuristic analogue of the breadth-first search algorithm. When finding the shortest route, the Dijkstra's algorithm considers the 'weights' of each segment of the route. However, as well as the breadth-first search algorithm, the Dijkstra's algorithm is very resource-intensive.

A* search algorithm is a modification of the Dijkstra's algorithm. It has the advantages of the depth-first search algorithm and, like the Dijkstra's algorithm, is heuristic, that is, it allows taking into account the cost of passing a route along each of its segments.

For the purposes of this study, we selected the A* search algorithm as the main algorithm. The basis of the A*search algorithm is the function of estimating the cost of passing through graph points. If the assessment function cannot be correctly selected, the search result may be an accessible but not optimal path. The A*search algorithm uses $f(n) = g(n) + h(n)$ as an evaluation function for expanding nodes options in the OPEN list. In this formula, $g(n)$ is an actual cost between the initial node and the current node n (the cost of the optimal path found), and $h(n)$ is an estimate of the cost of the optimal path between the current node n and the target node. The disadvantage of the algorithm is in too large search area, which leads to an exponential growth of the studied vertices with an increase in the distance to the ending point. It also affects the memory consumption (which also increases exponentially) and the complexity becomes polynomial when the heuristics satisfies the condition: $|h(n) - h^*(n)| \leq O(\log h^*(n))$ [4].

These deficiencies in the A* search algorithm may be fatal when using it for portable mobile robotic devices because miniature robots cannot carry powerful computing systems with a large amount of RAM. So, to bypass a simple obstacle using the A* search algorithm, 760 (92) points will need to be analyzed, and the path length will be only 89 points (Figure 1).

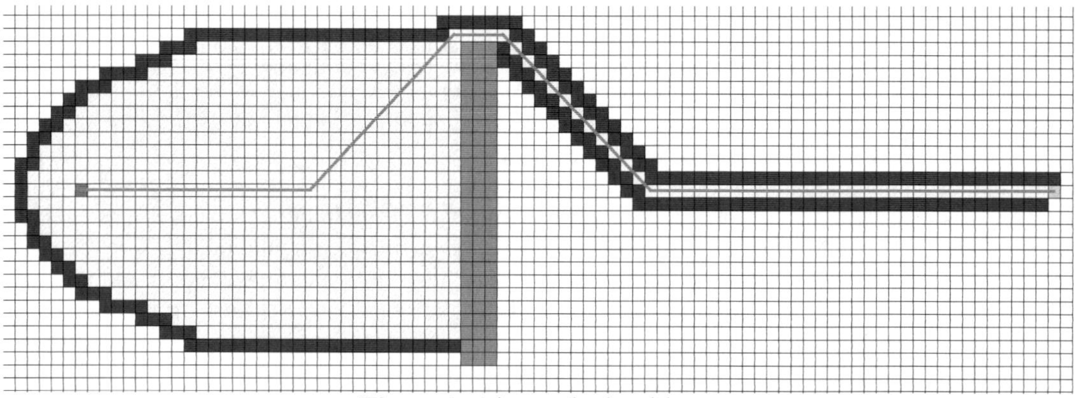

Figure 1. A* search algorithm.

For a 10×10 cm mobile robotic device, the length of the covered path will be only 8.9 meters. As the distance between the mobile device and the ending point of its route (and the number of obstacles in its path) increases, the complexity of calculating the trajectory of movement (as well as the amount of the required memory for storing data) increases exponentially.

3. The methods for optimization of the A* search algorithm.

There are several variations of the A*algorithm proposed for solving problems of exponential growth in the computation of the optimal path. These variations include iterative deepening A* (IDA*), memory-bounded A* (MA*), simplified MA (SMA*) and recursive best first search (RBFS). As a rule, all modifications are aimed at improving or changing the heuristic part of the A* algorithm. Although the heuristic component of the algorithm is the main and most important part of the pathfinding algorithm, optimizing only the heuristic part of the algorithm may not be enough to solve the optimization problems of autonomous mobile devices [5].

IOP Conf. Series: Materials Science and Engineering **516** (2019) 012054 doi:10.1088/1757-899X/516/1/012054

In dynamic and natural environments, it is required to make continuous calculations and adjustments to the route of movement in order to bypass unexpected obstacles, moving objects and targets. For example, to intercept a target moving fast along a complex unpredictable trajectory, it is required to recalculate the trajectory and correct the route faster than the target leaves the conditional point of its current location (when the size of the graph points of the A*algorithm is equal to the dimensions of the robot). Resolution of such problems in the usual way is not possible. However, as a rule, in real environments, the obstacles are more fragmented, and the interaction objects such as buildings, cars, or people are much larger than autonomous mobile devices, robots, and drones. Therefore, the distance between them can be significant and the dimensions of the passages between obstacles are many times greater than the dimensions of a mobile autonomous robotic device. This fact allows us to reduce significantly the number of required operations and the amount of RAM used to calculate and store the optimal route if in order to find the path, we use nested cycles with geometrically proportional graph nodes [6, 7].

4. Features of A*algorithm optimization

For the above-mentioned example of finding a path using the A* algorithm, we use nested cycles with geometrically proportional graph nodes (Figure 2).

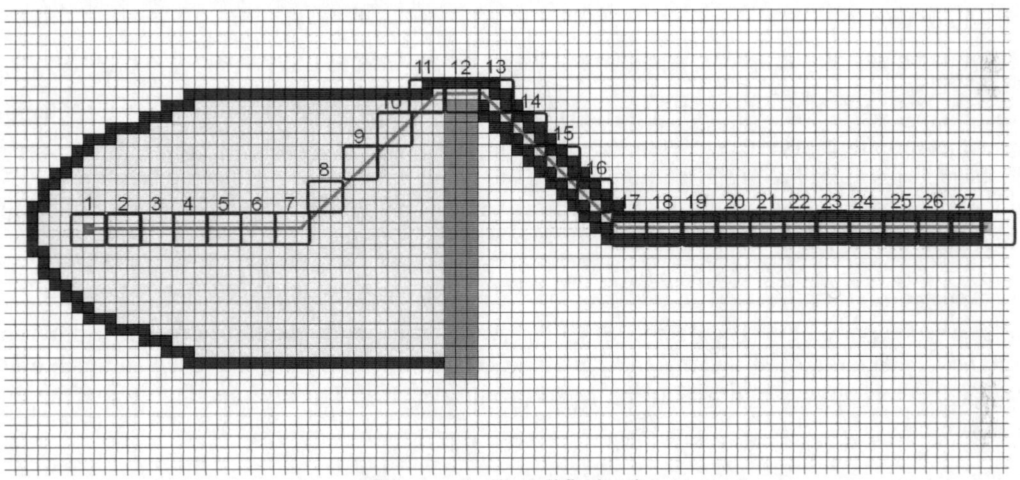

Figure 2. Simplified A*.

Figure 2 shows that in order to build a complete route, we need to perform 27 simplified pathfinding operations, using the A* algorithm and another 27 operations to search for a route until each next block of calculations. Table 1 shows these calculations.

Table 1. Calculating a path.

Step	Points checked			Step	Points checked		
	Main block	Additional block	Full solution		Main block	Additional block	Full solution
1	109	4	760	15	14	4	39
2	102	4	690	16	13	4	36
3	94	4	621	17	12	4	33
4	85	4	553	18	11	4	30
5	75	4	483	19	10	4	27
6	67	4	415	20	9	4	24
7	55	4	345	21	8	4	21
8	43	4	250	22	7	4	18
9	32	4	167	23	6	4	15
10	23	4	95	24	5	4	12

11	18	4	52	25	4	4	9
12	17	4	48	26	3	4	6
13	16	4	45	27	3	4	3
14	15	4	42				
Total	751	56	4566		105	52	273
			964				**4839**

Table 1 shows that the simplified A* algorithm (example shown in Figure 2) in case of the need to clarify the route periodically, needs to analyze only 964 points of the graph, and the maximum amount of required memory cells is 109. At the same time, using the A* algorithm (when clarifying the route at the same points and at the same intervals), it was required to analyze 4839 points of the graph and to use 760 memory cells to store data. For this case, the increase in performance using our approach to build a route was 4839/964 = 5.02 times, and the memory usage decreased by 760/109 = 6.97 times. Moreover, the efficiency will multiply increase in case of an increase in the distance to the final object, the frequency of route clarification and the degree of nesting of the algorithm. This will significantly reduce the requirements for the computing power of autonomous robotic devices.

Implementation of this approach requires substantial development and complication of the heuristic function of the algorithm, which may adversely affect performance, but the overall effectiveness of the solution will remain.

5. Conclusion

In this work, we were able to develop a fundamentally new approach to using the A* algorithm for solving the pathfinding problem in conditions of the need to continually adjust the route or track a moving target. This approach significantly reduces the computing power requirements of mobile robots and drones and allows scaling solutions immediately depending on the required accuracy of routing or the requirements for the speed of response to environmental changes.

Acknowledgments

The paper was supported by “The Tomsk State University competitiveness improvement programme” under grant (No 8.2.24.2018) and by the Russian Foundation for Basic Research (grant No 16-29-04388). The authors are grateful to Tatiana B. Rumyantseva from Tomsk State University for English language editing.

References

[1] Zheng-hong Hu, Jin Li. *2010 International Conference on Computational Aspects of Social Networks* 738–739

[2] Shashev D. V., Shidlovskiy S. V. 2017 *Journal of Physics Conference Series* **881** 19–26

[3] Aggarwal A., Bhalla J.S. 2013 *The Next Generation Information Technology Summit (4th International Conference)* 69–75

[4] Cheng Rao, Lianqing Yu. 2010 *The 2nd International Conference on Industrial Mechatronics and Automation* 221–224

[5] Miao Wang, Hanyu Lu *2012 International Conference on Industrial Control and Electronics Engineering* 1739–1742

[6] Shikhman M. V., Shidlovskiy S. V. 2017 *IOP Conference Series-Materials Science and Engineering* 363.

[7] Yao J., Zhang B., Zhou Q. 2009 *World Congress on Software Engineering* 515

IOP Conf. Series: Materials Science and Engineering **516** (2019) 012055 doi:10.1088/1757-899X/516/1/012055

Valve converter with steeply falling external characteristics

S Mendybayev, T Akimzhanov, S Zhumazhanov, Ye Sarsikeev

Saken Seifullin Kazakh Agrotechnical University, Astana, Kazakhstan
E-mail: sarsikeev@tpu.ru

Abstract. This article discusses the development of three-phase valve converters with a steeply external characteristic consisting of a parametric current source, to the outputs of which three-phase valve kits are connected. A parametric current source, made according to the "star" scheme, converts the initial three-phase EMF system into a three-phase current system; the resonant circuit of the parametric source improves the energy performance of the valve converter, it contributes to the growth of energy efficiency indicators. Three-phase converters with a steeply falling external characteristic are most appropriate in practice when using them as power sources for energy-intensive consumers of electrical technology, which, ultimately, leads to an increase in the quality of products.

1. Introduction

One of the most characteristic features of the current stage is the ever-increasing growth of technical requirements for sources and converters of electrical energy in terms of reliability and economy. Almost half of the electricity produced is consumed in the form of direct current or non-standard frequency current [1, 2]. In this case, most of the electrical energy conversion is performed by semiconductor valve converters (VC). Valve converters closely deal with the development of electric power industry and electromechanics. The complexity of the processes in energy systems required the widespread introduction of new, efficient circuit design solutions for the VC, based on the latest achievements of science and technology. Semiconductor converters (SC) are one of the main load elements of networks, their work significantly determines the mode of operation of networks [3]. Valve converters are one of the main nodes of many power sources. The most energy-intensive consumers of electrical energy are consumers of electronic technologies, which are arc furnaces, electrolysis of non-ferrous metals, etc. [4, 5]. For electrical installations, the most efficient power sources are valve converters with a steeply dipping external characteristic or converters with characteristics of current sources, providing a significant improvement in many technological indicators of the production process [6]. Existing current transducers have a low power factor (cosφ), especially with deep regulation [7, 8]. Since the use of valve devices impairs the quality of electrical energy, the development of reliable current converters is an important and urgent task. One of the ways to solve the problem is the use of current-parameter converters (CPC), having a high power factor close to unity. The CPC forms a new class of converters – a class of current converters. A new class of circuits required a new approach to analyzing the operation of a current converter (single-phase, multi-phase) for various types of loads.

IOP Conf. Series: Materials Science and Engineering **516** (2019) 012055 doi:10.1088/1757-899X/516/1/012055

2. Analysis of the work of current transformers for different types of loads

2.1. Active load (contact heating)

One of the types of active load of the converter valve is industrial contact heating, which is one of the methods of direct heating. Direct heating provides: increase in labor productivity; reduction of technological losses, increase the service life of production equipment, etc. The advantage of contact heating is the high rate of metal heating. However, industrial installations of contact heating have low energy indicators (efficiency, cosφ). In addition, there is no possibility of smooth control of the load current, without the use of additional auxiliary devices [9–11].

2.2. Active - inductive load (arc furnaces)

Arc furnaces are powerful consumers with low power factor ($\cos\varphi \approx 0{,}7 \div 0{,}8$). The technological mode of arc furnaces is characterized by frequent technological short circuits. Part of the arc furnaces, which include vacuum arc furnaces, operates on direct current. In this case, the furnace power source must contain an AC-DC converter. Vacuum arc furnaces are sensitive to mode changes. Therefore, there is an urgent need to stabilize the operating current of the furnaces and the possibility of programmatically controlling the technological current of the furnace [12, 13].

2.3. Load in the form of counter-emf

Industrial electrolysis cells are the load in the form of counter-emf. their volt-ampere characteristics (VAC) have a slight slope in the working area. The small slope of the VAC causes the need for automatic stabilization of the load current. This is due to the fact that fluctuations in the voltage of the network can lead to a significant change in the value of the technological mode of operation of the electrolyzers [14–16].

3. Three-phase CPC for various types of loads

Three-phase circuits of the CPC are used to supply powerful consumers. At the same time, the ripple of the current in the load is noticeably reduced, the heating intensity is increased due to an increase in the average value of the rectified current. The bridge circuit of the converter found the broadest application in powerful consumers with output voltages over one hundred and fifty volts. So, for example, when using a bridge circuit to power cells, as calculations show, the cost of a transformer is reduced by 12.5% in comparison with a six-phase zero circuit, the power factor rises by 0.01–0.05 [7].

3.1. Three-phase CPC for the active load

The circuit of one of the variants of a three-phase diode-thyristor CPC assembled by the bridge circuit and its time diagrams of operation is shown in Figure 1 and Figure 2. In the interval $t_0 - t_2$ of diodes *1 – 6*, an uncontrolled bridge is conducted in a cyclic mode: *5 6 1 – 6 1 2 – 1 2 3* etc. In this case, the positive load current mode is realized in the output circuit of the converter. An additional controllable set assembled on thyristors *1'*, *3'*, *5'* is introduced into the circuit to generate the load current. A control pulse is applied to the thyristor 1' of the additional gated set and the load current drops to zero at time t_2. The parametric current source is in short-circuit mode, which is provided by simultaneously operating gates of an uncontrolled bridge and an additional gated set. In order to control the pause in the load current, after the thyristor *1'*, the gate *3'* of the additional gated set is turned on after two cycles, etc. When the pause formation is complete (moment t_2, Figure 2), the gate pulses are not generated by thyristors of the additional gated set, and the positive load current mode is restored in the load (only uncontrolled diodes *1–6* operates).

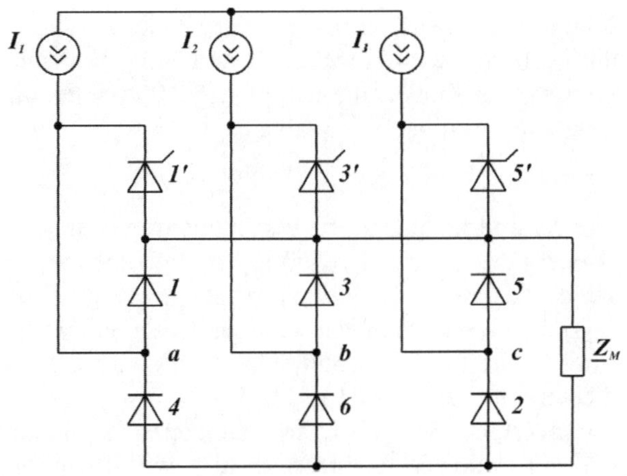

Figure 1. Diagram of a three-phase diode-thyristor CPC (I_1,I_2,I_3 – parametric current sources).

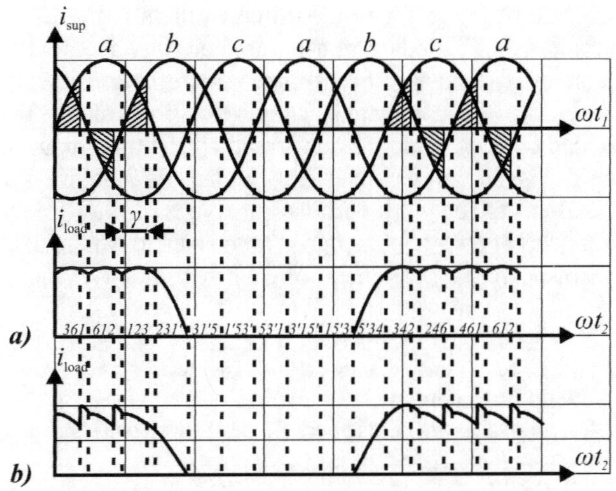

Figure 2. Diagram of a three-phase diode-thyristor CPC.

3.2. Three-phase CPC for active-inductive load

Current sources having steeply dipping external characteristics and providing high energy parameters (power factor, efficiency, etc.) are needed to supply arc furnaces. These requirements are fully met by three-phase current-to-voltage converters (CPC) with a natural current characteristic.

3.3. Three-phase CPC for loads in the form of counter-emf

The time diagrams of the load current of a three-phase current-measuring converter when operating on a counter-emf in the modes I_H + and $I_H = 0$ are shown in Figure 2*b*. The $I_H = 0$ mode for $Z_H = E_d$ is not typical for the CPC and given only for qualitative analysis in Figure 2*b*. A detailed analysis of the operation of single-phase and three-phase CPC when operating on the counter-emf was carried out in [7].

The implantation of current parametric converters for a series of non-ferrous metal electrolyzers allows stabilizing the current of the load (at the least by 1%) due to the resonance circuit of the parametric source.

The technical and economic comparison of converter substations intended for aluminum electrolyzers with the use of power units such as VAK, VAKEL, etc., controlled by saturation inductors, and current-to-voltage converters has shown: the use of the CPC yields an economic effect, due to increase the current output, with approximately the same capital costs and power losses [7].

4. Conclusion

The acceleration of the scientific and technical process imposes ever-increasing demands on sources and converters of electrical energy in terms of reliability and efficiency. The most energy-intensive consumers are consumers of electrical technology (arc furnaces, electrolysis of non-ferrous metals, etc.). As power sources of energy-intensive consumers, semiconductor converters with characteristics of current sources are widely used. The use of current sources allows for stabilization of the operating current, improves many indicators of the process [6].

Existing semiconductor converters (SC) with the characteristics of current sources, along with significant advantages, have a low power factor, especially with deep regulation. Since the main elements of the SC are thyristors, having a non-linear volt-ampere characteristic, this leads to a deterioration in the quality of electricity in industrial power grids. Therefore, the development and analysis of current converters is an important and urgent task.

One of the ways to solve the problem for powerful consumers is the use of valve converters with steeply falling external characteristic, called current-parameter converters (CPC).

The current-parameter converter has a natural current characteristic and is a parametric current source to the output of which valve kits are connected. The CPC allows using the latitudinal methods to smoothly adjust the output load current, which is confirmed by the time diagrams of their work given in the article. The CPC is a new class of valve converters – a class of current converters. The new class of converters required a new analysis of the work with various types of plug-in loads. The valve converters with a steeply falling external characteristic considered in the article can be used for the practical design of secondary power sources of powerful consumers of electrical energy.

References

[1] Nikulin A. D., Rodshtain L. S., Salnikov V. G. 1995 *Thyristor Conversion Technique in Non-Ferrous Metallurgy* (*Metallurgy Publ.:* Moscow)
[2] Svenchansky A. D., Evtyukova I. P., Katsevich L. S., Nekrasova N. M. 1982 *Electrotechnological industrial plants* (*Energoizdat Publ.:* Moscow)
[3] Akimzhanov T. B. *et al* 2014 *Proc. of the 9th International Forum on Strategic Technology (Coxs Bazar)* (Chittagong Univ Engn & Technol) 351–354
[4] Illahi F. *et al* 2018 *IEEE Transact. on power delivery* **33** 4 1727–1734
[5] Mendybayev S. *et al* 2018 *MATEC Web Conf.* **155**
[6] Akimzhanov T. B. *et al* 2018 *Proc. Conf. Materials Science and Engineering* **457** 012019
[7] Plotnikova I. V. *et al* 2016 *MATEC Web of Conferences* **79** 01019 doi: 10.1051 matecconf/20167901019
[8] Odnokopylov I. G. *et al* 2016 *IOP Conf. Ser.: Mater. Sci. Eng.* **189(1)** 012010 doi: 10.1088/1757-899X/189/1/012010
[9] Mendybayev S. A., Ivanova Ye. V. 2012 *Proc. of Int. Conf. Efficient and qualitative use of electricity.*
[10] Hocine L. *et al* 2009 *Energy* **349** 1207–1214
[11] Mendybayev S. A. *et al* 2016 *Proc. of Int. Conf. Technical and Mathematical Sciences*
[12] Maki S. *et al* 2002 *J Mater. Proces. Technol.* **125** 477–482
[13] Ozturk F. *et al* 2011 *Key Eng Mater* **479** 130–136
[14] Lucia O. *et al* 2014 *IEEE Transact. on indust. electron.* **61** 5 2509–2520
[15] Donsion M. P. *et al* 2004 *IEEE Transact. on power delivery* **19** 1 367–373
[16] Varadan S. *et al* 1996 *IEEE transact. on power delivery* **11** 3 1685–1691

IOP Conf. Series: Materials Science and Engineering **516** (2019) 012056 doi:10.1088/1757-899X/516/1/012056

Study of the effect of sugar concentration on the characteristic impedance of a magnetic coil

V P Yakubov , R M Makhmanazarov, A V Klokov, A S Zapasnoy, K V Zavyalova, A S Mironichev

Tomsk State University, Tomsk, Russia
E-mail: yvlp@mail.tsu.ru

Abstract. The work is devoted to the development of technology for non-invasive measurement of glucose in human blood. Modern blood glucose meters are discrete devices that do not fully provide a picture of the level of glucose in the blood. Some cases of hyperglycemia or hypoglycemia may not be recorded between measurements. In addition, finger piercing causes pain and subsequently leads to tissue damage. In this paper, the effect of glucose concentration in the human body on the complex resistance of an inductance coil, with the finger of a test person placed in it, is investigated.

1. Introduction

The incidence of diabetes is increasing every year. The number of people with diabetes over the past 35 years has increased 4 times. Now, more than 400 million people in the world have diabetes, and the prevalence of the disease continues to grow [1]. The World Health Organization estimates that by 2030, diabetes will be the seventh leading cause of death. Possible treatments include monitoring blood glucose through dietary regulation, oral medication or insulin administration, all of the methods presented have an adverse effect on daily life [2]. At present, patients with diabetes are advised to regularly check their blood glucose levels with a glucometer. This practice can help closely monitor blood glucose levels. Thus, diabetics and their doctors can get a clear picture of blood glucose levels to optimize therapy. This is an indicator for adjusting insulin dosage among diabetics who need daily insulin injections.

The incidence of diabetes is increasing every year. The number of people with diabetes over the past 35 years has increased 4 times. Now, more than 400 million people in the world have diabetes, and the prevalence of the disease continues to grow [1]. The World Health Organization estimates that by 2030, diabetes will be the seventh leading cause of death. Possible treatments include monitoring blood glucose through dietary regulation, oral medication or insulin administration, all of the methods presented have an adverse effect on daily life [2]. At present, patients with diabetes are advised to regularly check their blood glucose levels with a glucometer. This practice can help closely monitor blood glucose levels. Thus, diabetics and their doctors can get a clear picture of blood glucose levels to optimize therapy. This is an indicator for adjusting insulin dosage among diabetics who need daily insulin injections.

However, it is difficult to puncture the blood glucose level for diabetics who check their blood sugar levels several times a day [3]. Finger piercing causes pain and subsequently leads to tissue damage. A patient who has been suffering from diabetes for 20 years and who checks blood glucose three times a day will have his finger pricked more than 21,900 [4]. It also increases the risk of infection.

IOP Conf. Series: Materials Science and Engineering **516** (2019) 012056 doi:10.1088/1757-899X/516/1/012056

It is worth noting the minimally invasive methods for measuring glucose levels - spectroscopic. They are based on the integration of subcutaneous sensors that determine the concentration of glucose in the tissue fluid. However, the disadvantage of this technology is the need for frequent calibration and high susceptibility to biological fouling [2]. Among the technologies of non-invasive glucometer, one can also note: infrared and near-infrared; Raman spectroscopy; optical coherence tomography; fluorescence spectroscopy; bio-impedance spectroscopy and others [2].

The most advantageous is the bio-impedance spectroscopy method, which is based on measuring the impedance of a biological object in the frequency range [3]. One of the existing instruments for measuring sugar using this method is PENDRA (Pendragon medical). It is a bracelet, like a watch. With the help of an open oscillating LC circuit embedded in the bracelet, the measurement of the complex resistance is performed. Using the values of the complex resistance, the change in the concentration of glucose in the blood is calculated [3-4]. Given its design, it can be said that the device is not completely non-invasive. The contacts of the LC circuit, touching the skin surface, are exposed to external influences of temperature changes and contact with sweat, which adversely affect the operation of the device. To date, a sufficiently large number of other works promoting the theory and technique of non-invasive glucometer are already known [5–16].

Summing up a brief overview of existing non-invasive methods of glucometers, we can say that the simplest approach is based on measuring the impedance characteristics of a part of the human body with a developed blood system. Such a body could be, for example, a hand - body, almost always naked for inspection. In this paper, we propose to consider the possibility of non-contact spectral analysis of the impedance of a person's hand, more precisely of one finger, depending on the level of sugar in his blood.

2. The method of measuring the total complex resistance

Since all measurements of the impedance of any part of the human hand are significantly dependent on the contact of the measuring system with the skin and this gives large errors. Human skin is a complex protective shell, which even in one person is in constant change. Hair and fat cover do not give full contact. The slightest bias in the contact clamp significantly changes the measurement. In addition, it should be added that the blood does not flow across the surface of the arm and the electric current will have to make its way through the layers of muscles, ligaments, fat and other formations. Non-invasive method should penetrate to the carrier blood - vein, contactless. There are two options - electromagnetic wave or magnetic field. Leaving aside the possibility of using an electromagnetic wave, let us focus on the use of a variable magnetic field.

We will use the simplest measuring sensor - a coil, which is both a source and a receiver of a magnetic field (Figure 1). The measured variable impedance is the impedance that should respond to the appearance of a finger inside the coil.

a

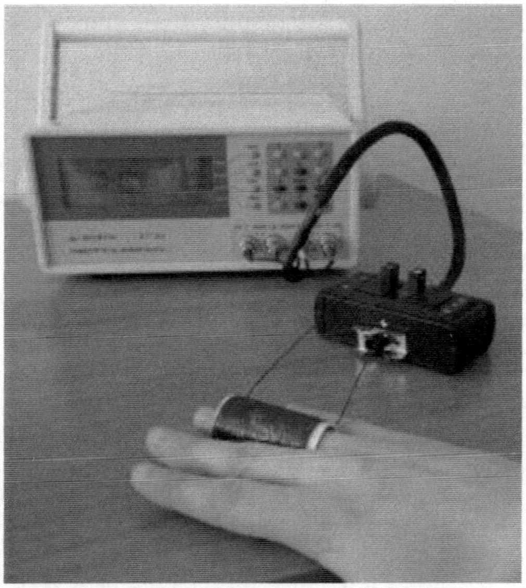

b

Figure 1. Measuring coil: in calibration mode (a) and in measurement mode (b).

Measurements are performed using the E7-20 instrument, which is a standard immittance meter in a wide band of tunable frequencies from 25 Hz up to 1 MHz. Immittance is a generalized concept for full (complex) resistance - impedance and full (complex) conductivity - admittance.

The principle of operation of this device is based on methods for measuring the parameters of electrical circuits, namely, bridge methods and the method associated with the use of Ohm's law relation to alternating current. The principle of operation of immittance bridge meters is based on the use of a measuring bridge, for balancing which the instrument contains sets of exemplary active and reactive (capacitive) resistances. Such devices can operate only at fixed frequencies. In the experiments, a device with a frequency range from 25 Hz up to 1 MHz was used (splitting the band into 17 frequencies: 25, 50, 60, 120, 200, 500, 1000, 2000, 5000, 10000, 50000, 100000, 20000, 500000, 1000000 Hz.)

An inductor with a winding diameter was designed and wound, so that any finger of the average person's hand can be placed into it. Subjects placed their fingers inside the coil and waited an average of 170 seconds while the instrument made measurements (10 seconds at each frequency). The principle of measurement of such immittance meters is based on the analysis of the passage of a test signal (usually sinusoidal) with a given frequency through the measured circuit, which has complex resistance. The voltage of the operating frequency from the internal generator is fed to the measured object. The voltage, current and phase shift between them is measured on a dedicated part of the circuit. The measured values are used to calculate the parameters of the circuits, in our case the complex resistance. Further, by analogy with the already well-established measurement methods in near-field diagnostics, we fix the offset of the measured parameters from the norm position (when there are no heterogeneities, the sugar level does not exceed the allowable values).

The measuring coil in the proposed use case consists of three elements connected in series: - the resistance of the copper wire from which the coil is made, - the inductance formed by the wound coil, and - the additional impedance introduced by the finger into the electrical circuit. Thus, the total impedance - immittance, measured by the device can be described by the ratio:

$$Z_x(f) = R + i2\pi f L + Z(f).$$

In the general case, of course, one should add inter-turn capacitance, but as the analysis showed, in the selected frequency range it is insignificant. In Figure 2 shows the result of measuring the coil

IOP Conf. Series: Materials Science and Engineering **516** (2019) 012056 doi:10.1088/1757-899X/516/1/012056

resistance in a wide frequency range. Shown here is the impedance modulus. This relationship is very well described by the first two terms in the above formula. The resistance of the wire is $R = 0.16723$ Om, $L = 3.12*10^{-5}$ Henry. These are small values.

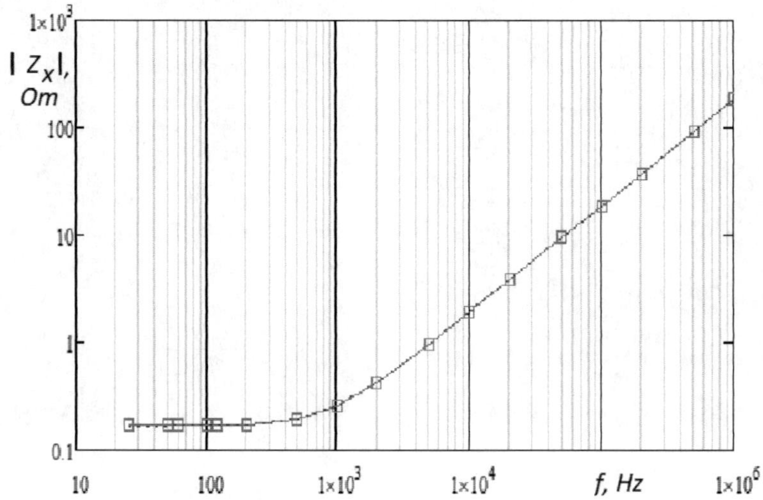

Figure 2. Modulus of impedance of the measuring coil in a wide frequency range.

3. The effect of sugar on the complex resistance

Immediately, we note that the effect of the finger is almost imperceptible on the background. To isolate this effect, it is necessary to perform a calibration for each measurement, which consisted in measuring the impedance in the absence of a finger $Z_0(f) = R + i2\pi f L$. This automatically takes into account the possibility, for example, of the thermal change in the ohmic resistance of the wire. In Figure 3 shows the dependence of

$$\delta = \left|\frac{Z_x(f)}{Z_0(f)}\right| - 1 = \left|\frac{Z(f)}{Z_0(f)}\right| - 1.$$

It can be seen, the influence of the finger as a whole is reduced to a few percent and falls on the frequency range up to 5 kHz.

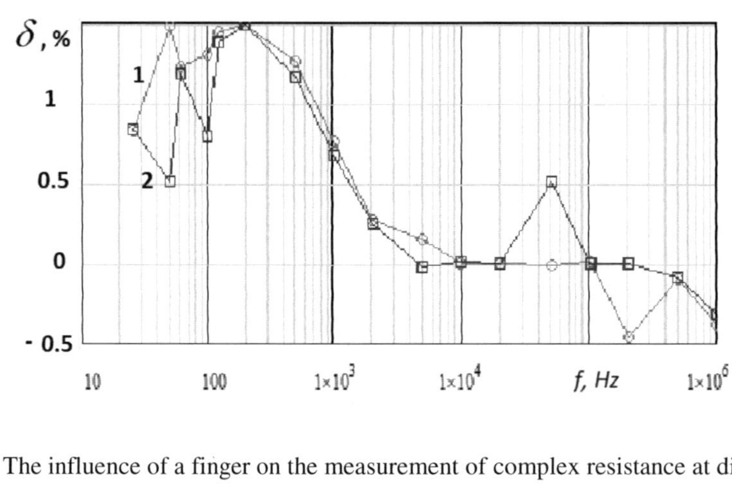

Figure 3. The influence of a finger on the measurement of complex resistance at different levels of sugar in the blood: 1 - sugar level 8.1 mmol / l and 2 - sugar level 5.7 mmol / l.

IOP Conf. Series: Materials Science and Engineering **516** (2019) 012056 doi:10.1088/1757-899X/516/1/012056

The level of sugar was controlled using an AccuCheck Performa type household glucose meter. All measurements were carried out in 2 stages. The first stage of measurement - connecting the coil, measuring the complex resistance of the circuit. This is a calibration. The next stage was carried out invasive measurement of glucose in human blood and record testimony. Blood was taken from the ring finger of the right hand. Next, the subject placed the ring finger of his right hand inside an inductor coil (Figure 1). Then, the measurement of the complex resistance of the system is made; at the end of the first two measurements, the participant in the experiment was offered food with a high sugar content. After the meal, the person was in a calm, unsettling, and inactive physical state for 30 minutes. Then the measurement procedure was repeated starting from the first stage.

The measurements performed showed that the influence of a finger makes a small contribution (less than 1-2%) to the total complex resistance. Sugar level gives an even smaller footprint. However, this trace can be enhanced by highlighting the parameter that most clearly reflects the physics of the processes. Let us explain the essence of these considerations.

An alternating magnetic field creates in the body of the finger a varying magnetic flux that permeates the blood vessels of the body of the finger. As a result, eddy currents are induced in the largest vessels, which develop in the cross section of the blood vessels. The blood containing a certain amount of free charge carriers begins to react, taking away energy from the "bulging" of its external magnetic field. Mobility of charges depends on the content of sugar dissolved in the blood. As a result, this interaction is reflected in the complex resistance of the blood. Here it is especially important that this interaction is of a reactive nature, since reactive eddy currents. As a result, the reactive component of the coil impedance will increase. For us, this means that the phase of the measured coil impedance will be the most informative.

In confirmation of the above reasoning in Figure 4 shows the detected relative changes in the phase of the coil's impedance, which are allocated to a relatively empty coil at each of the frequencies used. Here is the argument (phase) of the resistance of the empty coil and is the observed change.

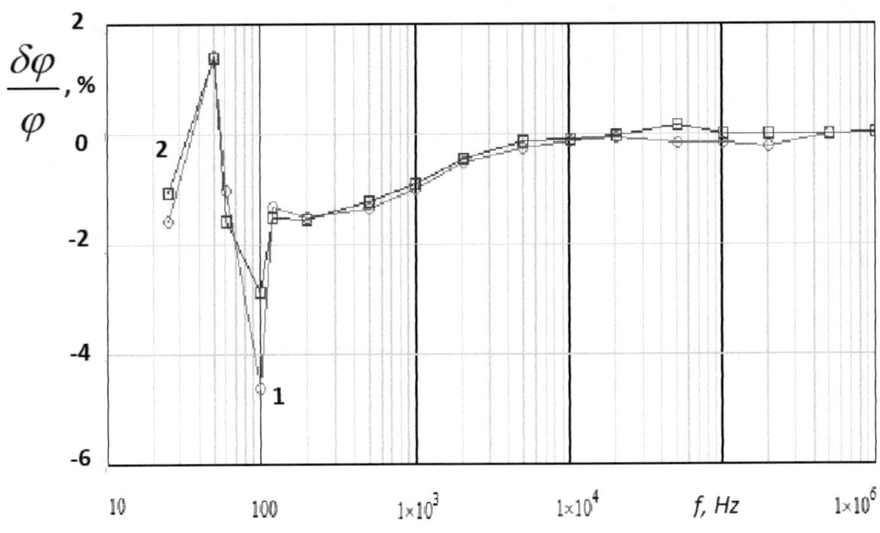

Figure 4. The change in the phase resistance of the measuring coil when inserting a finger at different levels of blood sugar: 1 - sugar level 8.1 mmol / l and 2 - sugar level 5.7 mmol / l.

It can be seen that the phase resistance of the measuring coil is the most sensitive value to the appearance of a finger in it and to the change in the sugar content in the blood. In the frequency range

IOP Conf. Series: Materials Science and Engineering **516** (2019) 012056 doi:10.1088/1757-899X/516/1/012056

of 25 Hz –5 kHz, even a small increase in the level of sugar from 5.7 to 8.1 mmol / l mmol / l causes a relative change of phase in units of%.

4. Summary

A study on the change in the complex resistance of the coil in the low-frequency region with a finger placed into it allowed, in a non-invasive way, to evaluate the increase in blood sugar levels. The most informative parameter is the phase resistance of the coil. The studies begun deserve wider continuation in world laboratories.

Acknowledgments

This work was supported by the grant of the Russian Science Foundation No. 18-75-10101.

References

[1] The official site of the World Health Organization Media Center URL: http://www.who.int/mediacentre/factsheets/fs312/en/
[2] The official website of MedicineNet, Determining the puncture of a finger URL: http://www.medicinenet.com/script/main/art.asp?articlekey=39540
[3] Gozani S. *et al* 1996 *Bulletin of the Harvard Institute of Neuroscience* (5) 44–56
[4] Caduff A. *et al* 2003 Biosensors and Bioelectronics **19** 209–217
[5] Beving H., Eriksson G. 1994 *Eur. J. Surg. Suppl.* **574** 87–89
[6] Fuchs K., Kaatze U. 2001 *J. Phys. Chem., Sect. B* **105** 2036–2042
[7] Grafenstein K., Duchna W. 1985 *Z. Gesamte Inn. Med.* **40** 589–591
[8] Grant E., Sheppard R., South G. 1978 *Dielectric Behavior of Biological Molecules in Solution* (Clarendon Press: Oxford)
[9] Hayashi Y. 2002 *J. Phys. D: App. Phys.* **35** 1–6
[10] Klonoff D. 1997 *Diabetes Care* **20** 433–443
[11] Koschinsky T., Heinemann L. 2001 *Diabetes Metab. Res. Rev.* **17** 113–123
[12] Marks V. 1996 *Clin. Chim. Acta* **251** 3–17
[13] Martinsen G., Grimnes S., Haug E.1999 *Skin Res. Technol.* **5** 179–181
[14] Martinsen G., Grimnes S., Sveen O. 1997 *Med. Biol. Eng. Comput.* **35** 172–176
[15] Tamada J., Garg S., Jovanovic L., Pitzer K., Fermi S., Potts R. 1999 *J. Am. Med. Assoc.* **282** 1839–1844
[16] Thennadil S., Rennert J., Wenzel B., Hazen K., Ruchti T., Block M. 2001 *Diabetes Technol. Ther.* **3** 357–365

IOP Conf. Series: Materials Science and Engineering **516** (2019) 012057 doi:10.1088/1757-899X/516/1/012057

Methods for robot localization on a map

M V Shikhman, S V Shidlovskiy

National Research Tomsk State University, Tomsk, Russia
E-mail: Shikhmar@gmail.com

Abstract. In many cases, to solve applied problems, the robot needs to know its real location, which is most often different from the data stored in the on-board system. For unmanned robotic devices, as well as for ground-based robots, it is most efficient to use local navigation algorithms, which consist in determining the coordinates of the device with respect to a certain starting point. The paper discusses localization algorithms provided that the map of the area is known in advance. Particular attention is paid to the Monte Carlo localization method because of several advantages. The paper presents an example of modeling the algorithm operation.

1. Introduction

The world has entered the digital era when new technologies are rapidly developing and changing the habitual way of life, new industries and professions are being formed, and new opportunities for development are appearing. Among the variety of areas in the digital industry, robotics is of particular interest. Robotic devices gradually penetrate into all spheres of human activity and have great prospects for further development. Many problems need to be solved to provide the subsequent evolution of such devices.

Mobile robots able to move independently in space are singled out the variety of robotic devices. For successful navigation, the robot must have a map of the area or skills to build it directly in the process of movement; pre-determine the route and be able to adjust it in the process of movement; control movement parameters to change its position; obtain environmental information and process it further; keep track of its location. In this paper, we consider methods for solving the latter problem, provided that the terrain map is known in advance. Localization is one of the most common tasks of robotic perception because knowing location of objects and the acting subject is the basis of any successful physical interaction.

2. The basic principles of building robot localization algorithms

The development of algorithms for robot localization consists of several areas of research including a mathematical and geometric description of navigation processes, computer technologies and technical devices that are necessary for implementing these processes. Gyroscopes, encoders, accelerometers, stereoscopic systems, laser range finders, and ultrasonic sonars can serve as technical devices for obtaining primary information about the environment, which is necessary for the operation of algorithms. Let us proceed to the description of the main algorithms used to localize the robot.

3. Iterative Closest Point

The most common algorithm for solving localization problems is the iterative closest point algorithm (ICP), which, in addition to determining the position of the robot, allows working with algorithms for finding the optimal route. The main idea of ICP is to combine images of the same object obtained from different angles but with common areas, the so-called overlapping areas. This is implemented by means of iterative minimization of the average distance between two clouds of points. To do this, first, an initial assessment of a rough conversion of one cloud to another is made followed by refinement during the minimization process. As a result, the algorithm finds the best transformation of one cloud to another for two given clouds of points P_1 and P_2 [1].

The task of minimizing the average distance between the clouds of points P_1 and P_2 may be represented as follows:

$$F = \frac{1}{N}\sum_{i=1}^{N} d\left(p_i^1; p_i^2\right) \longrightarrow \min$$

where $\left[p_i^1;\ p_i^2\right]_{i=1...N}$ is a set of pairs of closest points;

$d\left(p_i^1; p_i^2\right)$ is Euclidean distance between the closest points.

To calculate the distance between the point p_1 belonging to the cloud P_1 and the cloud P_2, you can use the so-called 'point-to-point distance' metrics. When using them, the distance between points p_1 and p_2 is minimized.

Another metric is 'point-to-plane distance', which minimizes the sum of squares of distances from p_i^1 to a plane S'_1 that is perpendicular to the plane S_1 at a point p_i^2 for all pairs of closest points.

In general, the iterative closest point algorithm includes the following steps.

- search for the closest pairs of points $\left[p_i^1;\ p_i^2\right]_{i=1...N}$;
- search for shift and rotation parameters, which reduce the F error by minimizing the distance or by the method of least squares;
- application of the found parameters for the point cloud P_1;
- steps 1-3 are repeated until the F error becomes less than a certain threshold value.

4. The method of planes detected in point clouds

Presentation of the initial information received from the sensors in the form of point clouds is not always appropriate because it is associated with high costs of computing resources. To solve this problem, representation of points in the form of elementary geometric elements (vertices, lines and planes) is often used. In this case, objects are characterized by much smaller dimension and noise level; and finding various relations between them becomes simpler. Thus, the first step in using this method is detection of linear objects and their parameters, which can be done using the segmentation method by combining points with close normals.

Then it is necessary to find a match in the database for each detected object. The pair plane is found based on the rotation and movement of the sensors defining the plane. To do this, we consider the global and local coordinate system. The local coordinate system is connected directly to the sensor and has a clear description. We also know the approximate position of this system relative to the global one, which is described by the linear displacement vector and the rotation matrix allowing us to directly determine the global plane equation. However, this step may find several paired planes or not find them at all; therefore, the next step is necessary.

To determine the parameters of the robot motion, you need to use at least three pairs of planes. For n pairs of planes, you can write a nonlinear system consisting of n equations, each of which includes values of the rotation angles relative to each plane of the three-dimensional surface and the vector of the linear displacement. These parameters can be found by optimization methods. As a result, they must satisfy all pairs of planes with a certain given accuracy, which allows discarding erroneously proposed planes [2].

5. Algorithms based on space splitting using an octree type structure

The main idea of these algorithms is to represent the entire surrounding space as a cube and then divide it into 8 parts (8 small cubes). Such division continues until reaching certain conditions, usually including the volume of the cube and the number of elements contained in it. Information about the environment is compared with the initial (stored in the database) through hierarchical access. The robot processes only visible nodes and the nodes included in them, sequentially from top to bottom. When the environment is divided into a certain number of cubes, the robot only works with the visible part of the space, that is, it checks the child nodes only when the main node is visible to the camera [3].

6. Kalman filter-based localization algorithms

To solve the robot localization problems, the Kalman filter is often used, which applies information about the laws of the robot's motion at the forecast stage and information from the robot sensors at the correction stage. The state of the system is the navigation parameters of the robot position (usually, its coordinates and shift parameters). Predicting the evolution of the state is based on odometric data and measurements allowing you to calculate the robot position in accordance with the map. The Kalman filter allows you to evaluate the state of the system based on a noisy prediction of its evolution and noisy measurements of this state [4].

The Kalman filter algorithm consists of the following steps:

- prediction of the current state based on estimates at the previous time step and taking into account the control parameters for the current step;
- prediction of observation upon displacement based on the observation model and the assessment of the state;
- shift of the robot in space;
- correction of the predicted state taking into account the error between the predicted and implemented observation;
- space shift calculation;
- extraction of reference points from the surrounding space, observation of the state

7. Approximate robot localization algorithm

When describing the method of planes and detected points in clouds, we have already discussed the reasonability of representing a cloud of points in the form of elementary geometric cells. This principle is applicable to approximate algorithms for the robot localization. In this case, we may present a map of terrain in the form of a certain geometric figure, most often a polygon.

At some initial time, the robot is at some point A, and there is some area of visibility S at its disposal. To determine its current location, the robot needs to compare the area of visibility S with all areas of the original map stored in the database. However, there may be several similar areas on the map (Figure 1). Thus, hypothetically, the robot can be at any of the points A_n.

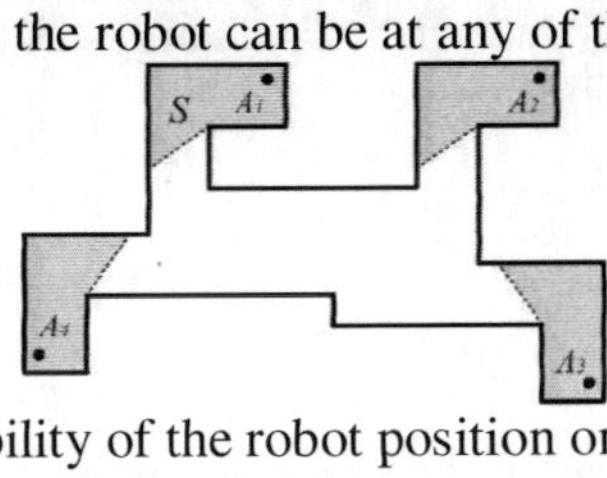

Figure 1. Variability of the robot position on the current map.

To determine exactly where the mobile robot is, it is most logical to use a probabilistic approach, which includes some kind of uncertainty of location. Thus, the probability of finding a robot at one or another point $P(A_n)$ is considered. During the process, the algorithm uses the methods of triangulation and construction of overlays [5].

8. Monte Carlo localization

In this case, to solve localization problems, the robot position is determined by the density of the distribution of probability of locating in a particular place approximated by a large set of particles. The particle is a hypothesis about the possible state of the robot at some point in time. The larger set of particles we have, the faster the program will find the right solution, that is the real position of the robot on the map.

The initial assumption of the algorithm is that there is usually a uniform random distribution of particles in the configuration space, that is, the robot does not have information about where it is located, and it is assumed that it can equally be at any point in space.

The algorithm includes two stages: prediction and correction. During the first stage, the robot motion from the previous time point to the current one and its angle of rotation are tracked. As a result, we obtain a preliminary set of approximating particles using a motion model. Here, the update of information about the location of each particle occurs. During the motion update, the robot predicts its new position based on the location of this trigger command, applying the simulated motion to each of the particles.

At the second stage, the search for paired planes is performed for each particle and there may exist several of them. To solve this uncertainty, a concept of 'particle weight' is introduced. The particle weight depends on how accurately the sensor readings coincide with the actual data and is calculated based on the sensors used. The final stage of correction is the formation of an approximating set of particles, in which a large particle belongs to the region with maximum weights [6].

8.1 Advantages of Monte Carlo localization

We should note that this method has several advantages. The particle filter, which is central to the Monte Carlo algorithm, can work with different probability distributions because it has a non-parametric representation. The time complexity of the particle filter is linear with respect to the number of particles, so you can find the optimal ratio of speed and accuracy. The implementation adapts to the available computing resources: the faster is the processor, the more particles can be generated and, therefore, we get a more accurate algorithm.

9. Implementation of the Monte Carlo localization

Based on the advantages of the Monte Carlo localization, the simulation was carried out specifically for this method. For this, we chose the following software: ROS (Robot Operating System), Gazebo software package, Rviz tool included in ROS, and MATLAB software package. In Gazebo, we used a TurtleBot robot simulator for simulation of the surrounding space. TurtleBot in Gazebo is a model of a real robotic device, which uses a Kinect sensor as a vision.

Then, using the Rviz tool, we created a map resulting from motion of the robot and receiving information from its sensor (Figure 2). Using the Matlab software package, the resulting map was transformed into a two-dimensional binary grid (Figure 3). Each cell in the grid has a value that represents true (1) or false (0) status of this cell.

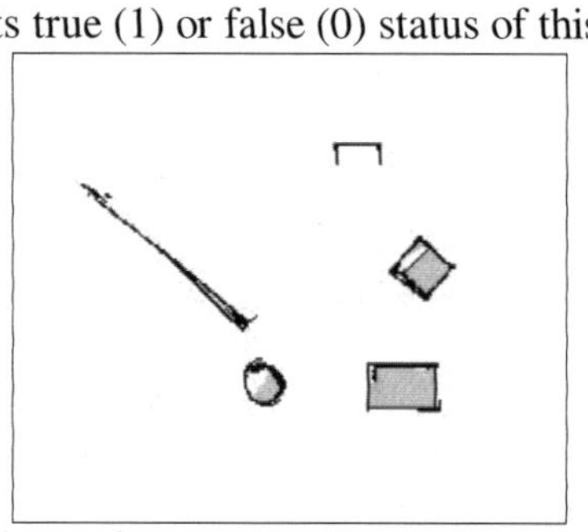

Figure 2. Result of building a map.

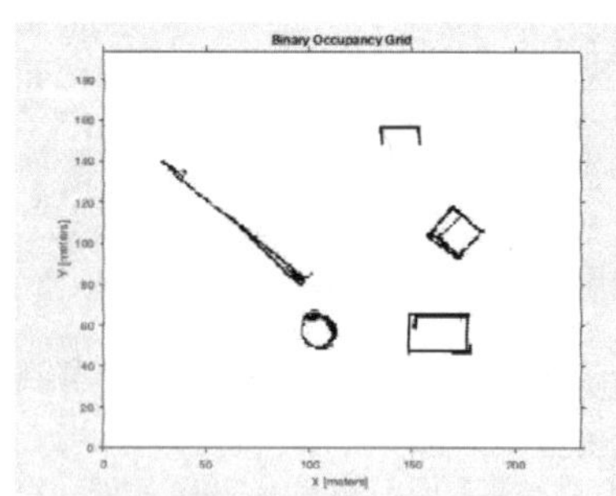

Figure 3. 2D binary grid.

IOP Conf. Series: Materials Science and Engineering **516** (2019) 012057 doi:10.1088/1757-899X/516/1/012057

The simulation of the Monte Carlo algorithm itself was carried out in the MatLab software environment. When applying this algorithm, there are two cases. First, when the approximate position of the robot is known and we just need to clarify it; and second, when the initial position of the robot is unknown. In this paper, we consider the second case because of its higher complexity. At the beginning, the algorithm assumes that the robot has an equal probability of being anywhere and generates uniformly distributed particles inside this space (Figure 4). The update occurs according to the algorithm and the current position of the robot is specified (Figure 5). As a result of the algorithm operation, the robot is localized (Figure 6).

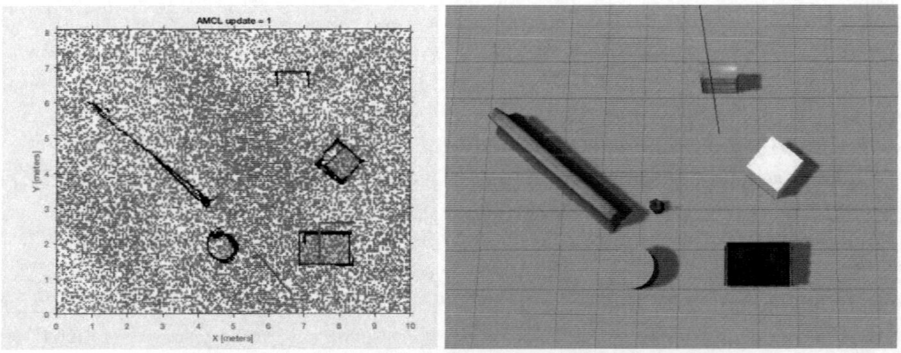

Figure 4. Start of the algorithm (third update).

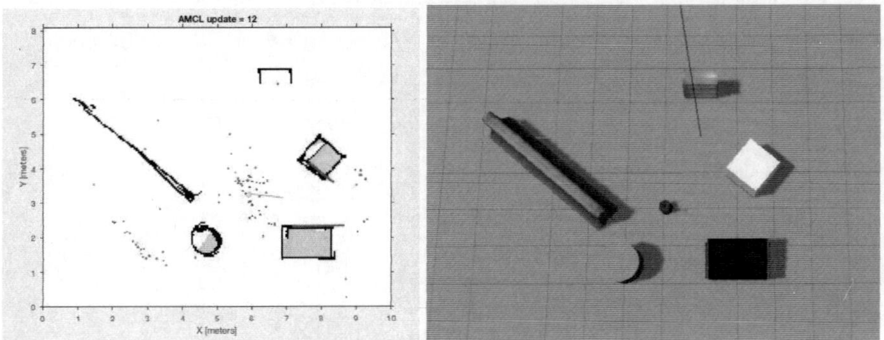

Figure 5. Algorithm operation (12th update).

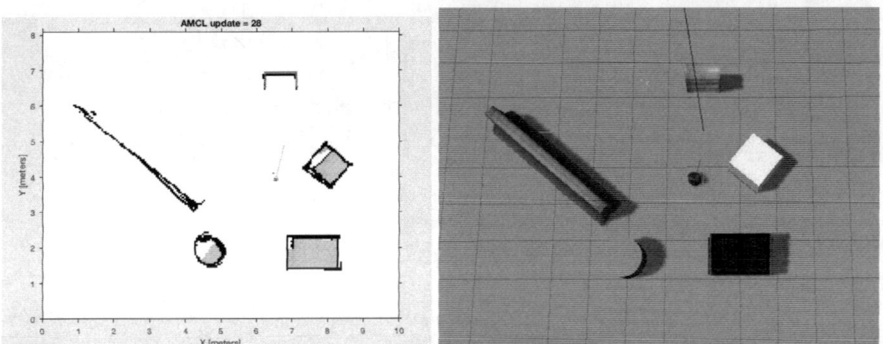

Figure 6. The end of the algorithm (28th update).

10. Conclusion

Monte Carlo localization allows specifying the current position of the robot when it moves and producing global localization in conditions of the absence of information about the robot location.

IOP Conf. Series: Materials Science and Engineering **516** (2019) 012057 doi:10.1088/1757-899X/516/1/012057

This algorithm has shown its effectiveness in solving the given problem, that is the robot localization provided that the terrain map is known in advance.

Acknowledgments
The research was carried out in Tomsk State University with the financial support of the Ministry of Education and Science of the Russian Federation, a unique project identifier: RFMEFI57817X0241. The authors are grateful to Tatiana B. Rumyantseva from Tomsk State University for English language editing.

References
[1] Borisov A. G., Gol S. A., Luksha S. S. 2013 *Vestnik RGRTU* **46 (4-3)** 35–42
[2] Kazmin V. N., Noskov V. P. 2015 *Proceedings of the Southern Federal University. Technical science* **8** 71–83
[3] Poslavskiy S. 2018 *Innovatika 2018* 118–120
[4] Gafurov O., Syryamkin V., Gafurov A. *et al.* 2012 *Telecommunications and Radio Engineering* **71**(17) 1565–1574 doi: 10.1615/TelecomRadEng.v71.i17.40
[5] Udovenko S., Sorokin A. 2014 *Information processing systems* **10** 248–254
[6] Dao Zuy Nam, Ivanovskiy S 2014 *Scientific Bulletin of NSTU* **55(2)** 109–121
[7] Sebastian Thrun *et al.* 2011 *Artificial Intelligence* **128** 99–141
[8] Kuznetsov D., Syryamkin V. 2015 *AIP Conference Proceedings* **1688** (040004) doi: 10.1063/1.4936037

IOP Conf. Series: Materials Science and Engineering **516** (2019) 012058 doi:10.1088/1757-899X/516/1/012058

Phase as the basis for wave vision

A P Kolandasamy, R M Makhmanazarov, I Yu Kuz'menko, T R Muksunov, V P Yakubov

Tomsk State University, Tomsk, Russia
E-mail: yvlp@mail.tsu.ru

Abstract. The work is devoted to the problem of obtaining images using radio and ultrasonic waves. These waves are not originally carriers of images, unless they carry a phase. In the simplest examples, it is theoretically and experimentally shown that waves with phase can be used to restore images of objects.

1. The method of synthetic aperture as the basis of vision systems

A person has no senses that would allow him to see the world around him in radio or acoustic radiation, and to obtain one, one has to use some technologically advanced devices, this motivates how vision systems arise, which allow using radars and sonar devices to orient in conditions of poor visibility. However, obtaining high-quality images is possible only when using the phase information contained in the radiation. Based on this so-called large aperture synthesis systems [1–10]. We explain its essence.

This can be done, for example, if the transmitter (T) and the receiver (R) are in direct contact, to obtain quadrature components of the scattered signal by the object, the direct heterodyning techniques has been used:

$$I = A\cos\varphi, \quad Q = A\sin\varphi. \tag{1}$$

Amplitude and phase carry all the information necessary for the implementation of the wave vision. It is enough to collect the signal quadrature distribution over a certain aperture (Figure 1). The resulting distribution of the complex amplitude ($E(\mathbf{r}) = I + iQ$) gives an image of the probed object by matched filtering with the hardware function of the system (HFS) $G(\mathbf{r})$, which is the response of the probing system to a single reflector. This filtering is performed by a simple calculation using fast integral convolution algorithms:

$$R(\mathbf{r}) = E(\mathbf{r}) * G^{*}(\mathbf{r}) \tag{2}$$

IOP Conf. Series: Materials Science and Engineering **516** (2019) 012058 doi:10.1088/1757-899X/516/1/012058

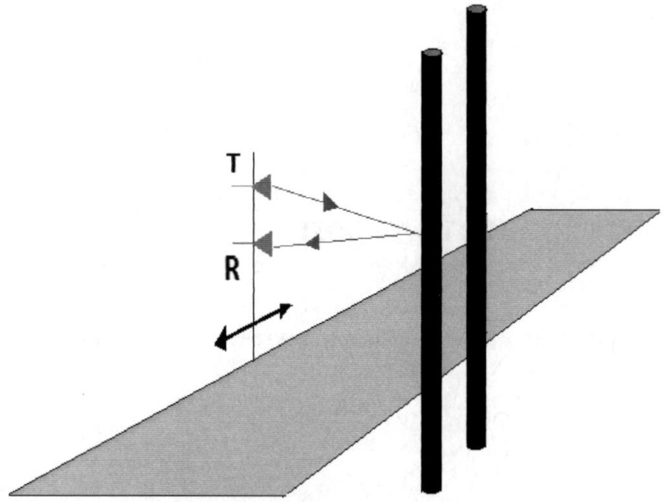

Figure 1. Typical measurement setup.

With a sufficiently large aperture of the probe system, the resulting distribution of the function is an image of the object being probed. Figure 2 shows an example of using the described procedure in the case where the reflector is a point. It is important to emphasize that at each observation point the reflected radiation is recorded taking into account the components of the total complex field, i.e. taking into account both amplitude and phase. This corresponds to coherent tomography or holography. From Figure 2 it can be seen that - the position of the reflector point is restored. Similarly, more complex images are restored.

a *b* *c*

Figure 2. Position of the sensing point *(a)*, the real part of the scattered field *(b)* and the reconstructed image of the point *(c)*.

Note that to restore the image is fundamentally important to measure the phase structure of the scattered field. From a physical point of view, obtaining an image is equivalent to focusing a field when a scattered field distributed in space (Figure 2*b*) is assembled at a point (Figure 2*c*). Only in this case, the observed interference pattern is compressed in this case to a point. This happens when the field is fully coherent when the phase of the field is written through quadrature's.

2. Other methods for measuring the quadrature's of the scattered field

In some cases, to measure the quadrature, it can be used not full, but partial, with a decrease in frequency to a level where direct digitization of the signal is already possible. This happens, for example, when using ultrasonic sounding at a frequency of $f_1 = 41$ kHz. If you select a heterodyne signal equal to $f_0 =$ 40 kHz, the difference frequency is equal to $\Delta f = f_1 - f_0 = 1$ kHz, which already makes it possible to use almost any analog-to-digital converters (ADC) for digitizing signals (Figure 3). This is important when creating multi-touch measuring systems, for example, when probing pavements.

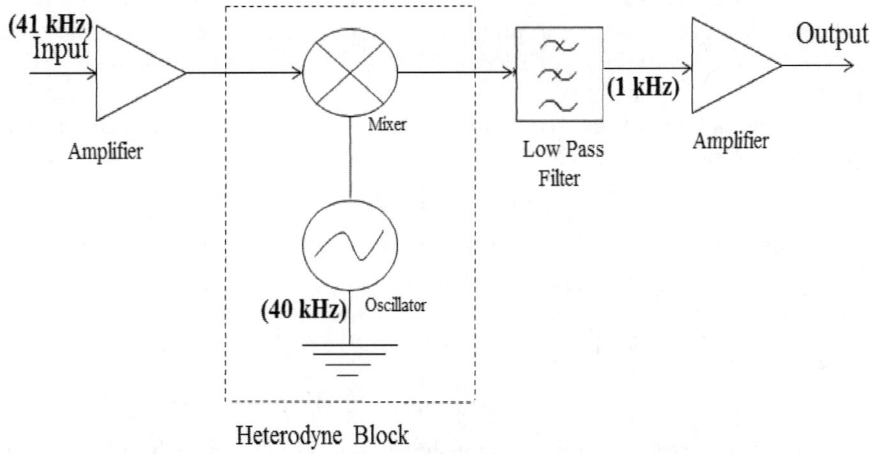

Figure 3. Receiver Schematic circuit Diagram.

After digitization, further selection of the complex amplitude or quadrature's of the signal is performed by digital heterodyning. The effectiveness of this procedure is illustrated in Figure 4, which shows the experimental result of reconstructing the image of a single pipe with a diameter of 1.5 cm. Here are the results of a 4-fold scan of the same pipe.

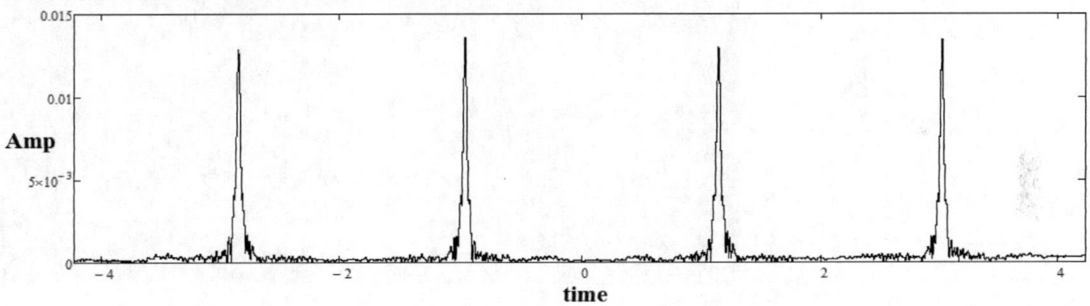

Figure 4. Image of 4 times sounding a single pipe (*time in sec*).

For comparison in Figure 5 shows a similar result of scanning two tubes located at a distance of 5.5 cm between them.

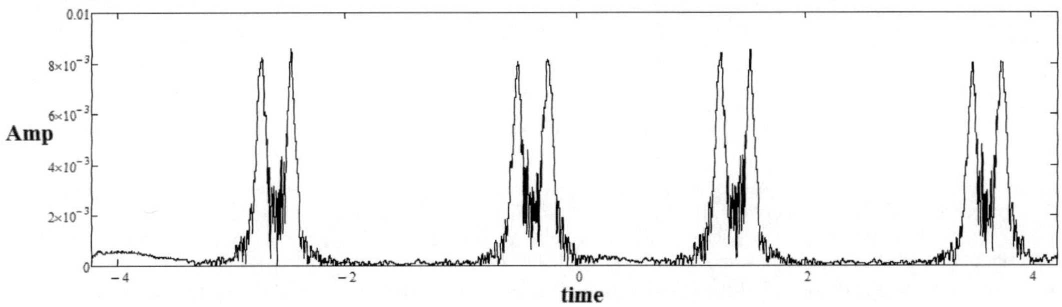

Figure 5. Image of 4 times sounding of two parallel tubes (*time in sec*).

Let us consider another variant of phase measurements, when, for one reason or another, the direct signal necessary for heterodyning at the receiving point is not available. This situation arises, for example, when using small unmanned aerial vehicles (UAVS) for sensing the earth, when placing on board a powerful coherent generator requires significant energy costs. A fairly simple solution is to use only one receiver, which registers only the amplitude of the signal reflected from the ground. At the same time, stealth and autonomy of the UAV is ensured. But what about the phase information necessary to obtain an image?

The output is when the radiation scattered by the earth is a mixture with external (reflective) coherent radiation. The intensity of the received radiation is estimated as

$$I = |E_0 + E_1|^2 = A_0^2 + A_1^2 + 2A_0A_1\cos(\varphi_1 - \varphi_0), \tag{3}$$

Where $\varphi = \varphi_1 - \varphi_0 = 2k|\mathbf{r_1} - \mathbf{r_0}|$ is the phase difference between the object and reference waves. It is assumed that $E_0 = A_0 \exp\{i\varphi_0\}$ - the reference wave, $E_1 = A_1 \exp\{i\varphi_1\}$ - the subject wave. We assume that the amplitudes of the waves are as $A_0 \approx A_1$. Then for a variable part of the intensity of the total field you can write

$$\Delta A^2 = I^2 - \langle I^2 \rangle = 2A_0A_1\cos(\varphi). \tag{4}$$

It is important that this expression, up to a constant factor, coincides with one of the quadrature's scattered field, which is necessary to reconstruct the image using the technology of synthesizing a large aperture. To illustrate the applicability of the proposed approach, we consider the case of three objects whose image should be obtained (Figure 6). It is seen that the image of all objects using the described procedure is surely restored.

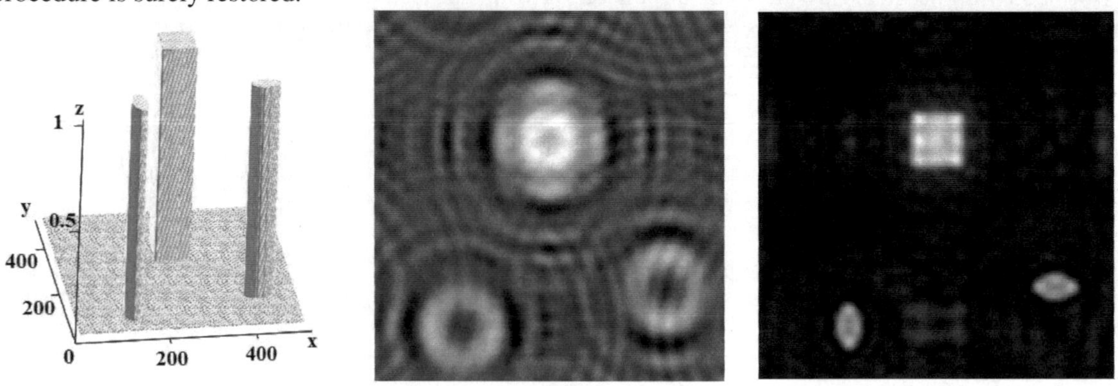

a *b* *c*

Figure 6. The position of the three objects *(a)*, the intensity distribution of the recorded interference field *(b)* and the reconstructed image of the objects *(c)*.

4. Summary

The restoration of images using wave radiation, one way or another, requires the preservation of phase information about the scattered field. Without using this information, focusing of images is impossible. Three different ways of extracting this information are considered; direct heterodyning, analog-digital heterodyning, and incoherent heterodyning mixed with a reference wave. In all cases, the images are well restored in the presence of phase information.

Acknowledgments

This work was supported by the grant of the Russian Science Foundation № 16-19-10272.

References

[1] Yakubov V.P. *et al* 2017 *Russian Physics Journal* **59**(12)1711–1717

IOP Conf. Series: Materials Science and Engineering **516** (2019) 012058 doi:10.1088/1757-899X/516/1/012058

[2] Yakubov V.P. *et al* 2018 *Russian Physics Journal*. 2018. Vol. **60**(11) 1905–1910
[3] Vaiman E.V. *et al* 2018 *Proceedings of the Fifteenth All-Russian Conference of Student Research Incubators* 163–166
[4] Makhmanazarov R. *et al* 2018 *Proceedings of the Fifteenth All-Russian Conference of Student Research Incubators* 178–181
[5] Yakubov V.P. *et al* 2016 *Inverse problems* (Publishing House of Scientific and Technology Literature:Tomsk)
[6] Yakubov V.P. 2006 *Electrodynamics* (Publishing house of the STL: Tomsk)
[7] Shipilov S. È. 2016 *Radio Wave Tomography: Achievements and Prospects* (Publishing House of Technology and Technology Literature:Tomsk)
[8] Yakubov V.P. *et al* 2018 *Russian Physics Journal* **61** (4) 624–634
[9] Yakubov V.P. 2018 *Semiconductor Nanotechnology: advances in information and energy processing. Cham* (Springer International Publishing: United States)
[10] 2019 *Electromagnetic and Acoustic Wave Tomography: Direct and Inverse Problems in Practical Applications* (CRC Press: Boca Raton)